Handbook of Experimental Pharmacology

Volume 225

More information about this series at
http://www.springer.com/series/164

Fabrizio Benedetti • Paul Enck •
Elisa Frisaldi • Manfred Schedlowski
Editors

Placebo

Springer

Editors

Fabrizio Benedetti
Neuroscience Department and
National Institute of Neuroscience Labs
University of Turin Medical School
Turin
Italy

Paul Enck
Department of Psychosomatic Medicine
University Hospitals Tuebingen
Tuebingen
Germany

Elisa Frisaldi
Neuroscience Department and
National Institute of Neuroscience Labs
University of Turin Medical School
Turin
Italy

Manfred Schedlowski
Institute of Medical Psychology and
Behavioral Immunobiology
University Hospital Essen
Essen
Germany

ISSN 0171-2004 ISSN 1865-0325 (electronic)
ISBN 978-3-662-52436-7 ISBN 978-3-662-44519-8 (eBook)
DOI 10.1007/978-3-662-44519-8
Springer Heidelberg New York Dordrecht London

Printed on acid-free paper

Springer is part of Springer Science+Business Media (www.springer.com)

Preface

A recent meeting on placebo and nocebo responses, sponsored by the Volkswagen Foundation, and held in Tuebingen (Germany) in January 2013, represented the starting point for inviting many scientists involved in experimental placebo and nocebo research to contribute to this volume by describing their work. Therefore, this volume presents the main lines of placebo research which are in progress and which will represent a challenge in the near future. Although this is not a comprehensive book on placebo and nocebo effects, we believe that a general overview of the ongoing studies may be useful to experimental pharmacologists, hopefully stimulating new avenues of debate and research.

Placebo is one of the most widespread words in the field of biomedical sciences. Until two decades ago, physicians and clinical scientists referred to this word when designing and interpreting clinical trials. In fact, placebo has always represented a comparator in the clinical trials setting, whereby the efficacy of a new treatment, be it pharmacological or not, has to be assessed. However, there still exists a semantic confusion within the scientific community in the use and meaning of the term placebo. On the one hand, placebo refers to an inert treatment, for example, a drug without any intrinsic pharmacological property. On the other hand, placebo effect, or response, refers to the therapeutic outcome following the administration of the inert treatment.

The still persisting confusion and misconception about the word placebo comes from the different meaning that this word has for the clinical trialist and the neuroscientist. In fact, the former is only interested in comparing the efficacy of a specific, e.g., pharmacological, intervention with a placebo treatment and to establish whether the drug is superior to the placebo. The clinical trialist is not interested in understanding whether the placebo-treated patients improve because of a spontaneous remission, a bias of the experimenter and/or patient, or different psychobiological factors. By contrast, the neuroscientist is interested in isolating the psychobiological components of the placebo response from the spontaneous fluctuations of the symptom, the patient's biased reports, and the experimenter's biased measurements. In this sense, the neuroscientist uses the placebo to probe several brain functions, ranging from endogenous pain modulation to anxiety mechanisms and from behavioral conditioning to social learning.

Nocebos and nocebo effects, on the other hand, are less studied and less understood, mainly due to many ethical constraints. In fact, nocebo is the evil twin of placebo, that is, a clinical worsening following placebo administration. In other words, expectations of adverse events or clinical worsening may lead to anticipatory anxiety which, in turn, may induce a real worsening.

Today placebo and nocebo effects are approached by means of modern biological tools that range from pharmacology to brain imaging and from genetics to single-neuron recordings in awake patients. Therefore, placebo and nocebo effects, or responses, are considered today psychobiological phenomena worthy of scientific inquiry, thus turning them from artifacts in clinical research into models for neuroscience. Besides these basic neurobiological insights, placebo research is also aimed both at exploring the possibility of exploiting placebo mechanisms in medical practice for the patient's benefit and at developing new clinical trial designs for the validation of new treatments.

Fabrizio Benedetti
Paul Enck
Elisa Frisaldi
Manfred Schedlowski

Contents

Part I Pain

Placebo and Nocebo Effects: An Introduction to Psychological and Biological Mechanisms 3
Jian Kong and Fabrizio Benedetti

Placebo, Nocebo, and Learning Mechanisms 17
Luana Colloca

A Meta-analysis of Brain Mechanisms of Placebo Analgesia: Consistent Findings and Unanswered Questions 37
Lauren Y. Atlas and Tor D. Wager

Placebo Analgesia: Cognition or Perception 71
Debora L. Morton, Wael El-Deredy, and Anthony K.P. Jones

Pain-Related Negative Emotions and Placebo Analgesia 81
Magne Arve Flaten

How Positive and Negative Expectations Shape the Experience of Visceral Pain 97
Sigrid Elsenbruch

Placebo Effects in Idiopathic and Neuropathic Pain Conditions 121
Lene Vase, Gitte Laue Petersen, and Karen Lund

Part II Emerging Models

Great Expectations: The Placebo Effect in Parkinson's Disease 139
Sarah Christine Lidstone

The Effects of Placebos and Nocebos on Physical Performance 149
Elisa Carlino, Alessandro Piedimonte, and Elisa Frisaldi

Learned Placebo Responses in Neuroendocrine and Immune Functions 159
Laura Wendt, Antje Albring, and Manfred Schedlowski

Placebo Responses on Cardiovascular, Gastrointestinal, and Respiratory Organ Functions 183
Karin Meissner

Placebo and Nocebo Effects in Itch and Pain 205
Andrea W.M. Evers, Danielle J.P. Bartels, and Antoinette I.M. van Laarhoven

Part III Clinical Practice and Clinical Trials

Clinical and Ethical Implications of Placebo Effects: Enhancing Patients' Benefits from Pain Treatment 217
Regine Klinger and Herta Flor

Traditional and Innovative Experimental and Clinical Trial Designs and Their Advantages and Pitfalls 237
Katja Weimer and Paul Enck

Lessons to be Learned from Placebo Arms in Psychopharmacology Trials 273
Bettina K. Doering, Winfried Rief, and Keith J. Petrie

The Emperor's New Drugs: Medication and Placebo in the Treatment of Depression 291
Irving Kirsch

Index 305

Part I

Pain

Placebo and Nocebo Effects: An Introduction to Psychological and Biological Mechanisms

Jian Kong and Fabrizio Benedetti

Contents

1 Unconscious Conditioning and Conscious Cognitive Expectancy 4
2 Brain Imaging Studies of Placebo and Nocebo Effects 6
3 Reward System and Placebo Effect 7
4 Endogenous Opioids and Placebo Analgesia 8
5 Do Placebo and Nocebo Share the Same Network? 9
6 Additive Effect of Placebo and Real Treatment 10
7 Predicting Placebo Responses 11
References 12

Abstract

Placebo and nocebo effects are essential components of medical practice and efficacy research, and can be regarded as a special case of context learning. A fundamental function of the central nervous system is to configure the way in which previous learned context becomes linked to corresponding responses. These responses could be either automatic procedures with little flexibility or highly adaptive procedures modified by associated contexts and consequences. Placebo and nocebo effects may represent a typical example of the combination of the two: conditioning effect, which is an inflexible, instinctual, and automatic response, and cognitive expectancy effect, which is a flexible adaptive response modified by prevailing conscious context. Given the fact that contextual learning originates in the brain, neuroimaging tools have been widely used to study

J. Kong (✉)
Psychiatry Department, Massachusetts General Hospital, Harvard Medical School, Building 120, 2nd street, Suite 101C, Charlestown, MA 02129, USA
e-mail: kongj@nmr.mgh.harvard.edu

F. Benedetti
Department of Neuroscience, University of Turin Medical School, National Institute of Neuroscience, Turin, Italy

F. Benedetti et al. (eds.), *Placebo*, Handbook of Experimental Pharmacology 225,
DOI 10.1007/978-3-662-44519-8_1

placebo and nocebo effects. In addition, pretest resting state fMRI may be a valuable biomarker to predict placebo responses.

Keywords
Placebo • Nocebo • Expectancy • Context learning • Consciousness • Unconsciousness • Resting state fMRI • Biomarker

1 Unconscious Conditioning and Conscious Cognitive Expectancy

A fundamental function of the central nervous system is to configure the way in which perceived information becomes linked to corresponding responses and meaningful experiences (Mesulam 1998; Lewis et al. 2009). These responses could be either automatic procedures with little flexibility or highly adaptive procedures modified by associated contexts and consequences.

The automatic process is fast and instinctual, but it restricts the range of events and tends to be inflexible. Inflexible bonds between incoming information and subsequent responses lead to instinctual and automatic behaviors that are resistant to change, even when faced by negative consequences. For instance, a study has shown that frogs whose optic nerve has been cut and allowed to fully regenerate after a 180° rotation of the eye can only snap at mud and moss on the ground when a fly was presented above the head (Mesulam 1998). Although human beings are less vulnerable to the emergence of such inflexible patterns, recent studies have shown that under certain circumstances, our actions can be initiated without conscious awareness of the goals to be attained or their motivating effect on our behavior (Custers and Aarts 2010). For instance, biased decisions may occur without conscious processing of contextual cues (Pessiglione et al. 2008). These findings imply that automatic instinctual processes still exist in human beings and influence our behavior.

The adaptive process, allowing for modifications and flexibility, exists only in advanced mammals. One of the most fundamental features of the human brain is that it does not passively analyze incoming information from the outside world; rather, it actively maintains ongoing representations and prior experiences, which can significantly sculpt neural responses to subsequent events (Mesulam 1998; Lewis et al. 2009).

Placebo and nocebo effects may represent a typical example of the combination of the two: conditioning effect, which is an inflexible, instinctual, and automatic response, and cognitive expectancy effect, which is a flexible adaptive response modified by prevailing conscious context. Taking our response to pain after placebo treatment as an example, the final effect may depend on both automatic conditioning responses and cognitive modulation based on previous knowledge and experience. The latter may modulate our pain experience during different states, including anticipation, pain experience, and posttreatment evaluation (Kong et al. 2007; Amanzio et al. 2013).

Benedetti et al. (2003) systematically investigated the relationship between the conditioning and cognitive expectancy in both healthy and patient populations. They found that verbally induced expectations of analgesia or hyperalgesia (in healthy subjects) and motor improvement or impairment (in Parkinsonian patients) completely antagonized the effects of a conditioning procedure. In addition, they also measured the effects of opposite verbal suggestions (i.e., verbal cues that suggest an opposite physiological response) on hormonal secretion. Results showed that verbally induced expectations of an increase or decrease in the level of growth hormone (GH) and cortisol did not have any effect on the secretion of these hormones. However, when a preconditioning procedure was performed with sumatriptan, a 5-HT1B/1D agonist that stimulates GH and inhibits cortisol secretion, a significant increase in GH and decrease in cortisol plasma concentrations were found after placebo administration, despite the opposite verbal suggestions were given. These findings suggests that placebo responses are mediated by conditioning when unconscious physiological functions, such as hormonal secretion, are involved, whereas they are mediated by expectation when conscious physiological processes, such as pain and motor performance, come into play, even though a conditioning procedure is performed.

Jensen and Karoly (2012) assessed whether a conditioning paradigm, using two similar, but not identical, facial cues during high and low pain, could induce placebo and nocebo responses with and without conscious awareness of the faces in two experiments in healthy subjects. The results showed significant placebo and nocebo effects using both clearly visible stimuli and nonconscious stimuli, indicating that even for conscious physiological processes, such as pain processing, the placebo and nocebo effects can operate without conscious awareness of the triggering cues. Interestingly, in this study, the data also showed that although the nocebo hyperalgesia effects evoked by both supraliminal and subliminal cues are comparable, placebo analgesia evoked by supraliminal cues is more robust than the placebo analgesia evoked by subliminal cues. We speculate that this may be associated with damage and risk avoidance of nocebo effect, which is important for survival. Thus, the unconscious, automatic response may have been more important for survival than the conscious, flexible responses, which may be a response that has developed during evolution.

Based on the results presented above, it seems that both placebo and nocebo effects involve unconscious, automatic conditioning and conscious cognitive modulation of expectation. The latter may be unique to human beings, and it can either enhance or eliminate the automatic conditioning effects depending on the specific context and outcome measurements. The resulting behavioral response is the combination of the two effects.

We believe that future studies should be focused on the following questions:

(1) What happens in the clinical setting? Is cognitive expectancy more important than unconscious conditioning, or vice versa? Both placebo and nocebo effects are clinical phenomena; the environment and context associated with each patient is

complex and unique; thus, we can imagine that there are multiple formulas/ proportions of different components. (2) What are the mechanisms of placebo and nocebo effects in the clinical setting? How can different conditioning experiences, particularly different pharmacological conditioning experiences, shape various pathways to produce placebo and nocebo effects (Amanzio and Benedetti 1999)? Are there any common mechanisms for different disorders?

2 Brain Imaging Studies of Placebo and Nocebo Effects

Given the fact that cognitive modulation/contextual learning originates in the brain, investigators started applying neuroimaging tools to study placebo and nocebo effects about one decade ago. With the development of advanced imaging techniques such as functional magnetic resonance imaging (fMRI), positron emission tomography (PET), and high-resolution EEG system, our understanding of placebo and nocebo effects has been greatly enhanced. Technical improvements in fMRI due to more powerful magnets, increasingly sophisticated imaging hardware, and in particular the development of new experimental paradigms and data analysis methods allow us to investigate neural events as dynamic processes within the brain and spinal cord. Both the spatial and temporal aspects of neural activity underlying placebo and nocebo effects can now be explored. Technical advances in PET imaging not only provide tools for investigating brain metabolism, blood flow changes, and other nonselective markers of neural activity, but also to investigate whole brain determinants of specific receptor-binding distributions in fully conscious humans. Such progress enables us to indirectly assess neurotransmitter changes associated with placebo analgesia. For instance, it allows us to indirectly measure the release of endogenous opioids in the brain.

Previous studies provide solid evidence that placebo treatment can significantly change outcome measurements. For instance, subjective pain intensity ratings have been shown to be reduced after placebo treatment in studies of placebo analgesia (Benedetti 2008; Benedetti et al. 2005; Kong et al. 2007). However, the mechanisms underlying these effects are far from apparent. A key question left unanswered concerns how and why the outcome measurements are reduced. Is it due to subjects' response bias or desire to please the experimenter, as suggested by some researchers (Hrobjartsson and Gotzsche 2001; Tedeschi et al. 1971; Cleophas 1995), or because biological changes happen after the placebo treatment, as for drug effects, or a combination of both?

Brain imaging tools hold the potential to resolve these questions. Using placebo treatment of pain as an example, previous studies suggest that brain regions including the thalamus, secondary somatosensory cortex (S2), insula, primary somatosensory cortex (S1), and anterior cingulate cortex (ACC) may be associated with the encoding of different pain intensity levels (Coghill et al. 1999; Bornhovd et al. 2002; Buchel et al. 2002; Alkire et al. 2004; Kong et al. 2006b, 2010a). Thus, we might reason that for post-placebo treatment, decreased brain activity in these regions indicates reduced incoming noxious information. In fact, under certain circumstances, brain imaging studies have begun to attest this very notion (Wager et al. 2004). This is just one example of how placebo treatment might be conceptualized through neurobiology. We believe that in order to fully understand placebo and nocebo phenomena, all contributions to outcome measurement changes should be properly acknowledged and investigated (Kong et al. 2007; Amanzio et al. 2013).

The literature on the placebo and nocebo effect is rich (see other related reviews for more details: Hrobjartsson and Gotzsche 2001; Hoffman et al. 2005; Kong et al. 2007; Finniss et al. 2010; Miller et al. 2009; Miller and Kaptchuk 2008; Barsky et al. 2002; Enck et al. 2008). In the next sections we will focus on several topics that may be helpful for understanding a complete picture of placebo treatment.

3 Reward System and Placebo Effect

Expectation is a critical component of all medical care and represents an important dimension of the "non-pharmacological component of pharmaceuticals" (Finniss et al. 2010; Benedetti 2008). In recent years, investigators have attempted to link positive expectancy to the reward system by reasoning that symptom reduction (decreased suffering) can be regarded as a special case of reward (de la Fuente-Fernandez et al. 2001, 2002; Scott et al. 2007, 2008; Leknes et al. 2011).

In an early study of placebo effects in patients with Parkinson's disease, investigators (de la Fuente-Fernandez et al. 2001, 2002) found that dopamine release increased in the ventral striatum (a key region in the reward system), suggesting the reward system association with expectations of improvement. In a subsequent study, Scott et al. (2007) found that dopamine release from the nucleus accumbens, observed during placebo administration, was related to its anticipated effects, perception-anticipation incongruity, and subsequent placebo effects. In a more recent morphometry study, Schweinhardt et al. (2009) found that dopamine-related traits can predict a substantial portion of the pain relief and individual gains from a sham treatment: the magnitude of placebo analgesia was correlated to gray matter density in the ventral striatum. In a more recent study, Yu and colleagues found that combining the resting state regional coherence at ventral striatum, Catechol-O-methyl transferase (COMT) and dopamine-related traits can be used to predict placebo response (Yu et al. 2014). The ventral striatum (nucleus accumbens) is also involved in placebo effects in anxiety (Petrovic et al. 2005), depression (Mayberg et al. 2002), psychotropic drug use (methylphenidate) (Volkow et al. 2003), and expectancy modulation of cue conditioning (Atlas et al. 2010). This involvement of the ventral striatum across multiple placebo conditions/modalities indicates that reward may represent a fundamental element of expectancy.

4 Endogenous Opioids and Placebo Analgesia

Placebo analgesia effect is one of the most robust and well-studied placebo effects. Although the dopamine reward system may play a role in placebo effects, most of the studies indicate the endogenous opioids and the pain descending modulatory network as the main mechanism in placebo analgesia (Zubieta et al. 2005; Scott et al. 2008; Wager et al. 2007; Eippert et al. 2009).

As the most studied pain modulatory mechanism for pain, the descending pain control system includes the periaqueductal gray (PAG), the rostral ventromedial medulla (RVM), frontal gyrus, anterior cingulate cortex, hypothalamus, and amygdala (Fields 2004; Kong et al. 2010b). PET studies have found significant μ-opioid binding potential changes after placebo treatment as compared with control condition (Zubieta et al. 2005; Scott et al. 2008; Wager et al. 2007), providing direct evidence of involvement of the endogenous opioid system.

In addition, fMRI studies (Wager et al. 2004; Price et al. 2007; Eippert et al. 2009) also found observable fMRI signal decreases in the brain's "pain matrix" as well as a decreased subjective pain experience, which provide direct evidence of the descending control system. Nevertheless, not all studies have found significant fMRI signal decrease in pain-related brain regions. In a previous study, Kong et al. (2006a) found significant fMRI signal increase in brain regions such as rostral ACC and anterior insula, but no significant fMRI signal decrease when comparing placebo treatment with controls, which suggest that other mechanisms such as emotion modulation may also be involved in placebo analgesia.

Despite some differences in previous studies, the results presented above may not be contradictory, as individual differences in detailed experimental paradigms may also underlie these differences. For instance, in our previous study (Kong et al. 2006a), the pain stimuli sequences after treatment were applied alternately between the placebo-treated side and untreated control side. Since opioid effects may be long lasting, as opioid activation can trigger naloxone insensitive analgesia that extends beyond the original opioid response (Atlas and Wager 2012), a method of applying stimuli on alternate sides (placebo and control) may not be sensitive to detect the opioid effects.

In addition, in three of the previous placebo analgesia studies (Wager et al. 2004; Price et al. 2007; Eippert et al. 2009) that reported an attenuation of brain activity in the pain matrix for reported analgesia, the authors used a relatively long duration of pain stimuli (20–30 s) as compared to the study by Kong et al. (2006a) that used pain stimuli lasting about 10 s. Their results confirm that the decreases in fMRI signal during the placebo condition (compared to the control condition) were mainly observed in a later phase of the pain stimulation. For example, Eippert et al. (2009) found a significant group-by-condition interaction during late pain (last 10 s), but not during early pain (first 10 s). Interestingly, they also found that naloxone could significantly reduce neural placebo effects, but only during late pain, not the early pain, implying that other mechanisms may be involved placebo analgesia. Indeed, a recent study shows that the endocannabinoid system could be involved under some circumstances (Benedetti et al. 2011).

In addition, one cue-expectancy study (Atlas et al. 2010) found that similar to placebo analgesia and nocebo hyperalgesia, significant fMRI signal decreases in pain-related brain regions and significant decreases in subjective pain intensity ratings were observed when subjects expected a low intensity pain as compared with control conditions. The rapid shifts of different predictive cues require any modulatory mechanism to be transient and reversible, due to the long-lasting effects of opioids in the brain; it is unlikely that endogenous opioid mechanisms are involved in cue modulation effects (Atlas and Wager 2012).

In summary, it appears that multiple brain mechanisms are involved in placebo analgesia, and that both subjective pain intensity rating reduction and fMRI signal reduction in pain-related brain region may happen without the involvement of endogenous opioids (Tracey 2010; Benedetti et al. 2011). Future research should focus on how to dissociate these different pain modulation mechanisms.

5 Do Placebo and Nocebo Share the Same Network?

A significant proportion of clinical improvement following different therapies, especially for subjective symptom outcomes, is directly attributable to placebo effects (Kaptchuk 2002), whereas a significant proportion of adverse events to medications is represented by nocebo effects (Amanzio et al. 2009; Barsky et al. 2002). Therefore, understanding similarities and differences in the mechanisms of placebo and nocebo effects represents an important challenge for future research.

Benedetti et al. (1997) found that nocebo hyperalgesia is mediated by cholecystokinin. In a brain imaging study, Kong et al. (2008) found that nocebo hyperalgesia may be predominantly produced through an affective-cognitive pain pathway (medial pain system), and the hippocampus may play an important role in this process. In a subsequent study that investigated the influence of expectancy on the analgesic effect of remifentanil, Bingel et al. (2011) also found that the positive expectancy effects were associated with activity in the endogenous pain modulatory system, and the negative expectancy effects with activity in the hippocampus, suggesting different mechanisms underlying placebo and nocebo effects. In a PET study, Scott et al. (2008) found an overlapping network including the anterior cingulate, orbitofrontal and insular cortex, nucleus accumbens, and amygdale, suggesting a similar mechanism of placebo analgesia and nocebo hyperalgesia. Future studies that directly compare the placebo and nocebo effects using the same cohort of subjects will help provide a general model of how different expectancies can modulate brain responses.

6 Additive Effect of Placebo and Real Treatment

Evidence-based medicine relies on the placebo-controlled randomized clinical trial (RCT) to distinguish the effects of an active (verum/genuine) pharmacological agent or procedure from the effects of a mimicking placebo treatment. Both for biomedicine and for complementary and alternative medicine (CAM), detecting this difference has become a challenge, especially for subjective outcomes.

The prevailing model for understanding verum-placebo differences has been the "additive model" (Kirsch 2000). This model presupposes that the placebo effect in the treatment and placebo arms of an RCT are of equivalent magnitude and that one can simply subtract the magnitude of the placebo response from the medication response to determine the presence (or absence) of verum effects. The possibility that under different conditions the pharmacological and placebo effects could act differentially in the two arms of a RCT is rejected a priori (Kleijnen et al. 1994; Kirsch 1999). Alternatively, it is possible that drug and placebo effects interact (Kirsch 1999). The possibility that placebo-induced expectancies might modify the drug effect (i.e., the difference between the drug response and the placebo response) has received insufficient attention. Recently, brain imaging has started shedding new light on this field.

Kong et al. (2009b) investigated how expectancy can modulate acupuncture treatment effects using a conditioning-like expectancy manipulation paradigm. The results indicate that expectancy can significantly modulate the analgesic effect of acupuncture treatment in both subjective pain rating changes and fMRI signal changes. In addition, it was found that, although verum acupuncture and sham acupuncture induced subjective reports of analgesia of equal magnitudes, fMRI analysis showed that (1) verum acupuncture produced greater fMRI signal decreases in pain-related brain regions during the application of calibrated heat pain stimuli on the right arm and (2) high expectancy produced greater fMRI signal changes in emotion-related brain regions.

Using a conditioning-like expectancy manipulation model, Bingel et al. (2011) investigated how different expectancies can modulate the analgesic effect of a potent opioid, remifentanil, using fMRI. The effect of a fixed concentration of the μ-opioid agonist remifentanil on constant heat pain was assessed under three experimental conditions using a within-subject design: with no expectation of analgesia, with expectancy of a positive analgesic effect, and with negative expectancy of analgesia (i.e., expectation of hyperalgesia or exacerbation of pain). Results showed that positive treatment expectancy can significantly enhance the analgesic benefit of remifentanil; at the same time, negative treatment expectancy abolished remifentanil analgesia effect. Brain imaging results also showed that the positive expectancy effects were associated with activity in the endogenous pain modulatory system, and the negative expectancy effects with activity in the hippocampus.

In another study, Atlas et al. (2012) also investigated the influence of expectancy on remifentanil. Two experiments were performed, and in both experiments remifentanil (or placebo treatment) was administered to all subjects during experimental thermal pain. Results showed that remifentanil and expectancy both reduced pain, but drug effects on pain reports and fMRI activity did not interact with expectancy. Regions associated with pain processing showed drug-induced modulation during both open (expected) and hidden (unexpected) administrations, with no differences in drug effects as a function of expectation. These findings suggest that remifentanil and placebo treatments both influence clinically relevant outcomes and operate without mutual interference.

In a more recent verbal suggestion balanced placebo study using a topical analgesic treatment (lidocaine), Schenk et al. (2014) investigated the interaction between lidocaine and expectancy with a clinical pain-related model (capsaicin pretreated skin) in healthy subjects. They found that active treatment can significantly reduce the pain rating as compared with placebo treatment, while the main expectancy effect (open administration compared with hidden administration) is not significant. However, unlike Atlas et al. (2012), they found a significant interaction between treatment and expectancy. In two lidocaine groups, open administration showed significantly greater pain rating reduction than hidden administration, but there was no significant difference between open and hidden administration of placebo treatment.

The inconsistent results may derive from different reasons: (1) different ways of manipulating expectancy, e.g., verbal suggestion vs. a more powerful conditioning-like procedure (Colloca et al. 2008; Wager et al. 2004; Kong et al. 2013b). (2) Different active treatment modalities (remifentanil, lidocaine, and acupuncture). Taken together, these results suggest that the effect of expectancy on treatment outcome may depend on the strength of the expectancy manipulation, the treatment modality, and the status of the participants (subjects vs. patients).

7 Predicting Placebo Responses

Given the high impact of the placebo response in medical practice and research, predicting placebo responses has always been an abstractive topic to placebo researchers. In recent years, investigators started applying imaging tools to predict placebo responses (Honey et al. 2008; Wager et al. 2011; Hashmi et al. 2012, 2014; Kong et al. 2013a; Yu et al. 2014). In a previous study, Wager et al. (2011) found that increased anticipatory activity in a frontoparietal network and decreases in a posterior insular/temporal network predicted placebo analgesia. During pain, decreases in limbic and paralimbic regions, not the pain-associated brain region, most strongly predicted placebo analgesia. These results indicate that enhancement of emotional appraisal circuits may be responsible for individual variation in placebo analgesia, rather than early suppression of nociceptive processing.

Ideally, a clinical applicable marker to predict placebo response should be a measurement applied before the treatment started. The pretest resting state functional connectivity has the potential to be used as a marker. In a previous study, Kong et al. (2013a, b) found that pretest resting state functional connectivity between the right frontoparietal network (as identified by independent component analysis) and the rostral anterior cingulate cortex/medial prefrontal cortex was positively associated with conditioning cue effects indicated by pain rating changes. In another study (Yu et al. 2014), it was found that the regional homogeneity (ReHo), an index of local neural coherence, in the ventral striatum, was significantly associated with conditioning effects on pain rating changes evoked by placebo visual cue. It was also found that the number of Met alleles at the COMT polymorphism was linearly correlated to the suppression of pain. These findings demonstrate the potential of combining resting state connectivity and genetic information to predict placebo effect. In the same study, personality was also found to represent a possible predictor. In another recent study (Hashmi et al. 2014), it was found that the efficiency of information transfer within local networks calculated with graph-theoretic measures (local efficiency and clustering coefficients) significantly predicted conditioned analgesia in older patients with knee Osteoarthritis.

Taking together, resting state functional connectivity holds the potential to predict placebo response. Nevertheless, we have to interpret the above results with caution, and independent replication of these studies is needed before we can draw solid conclusions.

Conclusions

In summary, both unconscious learning mechanisms and conscious expectancies can be involved in placebo and nocebo effects, particularly placebo analgesia and nocebo hyperalgesia. When expectancy is enhanced, it holds the potential to significantly enhance or overcome the real treatment effect. Appropriate application of the power of both learning and expectancy may provide new pathway to promote good therapeutic outcomes. In addition, brain imaging measurements may be applied as a potential marker to predict placebo responses, with profound implications for clinical trials.

References

Alkire MT, White NS, Hsieh R, Haier RJ (2004) Dissociable brain activation responses to 5-Hz electrical pain stimulation. Anesthesiology 100:939–946

Amanzio M, Benedetti F (1999) Neuropharmacological dissection of placebo analgesia: expectation-activated opioid systems versus conditioning-activated specific subsystems. J Neurosci 19(1):484–494

Amanzio M, Corazzini LL, Vase L, Benedetti F (2009) A systematic review of adverse events in placebo groups of anti-migraine clinical trials. Pain 146(3):261–269

Amanzio M, Benedetti F, Porro CA, Palermo S, Cauda F (2013) Activation likelihood estimation meta-analysis of brain correlates of placebo analgesia in human experimental pain. Hum Brain Mapp 34(3):738–752

Atlas LY, Wager TD (2012) How expectations shape pain. Neurosci Lett 520(2):140–148

Atlas LY, Bolger N, Lindquist MA, Wager TD (2010) Brain mediators of predictive cue effects on perceived pain. J Neurosci 30(39):12964–12977

Atlas LY, Whittington RA, Lindquist MA, Wielgosz J, Sonty N, Wager TD (2012) Dissociable influences of opiates and expectations on pain. J Neurosci 32(23):8053–8064

Barsky AJ, Saintfort R, Rogers MP, Borus JF (2002) Nonspecific medication side effects and the nocebo phenomenon. JAMA 287(5):622–627

Benedetti F (2008) Mechanisms of placebo and placebo-related effects across diseases and treatments. Annu Rev Pharmacol Toxicol 48:33–60

Benedetti F, Amanzio M, Casadio C, Oliaro A, Maggi G (1997) Blockade of nocebo hyperalgesia by the cholecystokinin antagonist proglumide. Pain 71(2):135–140

Benedetti F, Pollo A, Lopiano L, Lanotte M, Vighetti S, Rainero I (2003) Conscious expectation and unconscious conditioning in analgesic, motor, and hormonal placebo/nocebo responses. J Neurosci 23:4315–4323

Benedetti F, Mayberg HS, Wager TD, Stohler CS, Zubieta JK (2005) Neurobiological mechanisms of the placebo effect. J Neurosci 25(45):10390–10402

Benedetti F, Amanzio M, Rosato R, Blanchard C (2011) Non-opioid placebo analgesia is mediated by CB1 cannabinoid receptors. Nat Med 17:1228–1230

Bingel U, Wanigasekera V, Wiech K, Mhuircheartaigh RN, Lee MC, Ploner M et al (2011) The effect of treatment expectation on drug efficacy: imaging the analgesic benefit of the opioid remifentanil. Sci Transl Med 3(70):70ra14

Bornhovd K, Quante M, Glauche V, Bromm B, Weiller C, Buchel C (2002) Painful stimuli evoke different stimulus–response functions in the amygdala, prefrontal, insula and somatosensory cortex: a single-trial fMRI study. Brain 125:1326–1336

Buchel C, Bornhovd K, Quante M, Glauche V, Bromm B, Weiller C (2002) Dissociable neural responses related to pain intensity, stimulus intensity, and stimulus awareness within the anterior cingulate cortex: a parametric single-trial laser functional magnetic resonance imaging study. J Neurosci 22(3):970–976

Cleophas TJ (1995) The importance of placebo effects. JAMA 273:283

Coghill RC, Sang CN, Maisog JM, Iadorola MJ (1999) Pain intensity processing within the human brain: a bilateral, distributed mechanism. J Neurophysiol 82:1934–1943

Colloca L, Tinazzi M, Recchia S, Le Pera D, Fiaschi A, Benedetti F et al (2008) Learning potentiates neurophysiological and behavioral placebo analgesic responses. Pain 139(2):306–314

Custers R, Aarts H (2010) The unconscious will: how the pursuit of goals operates outside of conscious awareness. Science 329(5987):47–50

de la Fuente-Fernandez R, Schulzer M, Stoessl AJ (2002) The placebo effect in neurological disorders. Lancet Neurol 1:85–91

de la Fuente-Fernandez R, Ruth TJ, Sossi V, Schulzer M, Calne DB, Stoessl AJ (2001) Expectation and dopamine release: mechanism of the placebo effect in Parkinson's disease. Science 293:1164–1166

Eippert F, Bingel U, Schoell ED, Yacubian J, Klinger R, Lorenz J et al (2009) Activation of the opioidergic descending pain control system underlies placebo analgesia. Neuron 63(4):533–543

Enck P, Benedetti F, Schedlowski M (2008) New insights into the placebo and nocebo responses. Neuron 59(2):195–206

Fields H (2004) State-dependent opioid control of pain. Nat Rev Neurosci 5(7):565–575

Finniss DG, Kaptchuk TJ, Miller F, Benedetti F (2010) Biological, clinical, and ethical advances of placebo effects. Lancet 375(9715):686–695

Hashmi JA, Baria AT, Baliki MN, Huang L, Schnitzer TJ, Apkarian AV (2012) Brain networks predicting placebo analgesia in a clinical trial for chronic back pain. Pain 153(12):2393–2402. doi:10.1016/j.pain.2012.08.008

Hashmi JA, Kong J, Spaeth R, Khan S, Kaptchuk TJ, Gollub RG (2014) Functional network architecture predicts psychologically mediated analgesia related to treatment in chronic knee pain patients. J Neurosci 34(11):3924–3936

Hoffman GA, Harrington A, Fields HL (2005) Pain and the placebo: what we have learned. Perspect Biol Med 48(2):248–265

Honey GD, Corlett PR, Absalom AR, Lee M, Pomarol-Clotet E, Murray GK et al (2008) Individual differences in psychotic effects of ketamine are predicted by brain function measured under placebo. J Neurosci 28(25):6295–6303. doi:10.1523/JNEUROSCI.0910-08.2008
Hrobjartsson A, Gotzsche PC (2001) Is the placebo powerless? An analysis of clinical trials comparing placebo with no treatment. N Engl J Med 344(21):1594–1602
Jensen KB, Kaptchuk TJ, Kirsch I, Raicek J, Lindstrom KM, Berna C, Gollub RL, Ingvar M, Kong J (2012) Nonconscious activation of placebo and nocebo pain responses. Proc Natl Acad Sci USA 109(39):15959–15964
Kaptchuk TJ (2002) The placebo effect in alternative medicine: can the performance of a healing ritual have clinical significance? Ann Intern Med 136(11):817–825
Kirsch I (1999) How expectancies shape experience. American Psychological Association, Washington, DC
Kirsch I (2000) Are drug and placebo effects in depression additive? Biol Psychiatry 47(8):733–735
Kleijnen J, de Craen AJ, van Everdingen J, Krol L (1994) Placebo effect in double-blind clinical trials: a review of interactions with medications. Lancet 344(8933):1347–1349
Kong J, Gollub RL, Rosman IS, Webb JM, Vangel MG, Kirsch I et al (2006a) Brain activity associated with expectancy-enhanced placebo analgesia as measured by functional magnetic resonance imaging. J Neurosci 26(2):381–388
Kong J, White NS, Kwong KK, Vangel MG, Rosman IS, Gracely RH et al (2006b) Using fMRI to dissociate sensory encoding from cognitive evaluation of heat pain intensity. Hum Brain Mapp 27(8):715–721
Kong J, Kaptchuk TJ, Polich G, Kirsch I, Gollub RL (2007) Placebo analgesia: findings from brain imaging studies and emerging hypotheses. Rev Neurosci 18(3–4):173–190
Kong J, Gollub RL, Polich G, Kirsch I, Laviolette P, Vangel M et al (2008) A functional magnetic resonance imaging study on the neural mechanisms of hyperalgesic nocebo effect. J Neurosci 28(49):13354–13362
Kong J, Kaptachuk TJ, Polich G, Kirsch IV, Angel M, Zyloney C et al (2009a) Expectancy and treatment interactions: a dissociation between acupuncture analgesia and expectancy evoked placebo analgesia. Neuroimage 45:940–949, PMID: 19159691
Kong J, Kaptchuk TJ, Polich G, Kirsch I, Vangel M, Zyloney C et al (2009b) An fMRI study on the interaction and dissociation between expectation of pain relief and acupuncture treatment. Neuroimage 47(3):1066–1076, PMID: 19501656
Kong J, Loggia ML, Zyloney C, Tu P, Laviolette P, Gollub RL (2010a) Exploring the brain in pain: activations, deactivations and their relation. Pain 148:257–267, PMID: 20005043
Kong J, Tu PC, Zyloney C, Su TP (2010b) Intrinsic functional connectivity of the periaqueductal gray, a resting fMRI study. Behav Brain Res 211(2):215–219
Kong J, Jensen K, Loiotile R, Cheetham A, Wey HY, Tan T et al (2013a) Functional connectivity of frontoparietal network predicts cognitive modulation of pain. Pain 154(3):459–467
Kong J, Spaeth R, Cook A, Kirsch I, Claggett B, Vangel M et al (2013b) Are all placebo effects equal? Placebo pills, sham acupuncture, cue conditioning and their association. PLoS ONE 8 (7):e67485. doi:10.1371/journal.pone.0067485
Leknes S, Lee M, Berna C, Andersson J, Tracey I (2011) Relief as a reward: hedonic and neural responses to safety from pain. PLoS ONE 6(4):e17870
Lewis CM, Baldassarre A, Committeri G, Romani GL, Corbetta M (2009) Learning sculpts the spontaneous activity of the resting human brain. Proc Natl Acad Sci USA 106(41):17558–17563
Mayberg HS, Silva JA, Brannan SK, Tekell JL, Mahurin RK, McGinnis S et al (2002) The functional neuroanatomy of the placebo effect. Am J Psychiatry 159(5):728–737
Mesulam MM (1998) From sensation to cognition. Brain 121:1013–1052
Miller FG, Kaptchuk TJ (2008) The power of context: reconceptualizing the placebo effect. J R Soc Med 101(5):222–225

Miller FG, Colloca L, Kaptchuk TJ (2009) The placebo effect: illness and interpersonal healing. Perspect Biol Med 52(4):518–539

Pessiglione M, Petrovic P, Daunizeau J, Palminteri S, Dolan RJ, Frith CD (2008) Subliminal instrumental conditioning demonstrated in the human brain. Neuron 59(4):561–567

Petrovic P, Dietrich T, Fransson P, Andersson J, Carlsson K, Ingvar M (2005) Placebo in emotional processing–induced expectations of anxiety relief activate a generalized modulatory network. Neuron 46(6):957–969

Price DD, Craggs J, Verne GN, Perlstein WM, Robinson ME (2007) Placebo analgesia is accompanied by large reductions in pain-related brain activity in irritable bowel syndrome patients. Pain 127(1–2):63–72

Schenk LA, Sprenger C, Geuter S, Buchel C (2014) Expectation requires treatment to boost pain relief: an fMRI study. Pain 155(1):150–157. doi:10.1016/j.pain.2013.09.024

Schweinhardt P, Seminowicz DA, Jaeger E, Duncan GH, Bushnell MC (2009) The anatomy of the mesolimbic reward system: a link between personality and the placebo analgesic response. J Neurosci 29(15):4882–4887. doi:10.1523/jneurosci.5634-08.2009

Scott DJ, Stohler CS, Egnatuk CM, Wang H, Koeppe RA, Zubieta JK (2007) Individual differences in reward responding explain placebo-induced expectations and effects. Neuron 55(2):325–336

Scott DJ, Stohler CS, Egnatuk CM, Wang H, Koeppe RA, Zubieta JK (2008) Placebo and nocebo effects are defined by opposite opioid and dopaminergic responses. Arch Gen Psychiatry 65 (2):220–231

Tedeschi JT, Schlenker BR, Bonoma TV (1971) Cognitive dissonance: private ratiocination or public spectacle? Am Psychol 26:685–695

Tracey I (2010) Getting the pain you expect: mechanisms of placebo, nocebo and reappraisal effects in humans. Nat Med 16(11):1277–1283

Volkow ND, Wang GJ, Ma Y, Fowler JS, Zhu W, Maynard L et al (2003) Expectation enhances the regional brain metabolic and the reinforcing effects of stimulants in cocaine abusers. J Neurosci 23(36):11461–11468

Wager TD, Rilling JK, Smith EE, Sokolik A, Casey KL, Davidson RJ et al (2004) Placebo-induced changes in FMRI in the anticipation and experience of pain. Science 303:1162–1167

Wager TD, Scott DJ, Zubieta JK (2007) Placebo effects on human mu-opioid activity during pain. Proc Natl Acad Sci USA 104(26):11056–11061

Wager TD, Atlas LY, Leotti LA, Rilling JK (2011) Predicting individual differences in placebo analgesia: contributions of brain activity during anticipation and pain experience. J Neurosci 31(2):439–452

Yu R, Gollub RL, Vangel M, Kaptchuk T, Smoller JW, Kong J (2014) Placebo analgesia and reward processing: integrating genetics, personality, and intrinsic brain activity. Hum Brain Mapp. doi:10.1002/hbm.22496

Zubieta JK, Bueller JA, Jackson LR, Scott DJ, Xu Y, Koeppe RA et al (2005) Placebo effects mediated by endogenous opioid activity on mu-opioid receptors. J Neurosci 25(34):7754–7762

Miller FG, Colloca L, Kaptchuk TJ (2009) The placebo effect: illness and interpersonal healing. Perspect Biol Med 52(4):518–539

[illegible]

Placebo, Nocebo, and Learning Mechanisms

Luana Colloca

Contents

1 Introduction 18
2 Pharmacological Conditioning 18
 2.1 Pharmacological Conditioning and Placebo Responses in Animals 19
 2.2 Pharmacological Conditioning and Placebo Responses in Humans 21
3 Non-pharmacological Conditioning and Placebo Responses 24
4 Verbal Communication, Reserve Information, and Memories 26
 4.1 Beyond Direct Experience: Learning from Others 28
5 Conclusions 31
References 32

Abstract

Recent substantial laboratory and theoretical research hints for different learning mechanisms regulating the formation of placebo and nocebo responses. Moreover, psychological and biological variants may play a role as modulators of learning mechanisms underlying placebo and nocebo responses. In this chapter, we present pioneering and recent human and nonhuman research that has impressively increased our knowledge of learning mechanisms in the context of placebo and nocebo effects across different physiological processes and pathological conditions.

L. Colloca (✉)
National Institute of Mental Health (NIMH), National Institutes of Health (NIH), Building 10, Room 1C154, Bethesda, MD 20892-1156, USA

National Center for Complementary and Alternative Medicine (NCCAM), National Institutes of Health (NIH), Building 10, Room 1C154, Bethesda, MD 20892-1156, USA

Department of Bioethics, Clinical Center, National Institutes of Health, Building 10, Room 1C154, Bethesda, MD 20892-1156, USA
e-mail: luana.colloca@nih.gov

F. Benedetti et al. (eds.), *Placebo*, Handbook of Experimental Pharmacology 225,
DOI 10.1007/978-3-662-44519-8_2

Keywords
Conditioning • Expectation • Verbal suggestion • Observational learning • Drug-mimic effects • Vicarious hypoalgesia

1 Introduction

Behavioral and neurobiological placebo and nocebo responses are formed by processing verbal instructions, conditioning, and social cues including observations and complex interpersonal interactions (Colloca et al. 2013a, b; Colloca and Miller 2011b). Verbal communication through suggestions of benefits from a certain treatment via persuasive words can induce placebo responses (Amanzio and Benedetti 1999). Conversely, verbal suggestion of harm creates an opposite phenomenon, by invoking a nocebo response (Benedetti et al. 2007a; Colloca and Miller 2011c). The experience of varying degrees of benefit through prior pharmacological and non-pharmacological conditioning creates subsequent behavioral and neurobiological placebo and nocebo responses depending respectively upon the positive or negative effect of the treatment (Colloca and Benedetti 2006; Colloca et al. 2008a). Finally, observing and interacting with other persons play a role in the formation of placebo and nocebo responses (Colloca and Benedetti 2009; Vogtle et al. 2013). Placebo and nocebo responses are elicited without any practice and direct experience, which are essential aspects in optimizing learning capabilities and probably survival mechanisms. It is likely that verbal conditioning and social cues are processed by the brain to generate dynamically updated expectations that, in turn, shape different symptoms and neurobiological responses.

We describe central concepts and learning mechanisms underpinning the formation of placebo and nocebo responses, and suggest promising future laboratory investigations to help expand our knowledge and provide valuable evidence of the effectiveness of placebo and nocebo responses in contexts other than pain.

2 Pharmacological Conditioning

In this section, we present a series of studies that illustrate how different forms of learning impact placebo and nocebo responses in animals and humans.

Classical conditioning has been the prevalent paradigm to explain the genesis of placebo and nocebo responses in terms of learning principles and mechanisms. Therefore, we use the terms and concepts derived from Pavlov's classical experiments, demonstrating that dogs would salivate (conditioned response, CR) in response to a bell (conditioned stimulus, CS) that had previously been paired with the administration of food (unconditioned stimulus, US) (Pavlov 1927). These learned responses indicated that a ringing bell implied food, hence the salivary reaction upon hearing the bell. Similar to the conditioned stimulus of ringing a bell, visual, tactile, and gustatory stimuli associated with the efficacy of a medication can also become conditioned stimuli through their repeated association with the unconditioned stimuli in the form of different active medication. Placebos given along with the presentation of CS and subsequently the US elicit CRs that are similar to the response to medication (Ader 1987).

2.1 Pharmacological Conditioning and Placebo Responses in Animals

A pioneering study by Alvarez-Buylla and Carrasco-Zanini (1960) investigated hypoglycemic conditioning by using insulin for eight consecutive days and then replacing insulin with saline solution in dogs. There were no appreciable differences in the magnitude of the hypoglycemic response to insulin compared to those induced by the conditioning stimulus saline given along with a metronome's sound. Interestingly, the authors tested for the different CS components, extinction, and mechanisms underlying the conditioned hypoglycemia. After having established the CR, the injection alone did not produce any CR, while the auditory stimulus elicited a hypoglycemic effect which was as great as that produced by the combination of injection and sound. When tested for extinction, the conditioned hypoglycemia diminished progressively and was totally extinguished on the fifth day. The CR was also tested in alloxan diabetic dogs and depancreatized dogs, respectively. Both presented a CR suggesting that the conditioned reflex was not related to the disease or the pancreas activity (Alvarez-Buylla and Carrasco-Zanini 1960).

A few years later, Woods and colleagues extended these pioneering observations by varying the number of conditioning trials and the CS nature to define the optimal values for a conditioned hypoglycemic reflex (Woods et al. 1969). Rats were tested with and without a menthol cue. The menthol cue consisted in an odor of menthol through a gauze pack taped to the inside of the chambers where the animals were kept between blood drawings. When the menthol cue was used, the acquisition of the conditioning was more rapid, the CR larger, and the development of a detectable CR faster compared to the conditioning without menthol cue (Woods et al. 1969).

Another study entitled "Placebo effect in the rat" by R.J. Herrnstein demonstrated that a pharmacological conditioning with 14 administrations of scopolamine paired with sweetened milk was able to induce a placebo response following the presentation of the pure sweetened milk alone (Herrnstein 1962). Herrnstein was one of the first scientists who interpreted the effect of the pharmacological conditioning as a placebo response in which the presentation of the conditioned stimulus (e.g., the sweetened milk) caused a scopolamine-like alteration of behavior such as the decrease in rates of a lever-pressing task (Herrnstein 1962).

Other authors have also pursued this line of research providing proof of concepts for the area of placebo research across different domains. Notably, Robert Ader introduced the concept that the immune system can be conditioned with potential clinical benefits (Ader 1987; Ader et al. 1987, 1990, 1995). For example, Ader and Cohen used a schedule of pharmacological conditioning in which a novel saccharine-flavored solution was paired with the immunosuppressant, cyclophosphamide (Ader and Cohen 1982). The authors observed that merely giving a placebo such as saccharine solution following the administration of cyclophosphamide induced immunosuppression in rats. Interestingly, there was a dose–response effect: rats that received two doses of cyclophosphamide during the conditioning phase had greater conditioned immunosuppression responses than those which received one dose of cyclophosphamide, supporting the notion that the stronger the US effect, the more robust the CR. Ader and colleagues have also demonstrated that the antibody production can be conditioned using an antigen as an unconditioned stimulus of the immune system (Ader et al. 1993). Mice received repeated immunizations with keyhole limpet hemocyanin (KLH) paired with a gustatory conditioned stimulus. Subsequently, mice were reexposed to the gustatory stimulation alone and a conditioned enhancement of anti-KLH antibodies was found (Ader et al. 1993).

In a more recent experiment, Pacheco-López et al. conditioned rats with 0.2 % saccharin given just before the administration of the immunosuppressive drug cyclosporin A, which specifically inhibits calcineurin (Pacheco-Lopez et al. 2009). This experiment confirmed that the pharmacological properties of cyclosporin A could be elicited by the neutral stimulus behaviorally. Furthermore, the authors found that these effects were not limited to behaviors but impacted activity at the level of splenocytes. In fact, there was a change in the production of Th1-cytokine when the rats were reexposed to the saccharin alone. Therefore, the calcineurin activity in CD4 (+) T lymphocytes was identified as the intracellular target for inducing placebo immunosuppression after cyclosporin A exposure, suggesting that the use of placebos after a pharmacological conditioning triggers specific neurobiological pathways (Pacheco-Lopez et al. 2009).

More recently, Guo et al. investigated the effect of prior pharmacological opioid and non-opioid exposure in mice using a model of a hot-plate test (Guo et al. 2010). Conditioned cues were paired with either the opioid agonist morphine hydrochloride or non-opioid aspirin. Placebo analgesic responses evoked by morphine pharmacological conditioning were antagonized by naloxone suggesting that the opioidergic system mediates this effect. By contrast, after aspirin conditioning, the placebo responses were not blocked by naloxone indicating that the substance used during the conditioning phase triggers the underlying systems leading to a specific effect (Guo et al. 2010). In another study, the authors investigated the relation between receptors at the level of rACC and placebo analgesia finding that rACC is the key brain region involved in opioid-mediated placebo analgesia with a determinant role of μ-opioid receptors (Zhang et al. 2013). Placebo analgesia has an effect that is transferable to other domains. After being conditioned with 10 mg of morphine in a model of pharmacologically induced placebo analgesia, plasma levels of corticosterone and ACTH were reduced and the effect produced significant changes to stress in a behavioral despair test (Guo et al. 2011).

Caution is urged in generalizing this knowledge. Pre-drug cues can also elicit conditioned compensatory responses (CCRs) that are opposite in direction to the US when tolerance, a decrease response to a drug within the course of administrations, is present. An early study by Subkov and Zilov (1937) showed that dogs treated with epinephrine every few days presented tachycardia that decreased over time developing tolerance. On a final test, epinephrine was replaced by inert Ringer's solution and an opposite bradycardic response was observed. Many other studies have shown that when tolerance occurs, pre-drug cues can elicit paradoxical CCRs on pharmacological tolerance likewise because pharmacological stimulations initiate adaptive responses that compensate for the primary drug effect (Siegel et al. 2000).

2.2 Pharmacological Conditioning and Placebo Responses in Humans

The above-described studies in animals have been partially repeated in human patients with immune disorders. Based on these findings, the pharmacological conditioning of the immune system appears to be an important result because it is suggestive of potential influences of conditioned placebo responses during the course of specific symptoms and the response to a pharmacological immune therapy. Importantly, Ader and colleagues have attempted to provide proof-of-concept evidence that a schedule of pharmacological reinforcement with immunosoppressors associated with placebos actually works in maintaining good clinical outcomes in patients suffering from immune disorders. For example, a child with lupus erythematosus was treated with cyclophosphamide given in association with a taste and smell beverage (Olness and Ader 1992). Remarkably, successful clinical outcomes were obtained by using taste and smell beverages alone on half of the monthly chemotherapeutic sessions. In another study, multiple sclerosis patients received four intravenous treatments with cyclophosphamide in association with anise-flavored syrup. Peripheral leukocyte count was assessed following the syrup alone, and eight out of ten patients displayed decreased peripheral leukocytes, an effect that mimicked that of cyclophosphamide (Giang et al. 1996).

Gobel et al. performed a similar experiment in which healthy subjects received cyclosporin A along with a strawberry-flavored milk drink (Goebel et al. 2002). The effects of conditioned immunosuppression were assessed by measuring interleukin-2 (IL-2) and interferon gamma (IFN-gamma) mRNA expression, in vitro release of IL-2 and IFN-gamma, and lymphocyte proliferation. A placebo given with the flavored drink significantly suppressed the immune functions in terms of interleukin-2 (IL-2) and interferon gamma (IFN-gamma) mRNA expression, in vitro release of IL-2 and IFN-gamma, as well as lymphocyte proliferation, revealing for the first time the mechanisms underlying conditioned immune responses (Goebel et al. 2002).

Conditioned placebo responses have also been demonstrated in conditions other than the immune system in human and animal experimental settings. Benedetti et al. (2007a, b) demonstrated that a pharmacological conditioning with morphine induced robust placebo analgesic responses when morphine is replaced with a placebo (Benedetti et al. 2007b). Morphine was given twice at intervals of 1 week. The placebo without prior morphine conditioning induced a small but significant increase in pain tolerability, which indicates smaller effects when a placebo is given for the first time compared with its administration after pharmacological conditioning (Benedetti et al. 2007b). Amanzio and Benedetti had also shown that the administration of morphine for two consecutive days produced substantial placebo responses when the placebo is given on the third day (Amanzio and Benedetti 1999). Therefore, it is important to note that different schedules of pharmacological conditioning influenced elicited morphine-like effects, and that these effects last at least in a range of days and weeks. Interestingly, these observations suggest that a pharmacological conditioning procedure creates a memory of the learned response that can be re-evoked over time.

The effects of conditioning have been explored using other drugs such as the serotonin agonist of the $5\text{-HT}_{1B/1D}$ receptors, sumatriptan, which stimulates growth hormone (GH) and inhibits cortisol secretion (Benedetti et al. 2003). The administration of a placebo after the repetitive administration of sumatriptan produced similar hormonal responses. Indeed, the placebo-induced GH increases and cortisol decreases (Benedetti et al. 2003).

Some additional human studies have adopted a pharmacological conditioning with drugs such as the dopamine agonist, apomorphine (Benedetti et al. 2004, 2009). A subcutaneous placebo was given after three repetitive subcutaneous administration of the dopaminergic agonist, apomorphine, to explore conditioned placebo responses at the level of single neurons in patients suffering from Parkinson's disease who underwent surgical implantation of electrodes for high-frequency deep brain stimulation. Notably, patients who showed a clear-cut conditioned placebo response, depicted clinically by a significant decrease of arm rigidity and subjective reports of well-being, presented a significant decrease of the neuronal discharge recorded at the level of the subthalamic region (Fig. 1). Nonresponders showed no differences in clinical assessment of rigidity, self-reports, and neuronal discharge characteristics. This study was the first one documenting a pharmacologically induced conditioned effect at the level of specific neuronal populations in Parkinson patients (Benedetti et al. 2004). The CR produced by the administration of the placebo induced effects similar to the neural patterns of activity elicited by apomorphine (Levy et al. 2001; Stefani et al. 2002). It remains obscure why only some patients respond to the pharmacological conditioning procedures.

Overall, these studies suggest that learned placebo responses following the exposure to drugs represent specific effects depending on the kind of drug exposure that is originally performed. These responses can be potentially relevant for clinical practice if we understand the underpinning mechanisms. Conditioned drug effects can be therapeutically exploited in routine clinical practice by integrating placebos in schedules of reinforcement, so that conditioned stimuli acquire properties and characteristics of USs. These effects, if generalizable, may become part of the

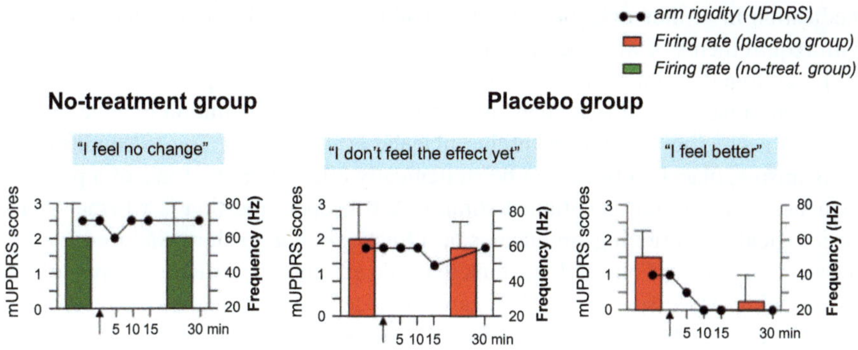

Fig. 1 Placebo responses at the level of single neuronal activity. After a pharmacological conditioning with apomorphine, a placebo was given and variations in frequency of discharge of subthalamic single neuronal activity, report of self-being, and rigidity scores were assessed. Three representative patients with Parkinson Disease are depicted. The *first graph* represents the neurophysiological, clinical, and subjective responses for a patient assigned to the natural history group. No changes were observed for all the measures in the first graph. The *second and third graphs* show the responses measured from a placebo nonresponder and a placebo responder, respectively. No changes were observed in patients who were nonresponders. In contrast, those who responded to the placebo given after the pharmacological conditioning with apomorphine presented a change in the neurophysiological, clinical, and subjective outcomes [Data from Benedetti et al. (2004)]

pharmacotherapeutic protocol preserving therapeutic benefits while costs and side effects are likely reduced (Colloca and Miller 2011a).

In line with these considerations, a recent clinical trial showed that a schedule of conditioning with corticosteroids was effective in reducing the relapse of symptoms in patients with psoriasis (Ader et al. 2010). Patients with mild-to-moderate psoriasis received medication that was followed by unconditioned effects of the drug (100 % reinforcement schedule), or placebo medication that was never reinforced by the active medication. Indeed, the results were clinically comparable to the reduction in symptoms induced by a full dose of corticosteroids (Ader et al. 2010).

Recent research in children with Attention Deficit Hyperactivity Disorder indicates that placebo effects may have potential therapeutic applications (Sandler et al. 2008, 2010; Sandler and Bodfish 2008). Children were randomly assigned to 1 of 3 schedules of 8-week treatments: (1) reduction of amphetamine dose by pairing drug with placebos; (2) reduction of amphetamine without placebo substitution; or (3) full dose of amphetamine treatment. Children in arm 1 received an open placebo pill paired with 50 % reduced dose of amphetamine. The same reduction of treatment was performed in arm 2 but without placebos as cue (control group). Pairing a conditioned stimulus with amphetamines produced conditioned placebo responses that allowed children with ADHD to be treated effectively with a lower dose of psychostimulant medication. The placebo treatment was overtly described to both parents and children transparently (Sandler and Bodfish 2008). Parents and children were informed that placebos consisted of a pill with no medication in it, thus overcoming the ethical problem of deception and consistent with requirements of informed consent.

Taken together, these studies in patients and research with placebos given after active pharmacological treatment suggest that placebo substitution may be understood as a specific way for promoting placebo effects. According to conditioning mechanisms, placebo effects can be strategically elicited on the basis of a planned sequence of drug and conditioned stimuli. A still open question is whether pharmacological conditioning produces side effects similar to those induced by the active treatment. It is plausible to think that side effects can be elicited as part of the conditioning processes. With this regard, Benedetti and colleagues used repeated administrations of analgesic doses of buprenorphine in postoperative patients, a treatment that produces a mild reduction of ventilation, to study the role of pharmacological conditioning on side effects. Placebos given after repetitive administration of buprenorphine produced mild reduction of ventilation mimicking the buprenophine respiratory depressant response (Benedetti et al. 1998). This effect was reversible by the administration of naloxone, indicating the release of endogenous opioids that can account for the reduction in ventilation (Benedetti et al. 1999). Thus, conditioned placebo effects may expand to adverse events and this possibility deserves further investigation.

3 Non-pharmacological Conditioning and Placebo Responses

Potentially, any CS-USs can induce strong placebo responses and the driving force for these effects is represented by the experience of efficacy and mastery induced by the USs during the conditioning phases. Based on this concept, simulation of efficacious treatments, such as surreptitiously reducing the intensity of painful stimulations delivered after a placebo cream, has been extensively used to produce models of studying placebo responses in various laboratory environments (Reiss 1980).

In a pioneering study, Price et al. used painful stimuli and a placebo cream to study placebo analgesia in healthy subjects (Price et al. 1999). The testing subjects were randomized to three experimental conditions receiving either a strong placebo (A), a weak placebo (B), or a control agent (C). The authors manipulated the intensity of the painful stimulation by decreasing it to 67 % in condition A and 17 % in condition B. No reduction was performed under condition C serving as control. Therefore, the placebo analgesic responses were contrasted with the experience of relief given during the conditioning phase. Those who received the strong placebo experienced the largest placebo analgesic response when a control level of pain was delivered. Conversely a lower placebo analgesic response was observed in condition B in which subjects were conditioned with small pain reduction (Price et al. 1999). The findings indicate that previous exposure to distinct intensities of the US determined the magnitude of the placebo effect.

Fig. 2 Relationship between the number of conditioning trials and placebo and nocebo responses. *On the left*, runs of short and long conditioning schedules (acquisition phase) with high, medium, and low of pain are shown. Each *square* corresponds respectively to 10 (short conditioning) and 40 (long conditioning) trials of conditioning. *On the right*, placebo and nocebo responses are shown (testing phase). A short conditioning elicited modest placebo (*green*) and nocebo (*red*) responses as compared to a control condition (*yellow*). These effects showed extinction over time. By contrast a longer schedule of conditioning induced both robust placebo and nocebo responses that lasted over the entire experimental session [Data from Colloca et al. (2010)]

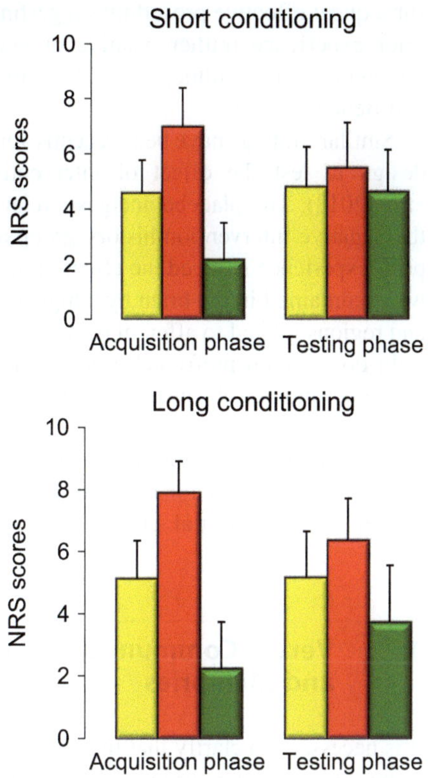

Notably, a recent study showed that the number of CS-US pairings impacts placebo responses. Colloca et al. used different schedules of full conditioning in which 10 vs. 40 CS-US pairings were delivered during the conditioning phase. Interestingly, there was a net relation between the magnitude of placebo and nocebo responses and the number of trials used for the conditioning (Colloca et al. 2010). The increase in number of associations during the conditioning resulted in robust placebo and nocebo responses that persisted over the entire experimental session as depicted in Fig. 2 (Colloca et al. 2010).

Research has also shown that prior experiences via conditioning impact placebo responsiveness (Colloca and Benedetti 2006; Kessner et al. 2013). For example, a positive, full-conditioning procedure induces robust analgesic responses of a subsequent placebo, but the identical procedure performed after an ineffective experience does not significantly impact the formation of placebo effects (Colloca and Benedetti 2006). The simulated effective intervention induced by reducing the intensity of painful stimulations induced robust analgesic responses in Group 1. A second group of subjects in the same study underwent a simulation of ineffective intervention with no reduction of intensity of painful stimulation, and after a time lag of 4–7 days, received the same effective manipulation as Group 1. As a result, the prior experience of ineffectiveness negatively impacted the effects of the subsequent effective procedure suggesting that placebo analgesia is finely tuned by prior experience (either positive or negative), and that the effect of an initial intervention may influence the formation of future placebo responses (Colloca and Benedetti 2006).

Similar findings have been recently reported by Kessner et al. who used the same design to test the effect of intervention history in an fMRI study (Kessner et al. 2013). The placebo analgesia related to the tested intervention was lower in the negative intervention history group as compared to the positive. The negative prior experience reduced the effect of the following positive one and this reduction was maintained in the brain by a higher activation of the bilateral posterior insulae and regions related to afferent nociceptive processing, and a lower activation of the right dorsolateral prefrontal cortex that is also involved in nociceptive inhibition processes and placebo analgesia. The above and many other similar studies indicate that conditioning via pharmacological or biologically significant prior exposures is a key modulating factor of the placebo effect owing to the fact that learning mechanisms account for a wealth of behavioral and clinical placebo and nocebo responses (Kessner et al. 2013).

4 Verbal Communication, Reserve Information, and Memories

It is necessary to clarify that the ability of one stimulus (CS) to evoke the original response by prior pairing with the US may only partially explain conditioned response in humans. Humans learn to anticipate relationships among events so that they can represent their own environment via verbal suggestions and observation. Therefore, while pairing and contiguity are determinant components, learning depends strongly on both the information that the CS provides about the US and the acquired awareness of a relation among events (Colloca and Miller 2011b; Kirsch 1985; Rescorla 1988a, b). This concept is well illustrated by studies focusing on the interactions of verbal suggestions and conditioned placebo effects.

In an earlier study, Voudouris and colleagues tested the effects of verbal suggestions and conditioning procedures (Voudouris et al. 1990). Healthy subjects underwent an iontophoretic pain stimulation attending four sessions during four consecutive days. During the first session, half the subjects were told that a topical cream was a powerful painkiller and would provide pain relief and the other half was told that the cream was a placebo. During the second session, half of the subjects received a cream (placebo) and the other half were given none. In the third session, half the subjects were conditioned by surreptitiously reducing the pain intensity after the application of placebo cream. The other half received the same pain stimulus. Thus, Group 1 received a combination of verbal suggestions and conditioning manipulation; Group 2 received verbal suggestions alone; Group 3 received conditioning alone; and Group 4 represented the control group. There was an enhancement of placebo responses in both Groups 1 and 3, but conditioning was effective in eliciting placebo analgesia with and without verbal suggestions (Voudouris et al. 1990).

When studied at the level of both N1 and biphasic N2-P2 components of scalp laser-evoked potentials (LEPs), verbal suggestions and conditioning clearly show that conditioning modulates placebo analgesia (Colloca et al. 2008b; Wager et al. 2006). N1 is generated in the second somatosensory area, while N2-P2 is a biphasic negative–positive complex obtained at the vertex which originates in the bilateral operculo-insular areas and in the cingulate gyrus. It was observed that verbal suggestions induced modest LEP changes occurring without subjective perception of pain reduction, whilst N2-P2 amplitude reductions induced by the conditioning, were robust and occurred along with a subjective self-report of pain relief (Colloca et al. 2008b).

Recently, Fiorio and colleagues showed that while a conditioning manipulation influences tactile perception and the late components (N140 and P200) of the somatosensory evoked potentials (SEPs) (Fiorio et al. 2012), verbal suggestions alone did not change SEPs (Fiorio et al. 2014).

While it is clear that conditioning is the most effective procedure to elicit a placebo response, it is interesting to note that reverse verbal suggestions communicating conflicting and opposite information about the US can influence clinical outcomes and behaviors (Chung et al. 2007; Flaten et al. 1999; Luparello et al. 1970).

Luparello and coworkers reported significant increases in airway resistance in nearly half the asthmatic patients under investigation when they inhaled a nebulized saline solution along with the information that it was an allergen with irritant properties. Interestingly, these patients reversed their airway obstruction by inhaling the same substance presented as a medicine with beneficial effects on asthma. Similarly, the effects of the bronchoconstrictor carbachol were higher when it was administered along with the information that it was a bronchoconstrictor than when subjects were told it was a bronchodilator (Luparello et al. 1970).

Different outcomes were found in healthy participants who were given decaffeinated coffee under two different verbal suggestions: participants in Group 1 were told that they would receive either regular or decaffeinated coffee according to a double-blind design, while participants in Group 2 received decaffeinated coffee presented as real coffee. Placebo responses were higher in Group 2 rather than Group 1, suggesting that verbal suggestions may shape perception and sensation (Kirsch and Weixel 1988). Moreover, Flaten et al. showed that carisoprodol, a centrally acting muscle relaxant, resulted in opposite outcomes, either relaxant or stimulant, depending on the interaction of verbal suggestions and given drug, suggesting that instructional learning can strongly shape experiences based on a priori expectations (Flaten et al. 1999).

Communication can influence experience with negative outcomes (Colloca and Finniss 2012). Healthy participants were alerted to the hyperalgesic effect of a treatment perceived pain despite the intensity of stimulation was ranging from no-painful to low painful levels (Colloca et al. 2008a). Negative suggestions produced allodynic effects, whereby non-painful tactile stimuli become painful.

In addition, low-intensity painful stimuli were perceived as high-intensity stimuli after negative verbal suggestion, with or without preconditioning, indicating that nocebos can also induce hyperalgesic effects, whereby low-intensity painful stimuli are perceived as high-intensity stimuli (Colloca et al. 2008a). Rodriguez-Raecke et al. showed that contextual information given once at the beginning of the investigation indicating that repeated painful stimulations over several days would increase pain sensation from day to day, impacted pain perception over 8- and 90-day periods with brain changes at the level of the insula (Rodriguez-Raecke et al. 2010).

4.1 Beyond Direct Experience: Learning from Others

Beyond firsthand experience, humans and animals learn by observing others in the absence of any direct reinforcement. Colloca and Benedetti first demonstrated that placebo analgesic effects could be elicited by observing the experience of another person (a demonstrator) who was carefully trained to simulate the analgesic experience (Colloca and Benedetti 2009). In the experiment, two silver chloride electrodes were applied to the back of the nondominant hand and a sham electrode was pasted above the subject's middle finger while a set of painful and non-painful stimuli were delivered. The demonstrator rated audibly the painful stimuli that were paired to a red light and the non-painful stimuli paired to a green light and the simulation of efficacious treatment. The experimental subjects paid attention to the entire session and at the end of this observational phase were asked to undergo a similar experimental session. However, the stimulus intensities were set at their painful level for both the green and the red stimuli. Interestingly, all the green painful stimuli were deemed less painful compared to the red-associated stimuli, indicating that observing a beneficial treatment in another person elicited placebo analgesia. The observed effects were stable over the entire experimental session (a total of 18 stimuli), showing no extinction and indicating implicit acquisition and retention of behavioral output. The effect size of observationally induced placebo analgesic responses was comparable to those induced by direct prior experience of analgesia via a conditioning schedule. The information drawn from observational learning may have established a self-projection into the future outcome boosting expectation of analgesia. The higher observationally induced placebo responses were reported by those subjects who had higher empathy scores suggesting that empathy might predict placebo analgesia elicited by observational learning (Colloca and Benedetti 2009).

We have further studied observationally induced placebo analgesia by looking at different components such as the live interaction with a demonstrator as compared to merely observing a video (Hunter et al. 2014). Testing subjects were randomized to watch either the video of the demonstrator or the same live demonstrator showing an analgesic benefit following the presentation of the green light. The subjects then received the same set of painful stimuli after the brief presentation of either a red or green light. The live face-to-face observation vs. a video replay induced similar

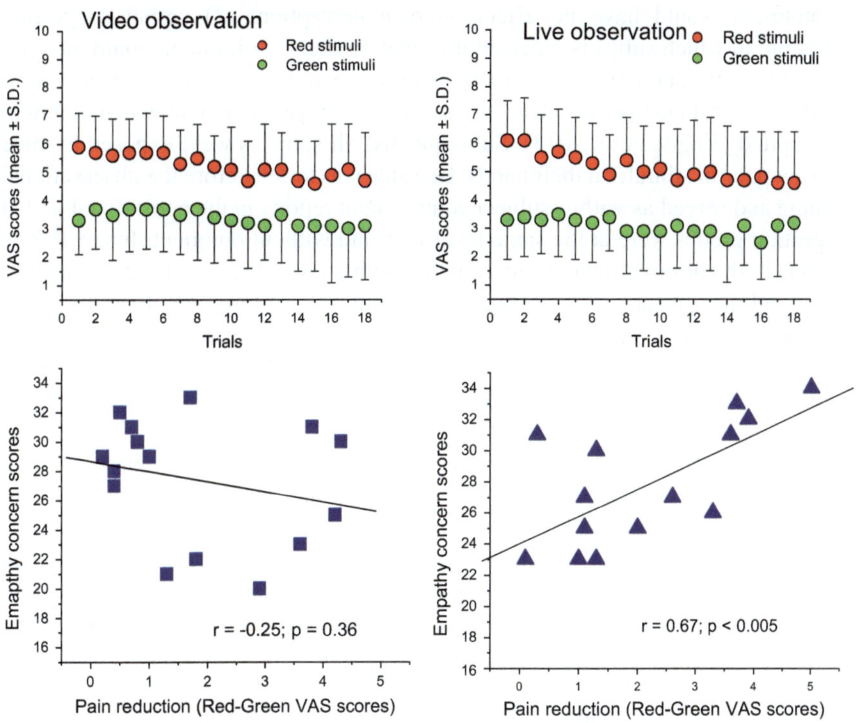

Fig. 3 Observationally induced placebo analgesia and empathy. *At the top*, placebo analgesic scores induced by video and live face-to-face observation are depicted. Placebo analgesia was similarly induced by observing a video or a live demonstrator. *At the bottom*, correlations with empathetic scores are shown. A positive correlation with empathy was found for the live face-to-face observation only [Data from Hunter et al. (2014)]

placebo analgesic effects in terms of magnitude emphasizing that observation conveys potential cues to induce expectations of benefit and activate specific mechanisms independently of the social interactions. However, empathy strongly correlated with placebo analgesic responses in the live observation group only, but not in the video replay group (Fig. 3) (Hunter et al. 2014). These findings suggest that observation induces placebo analgesia and that empathy may facilitate these effects when live interactions are involved but without being a driving factor. Two recent studies confirming and extending the findings on vicarious learning have adopted during the observational phase, either a video reply (Vogtle et al. 2013) or live demonstrators (Swider and Babel 2013). Observationally induced changes in pain were correlated with the empathy scores only when live demonstrators were involved in the experimental settings, confirming that empathy predicts these effects when interpersonal interactions are involved. It is worth noting that the effect of observation and modeling applies to nocebo effect as well.

Vögtle et al. have studied young women, randomly assigning them to one of three conditions: (1) control condition in which subjects received information that an ointment would have no effect on pain perception; (2) verbal suggestion condition, in which subjects received information that the ointment would increase pain sensitivity; and (3) observational learning condition, in which subjects were asked to watch a video in which a demonstrator displayed more pain when ointment was applied (Vogtle et al. 2013). Subsequently, all subjects were exposed to three pressure painful stimuli on their hands. One side was tested before the observational learning and served as within-subject control. Pain reports in the control and verbal suggestion groups were at the same level with and without ointment. Interestingly, subjects in the observational group reported higher pain after watching the demonstrator and these responses were higher than in the control group with and without ointment (Vogtle et al. 2013). The nocebo responses induced by observational learning correlated with pain catastrophizing scores, indicating the importance of studying the mechanisms underlying observational learning, psychological traits, and nocebo hyperalgesia (Vogtle et al. 2013).

Gender effects influence the magnitude of nocebo induced by observational learning (Swider and Babel 2013). Subjects (men and women) were assigned to observational experimental groups in which either a male or a woman was respectively observed. Subjects rated red-associated stimuli as more painful than the ratings of subjects from control groups who did not observe a demonstrator before receiving the same pain stimuli. Also, regardless of the sex of the subject, nocebo hyperalgesia was greater after a male demonstrator was observed (Swider and Babel 2013).

It has been also recently reported that observation may trigger nocebo mass psychogenic illness (Mazzoni et al. 2010). Healthy subjects were invited to self-administer an intranasal product containing a suspected environmental toxin, which can cause headache, nausea, itchy skin, and drowsiness. Half of the subjects observed an actor who inhaled the product. Those who had observed the actor displaying signs of illness reported a significant increase of the four described symptoms, suggesting that observational learning is likely involved in mass psychogenic illnesses (Mazzoni et al. 2010). Interestingly, empathic stress responses modulated the HPA-axis activity and such a modulation is shaped by the familiarity between observer and target (partners vs. strangers), and the modality of observation (real-life vs. virtual). The exposure to a psychosocial stressor induced in the observer (26 %) physiologically significant cortisol increases. This effect was larger in intimate observer-target dyads (40 %) and during the real-life representation of the stressor (30 %) (Engert et al. 2014).

One may argue that these self-reported scores represent biases generated by the subjects' wishes to please the researcher or fit in with the perceived experimental proposition (Hrobjartsson et al. 2011). However, the experimental settings include control groups (e.g., verbal suggestion and natural history groups) that have received the same instruction about what to expect, and there was no analgesic or hyperalgesic response, indicating that biases are unlikely to account for the difference in the placebo and nocebo effects found in observational learning models. Observation of the demonstrator's benefit may have acted as a US, indicating possible commonalities between observational learning and classical conditioning.

Attempts to analyze observational learning within an associative learning framework have been made for aversive and fear models. In rats, observational aversive learning fails to show blocking, latent inhibition, and overshadowing that are three characteristics of classical conditioning (Galef and Durlach 1993). By contrast, studies in humans have found that observational aversive learning is characterized by features of classing conditioning including overshadowing and blocking (Lanzetta and Orr 1980). We can speculate that humans alter and adapt their behaviors, due to their ability to use symbols, thus setting them apart from the limited stimulus–response world of animals. Further behavioral and brain imaging studies are needed to illustrate the mechanisms involved in the observationally induced placebo and nocebo phenomena.

5 Conclusions

Aspects of conditioning, instructional, and observational learning are likely to combine promoting expectations of benefits and anticipations of negative outcomes (e.g., increase of pain). Expectations are central to the formation of placebo and nocebo responses, are influenced by emotions, and are dynamically shaped by the prior experiences and likelihood of positive or negative outcomes (Colloca and Miller 2011b; Kirsch 1985).

Expectations can be induced explicitly by suggestions of positive or negative outcomes and implicitly by individual previous experience. It is imperative to keep away from any strict dichotomy between conditioning and expectation mechanisms, as the former involves information processing by which a subject expects a future event, which may or may not be conscious. Conversely, expectations formed on the basis of instructions are often associated with unconscious prior experience and thus involving different grades of awareness.

When a perception, such as pain relief, is consciously accessible, verbal instructions become a crucial modulator of placebo effects. By contrast, conditioned placebo responses are shaped by unconscious conditioning but are not affected by verbal instructions and such an event cannot be experienced and perceived by human cognition (e.g., changes in cortisol levels).

If learning mechanisms are understood as processes generating expectations and conditioned responses in humans and animals without being mediated by consciousness, it follows that expectations are not necessarily conscious (Colloca and Miller 2011b). However, it is reasonable to assume that by and large, the closer the phylogenetic distance to human, the larger the role of cognition and emotions. Conscious and unconscious expectations in forming placebo responses are partially an open question and deserve further investigation.

In conclusion, this chapter has explored a wealth of research serving to elucidate the mechanisms responsible for activating learning mechanisms and placebo and nocebo responses. In particular, learning mechanisms have been demonstrated to be a key mediator of expectations and placebo and nocebo responses. We formally systemized here a large body of evidence, integrating behavioral and neurobiological literature and reframing the placebo effect as a complex emotional and learning phenomenon. This approach has the potential to guide future research opening new avenue in placebo and nocebo investigation. Viewing the placebo effect via a learning perspective will endorse a better knowledge of the phenomenon also in health care. In fact, the ramifications of such approach are of paramount importance to the study of symptom management, given the potential capacity of the placebo and nocebo responses in affecting clinical outcomes across different pathological conditions.

Acknowledgments This research is supported by the Intramural Research Program of the Clinical Center, NIH, the National Center for Complementary and Alternative Medicine (NCCAM), and the National Institute of Mental Health (NIMH).

Conflict of Interest There are no conflicts of interest.

References

Ader R (1987) Conditioned immune responses: adrenocortical influences. Prog Brain Res 72:79–90

Ader R, Cohen N (1982) Behaviorally conditioned immunosuppression and murine systemic lupus erythematosus. Science 215:1534–1536

Ader R, Grota LJ, Cohen N (1987) Conditioning phenomena and immune function. Ann N Y Acad Sci 496:532–544

Ader R, Felten D, Cohen N (1990) Interactions between the brain and the immune system. Annu Rev Pharmacol Toxicol 30:561–602

Ader R, Kelly K, Moynihan JA, Grota LJ, Cohen N (1993) Conditioned enhancement of antibody production using antigen as the unconditioned stimulus. Brain Behav Immun 7:334–343

Ader R, Cohen N, Felten D (1995) Psychoneuroimmunology: interactions between the nervous system and the immune system. Lancet 345:99–103

Ader R, Mercurio MG, Walton J, James D, Davis M, Ojha V, Kimball AB, Fiorentino D (2010) Conditioned pharmacotherapeutic effects: a preliminary study. Psychosom Med 72:192–197

Alvarez-Buylla R, Carrasco-Zanini J (1960) A conditioned reflex which reproduces the hypoglycemic effect of insulin. Acta Physiol Lat Am 10:153–158

Amanzio M, Benedetti F (1999) Neuropharmacological dissection of placebo analgesia: expectation-activated opioid systems versus conditioning-activated specific subsystems. J Neurosci 19:484–494

Benedetti F, Amanzio M, Baldi S, Casadio C, Cavallo A, Mancuso M, Ruffini E, Oliaro A, Maggi G (1998) The specific effects of prior opioid exposure on placebo analgesia and placebo respiratory depression. Pain 75:313–319

Benedetti F, Amanzio M, Baldi S, Casadio C, Maggi G (1999) Inducing placebo respiratory depressant responses in humans via opioid receptors. Eur J Neurosci 11:625–631

Benedetti F, Pollo A, Lopiano L, Lanotte M, Vighetti S, Rainero I (2003) Conscious expectation and unconscious conditioning in analgesic, motor, and hormonal placebo/nocebo responses. J Neurosci 23:4315–4323

Benedetti F, Colloca L, Torre E, Lanotte M, Melcarne A, Pesare M, Bergamasco B, Lopiano L (2004) Placebo-responsive Parkinson patients show decreased activity in single neurons of subthalamic nucleus. Nat Neurosci 7:587–588

Benedetti F, Lanotte M, Lopiano L, Colloca L (2007a) When words are painful: unraveling the mechanisms of the nocebo effect. Neuroscience 147:260–271

Benedetti F, Pollo A, Colloca L (2007b) Opioid-mediated placebo responses boost pain endurance and physical performance: is it doping in sport competitions? J Neurosci 27:11934–11939
Benedetti F, Lanotte M, Colloca L, Ducati A, Zibetti M, Lopiano L (2009) Electrophysiological properties of thalamic, subthalamic and nigral neurons during the anti-parkinsonian placebo response. J Physiol 587:3869–3883
Chung SK, Price DD, Verne GN, Robinson ME (2007) Revelation of a personal placebo response: its effects on mood, attitudes and future placebo responding. Pain 132:281–288
Colloca L, Benedetti F (2006) How prior experience shapes placebo analgesia. Pain 124:126–133
Colloca L, Benedetti F (2009) Placebo analgesia induced by social observational learning. Pain 144:28–34
Colloca L, Finniss D (2012) Nocebo effects, patient-clinician communication, and therapeutic outcomes. JAMA 307:567–568
Colloca L, Miller FG (2011a) Harnessing the placebo effect: the need for translational research. Philos Trans R Soc Lond B Biol Sci 366:1922–1930
Colloca L, Miller FG (2011b) How placebo responses are formed: a learning perspective. Philos Trans R Soc Lond B Biol Sci 366:1859–1869
Colloca L, Miller FG (2011c) The nocebo effect and its relevance for clinical practice. Psychosom Med 73:598–603
Colloca L, Sigaudo M, Benedetti F (2008a) The role of learning in nocebo and placebo effects. Pain 136:211–218
Colloca L, Tinazzi M, Recchia S, le Pera D, Fiaschi A, Benedetti F, Valeriani M (2008b) Learning potentiates neurophysiological and behavioral placebo analgesic responses. Pain 139:306–314
Colloca L, Petrovic P, Wager TD, Ingvar M, Benedetti F (2010) How the number of learning trials affects placebo and nocebo responses. Pain 151:430–439
Colloca L, Flaten MA, Meissner K (2013a) Placebo and pain: from bench to bedside. Elsevier, Oxford
Colloca L, Klinger R, Flor H, Bingel U (2013b) Placebo analgesia: psychological and neurobiological mechanisms. Pain 154:511–514
Engert VP, Franziska P, Miller R, Kirschbaum C, Singer T (2014) Cortisol increase in empathic stress is modulated by social closeness and observation modality. Psychoneuroendocrinology 45:192–201
Fiorio M, Recchia S, Corra F, Simonetto S, Garcia-Larrea L, Tinazzi M (2012) Enhancing non-noxious perception: behavioural and neurophysiological correlates of a placebo-like manipulation. Neuroscience 217:96–104
Fiorio M, Recchia S, Corra F, Tinazzi M (2014) Behavioural and neurophysiological investigation of the influence of verbal suggestion on tactile perception. Neuroscience 258:332–339
Flaten MA, Simonsen T, Olsen H (1999) Drug-related information generates placebo and nocebo responses that modify the drug response. Psychosom Med 61:250–255
Galef BG Jr, Durlach PJ (1993) Absence of blocking, overshadowing, and latent inhibition in social enhancement of food preferences. Anim Learn Behav 21:214–220
Giang DW, Goodman AD, Schiffer RB, Mattson DH, Petrie M, Cohen N, Ader R (1996) Conditioning of cyclophosphamide-induced leukopenia in humans. J Neuropsychiatry Clin Neurosci 8:194–201
Goebel MU, Trebst AE, Steiner J, Xie YF, Exton MS, Frede S, Canbay AE, Michel MC, Heemann U, Schedlowski M (2002) Behavioral conditioning of immunosuppression is possible in humans. FASEB J 16:1869–1873
Guo JY, Wang JY, Luo F (2010) Dissection of placebo analgesia in mice: the conditions for activation of opioid and non-opioid systems. J Psychopharmacol 24:1561–1567
Guo JY, Yuan XY, Sui F, Zhang WC, Wang JY, Luo F, Luo J (2011) Placebo analgesia affects the behavioral despair tests and hormonal secretions in mice. Psychopharmacology (Berl) 217:83–90
Herrnstein RJ (1962) Placebo effect in the rat. Science 138:677–678

Hrobjartsson A, Kaptchuk TJ, Miller FG (2011) Placebo effect studies are susceptible to response bias and to other types of biases. J Clin Epidemiol 64:1223–1229
Hunter T, Siess F, Colloca L (2014) Socially-induced placebo analgesia: a comparison of a pre-recorded versus live face-to-face observation. Eur J Pain 18(7):914–922
Kessner S, Wiech K, Forkmann K, Ploner M, Bingel U (2013) The effect of treatment history on therapeutic outcome: an experimental approach. JAMA Int Med 1–2
Kirsch I (1985) Response expectancy as a determinant of experience and behavior. Am Psychol 40:1189–1202
Kirsch I, Weixel LJ (1988) Double-blind versus deceptive administration of a placebo. Behav Neurosci 102:319–323
Lanzetta JT, Orr SP (1980) Influence of facial expressions on the classical conditioning of fear. J Pers Soc Psychol 39:1081–1087
Levy R, Dostrovsky JO, Lang AE, Sime E, Hutchison WD, Lozano AM (2001) Effects of apomorphine on subthalamic nucleus and globus pallidus internus neurons in patients with Parkinson's disease. J Neurophysiol 86:249–260
Luparello TJ, Leist N, Lourie CH, Sweet P (1970) The interaction of psychologic stimuli and pharmacologic agents on airway reactivity in asthmatic subjects. Psychosom Med 32:509–513
Mazzoni G, Foan L, Hyland ME, Kirsch I (2010) The effects of observation and gender on psychogenic symptoms. Health Psychol 29:181–185
Olness K, Ader R (1992) Conditioning as an adjunct in the pharmacotherapy of lupus erythematosus. J Dev Behav Pediatr 13:124–125
Pacheco-Lopez G, Riether C, Doenlen R, Engler H, Niemi MB, Engler A, Kavelaars A, Heijnen CJ, Schedlowski M (2009) Calcineurin inhibition in splenocytes induced by pavlovian conditioning. FASEB J 23:1161–1167
Pavlov IP (1927) Conditioned reflexes: an investigation of the physiological activity of the cerebral cortex. Oxford University Press, London
Price DD, Milling LS, Kirsch I, Duff A, Montgomery GH, Nicholls SS (1999) An analysis of factors that contribute to the magnitude of placebo analgesia in an experimental paradigm. Pain 83:147–156
Reiss R (1980) Pavlovian conditioning and human fear: an expectancy model. Behav Ther 11:380–396
Rescorla RA (1988a) Behavioral studies of Pavlovian conditioning. Annu Rev Neurosci 11:329–352
Rescorla RA (1988b) Pavlovian conditioning. It's not what you think it is. Am Psychol 43:151–160
Rodriguez-Raecke R, Doganci B, Breimhorst M, Stankewitz A, Buchel C, Birklein F, May A (2010) Insular cortex activity is associated with effects of negative expectation on nociceptive long-term habituation. J Neurosci 30:11363–11368
Sandler AD, Bodfish JW (2008) Open-label use of placebos in the treatment of ADHD: a pilot study. Child Care Health Dev 34:104–110
Sandler A, Glesne C, Geller G (2008) Children's and parents' perspectives on open-label use of placebos in the treatment of ADHD. Child Care Health Dev 34:111–120
Sandler AD, Glesne CE, Bodfish JW (2010) Conditioned placebo dose reduction: a new treatment in attention-deficit hyperactivity disorder? J Dev Behav Pediatr 31:369–375
Siegel S, Baptista MA, Kim JA, Mcdonald RV, Eise-Kelly L (2000) Pavlovian psychopharmacology: the associative basis of tolerance. Exp Clin Psychopharmacol 8:276–293
Stefani A, Bassi A, Mazzone P, Pierantozzi M, Gattoni G, Altibrandi MG, Giacomini P, Peppe A, Bernardi G, Stanzione P (2002) Subdyskinetic apomorphine responses in globus pallidus and subthalamus of parkinsonian patients: lack of clear evidence for the 'indirect pathway'. Clin Neurophysiol 113:91–100
Subkov AAZ, Zilov GN (1937) The role of conditioned reflex adaption in the origin of hyperergic reactions. Bull Biol Med Exp 4:294–296

Swider K, Babel P (2013) The effect of the sex of a model on nocebo hyperalgesia induced by social observational learning. Pain 154:1312–1317

Vogtle E, Barke A, Kroner-Herwig B (2013) Nocebo hyperalgesia induced by social observational learning. Pain 154:1427–1433

Voudouris NJ, Peck CL, Coleman G (1990) The role of conditioning and verbal expectancy in the placebo response. Pain 43:121–128

Wager TD, Matre D, Casey KL (2006) Placebo effects in laser-evoked pain potentials. Brain Behav Immun 20:219–230

Woods SC, Makous W, Hutton RA (1969) Temporal parameters of conditioned hypoglycemia. J Comp Physiol Psychol 69:301–307

Zhang RR, Zhang WC, Wang JY, Guo JY (2013) The opioid placebo analgesia is mediated exclusively through mu-opioid receptor in rat. Int J Neuropsychopharmacol 16:849–856

A Meta-analysis of Brain Mechanisms of Placebo Analgesia: Consistent Findings and Unanswered Questions

Lauren Y. Atlas and Tor D. Wager

Contents

1 Biological Mechanisms of Placebo Analgesia 39
2 Advantages of Meta-analyses of Expectancy-Based Pain Modulation 40
3 Methods 42
3.1 Study Selection and Coordinate Identification 42
3.2 Analysis 48
4 Results 49
4.1 Meta-analyses 1 and 2: Expectancy-Based Pain Modulation 49
4.1.1 Expectancy-Induced Reductions During Noxious Stimulation 49
4.1.2 Expectancy-Induced Increases During Noxious Stimulation 49
4.2 Meta-analyses 3 to 6: Stimulus Expectancies Versus Treatment Expectancies 49
4.2.1 Placebo-Induced Reductions During Noxious Stimulation 49
4.2.2 Stimulus Expectancy-Induced Reductions During Noxious Stimulation 50
4.2.3 Placebo-Induced Increases During Noxious Stimulation 51
4.2.4 Stimulus Expectancy-Induced Increases During Noxious Stimulation 52
4.2.5 Treatment Expectancy Effects Versus Stimulus Expectancy Effects 52
5 Discussion 57
5.1 Placebo Effects on Brain Regions Traditionally Associated with Pain Processing . 58
5.2 Reductions During Pain in the Amygdala 60
5.3 Modulatory Mechanisms 60
5.4 Treatment Expectancies Versus Stimulus Expectancies 61
5.5 Relationship to Prior Meta-analyses of Placebo Analgesia 61
5.6 Unanswered Questions: Extending the Meta-analysis 62
References 64

L.Y. Atlas (✉)
National Center for Complementary and Alternative Medicine, National Institutes of Health, 10 Center Drive, Rm 4-1741, Bethesda, MD 20892
e-mail: lauren.atlas@nih.gov

T.D. Wager
Departments of Psychology and Neuroscience, University of Colorado, Boulder, CO, USA

F. Benedetti et al. (eds.), *Placebo*, Handbook of Experimental Pharmacology 225,
DOI 10.1007/978-3-662-44519-8_3

Abstract

Placebo treatments reliably reduce pain in the clinic and in the lab. Because pain is a subjective experience, it has been difficult to determine whether placebo analgesia is clinically relevant. Neuroimaging studies of placebo analgesia provide objective evidence of placebo-induced changes in brain processing and allow researchers to isolate the mechanisms underlying placebo-based pain reduction. We conducted formal meta-analyses of 25 neuroimaging studies of placebo analgesia and expectancy-based pain modulation. Results revealed that placebo effects and expectations for reduced pain elicit reliable reductions in activation during noxious stimulation in regions often associated with pain processing, including the dorsal anterior cingulate, thalamus, and insula. In addition, we observed consistent reductions during painful stimulation in the amygdala and striatum, regions implicated widely in studies of affect and valuation. This suggests that placebo effects are strongest on brain regions traditionally associated with not only pain, but also emotion and value more generally. Other brain regions showed reliable increases in activation with expectations for reduced pain. These included the prefrontal cortex (including dorsolateral, ventromedial, and orbitofrontal cortices), the midbrain surrounding the periaqueductal gray, and the rostral anterior cingulate. We discuss implications of these findings as well as how future studies can expand our understanding of the precise functional contributions of the brain systems identified here.

Keywords

Placebo effect • Placebo response • Expectancy • Pain • fMRI • PET • Opioid • Prefrontal cortex • Periaqueductal gray • Amygdala • Meta-analysis • MKDA • Neuroimaging

For decades, the public and scientific community have been well aware of the "powerful placebo effect" (Beecher 1955). However, many scientists and laypeople alike still think placebo effects represent false improvement, or changes in subjective reports without "real" (viz., clinically or functionally meaningful) changes in objective symptoms. Placebo analgesia, or placebo-based pain reduction, provides a particularly unique challenge to researchers seeking to determine whether placebos cause *functionally and/or neurobiologically significant* changes, as pain itself is subjective and psychological. When someone says she is in pain, how do we evaluate whether she is telling the truth? To do this, we need reliable biological markers linked to pain processing. Brain imaging techniques have provided a powerful way to assess placebo effects, and to understand how they influence pain reports. Today, researchers can conduct carefully controlled studies of placebo-related changes in the brain, and test whether placebos cause changes in early nociceptive pathways (Eippert et al. 2009b; Geuter and Buchel 2013), understand the neurochemical bases underlying placebo effects (Scott et al. 2008; Wager et al. 2007b; Zubieta et al. 2005), and determine the brain changes that are associated with placebo-induced changes in subjective pain (Wager et al. 2011).

To date, over 40 neuroimaging studies have been published on placebo effects. Here, we provide a summary of the most consistent findings across studies in relation to theories of the biological causes and effects of placebo treatment. We present a formal meta-analysis of 25 studies that measured placebo effects and related expectancy effects on brain responses using functional magnetic resonance imaging (fMRI) and positron emission tomography (PET). We report the brain regions that show consistent placebo-induced reductions in pain-related processing during noxious stimulation, which provides information on how placebos affect the systems thought to generate and regulate pain, and may provide clues about how psychological context informs the construction of pain in the central nervous system. We also summarize brain circuits that show increases in activation with placebo treatment, which can inform us both about pain-modulatory mechanisms and about the neurobiological underpinnings of expectations and beliefs more broadly. We then discuss limitations in our current knowledge and how to address some of the outstanding questions in future work.

1 Biological Mechanisms of Placebo Analgesia

The first evidence that placebo analgesia depends on biological mechanisms was published in 1978. Levine et al. (1978) showed that the opioid antagonist naloxone abolished placebo effects on pain, suggesting that placebo analgesia depends on endogenous opioid release. In 2002, Petrovic et al. (2002) published the first neuroimaging study of placebo analgesia. Using PET imaging, they compared placebo analgesia with opioid analgesia produced by the μ-opioid agonist remifentanil. They showed that the effects of endogenous placebo-based opioids and exogenous drug-based opioids overlapped during pain processing: Both caused increases in glucose metabolism in the same brain region, the rostral anterior cingulate cortex (rACC). The first fMRI study of placebo analgesia was conducted in 2004 (Wager et al. 2004). This study showed that placebo administration caused increases during pain anticipation in the lateral and medial prefrontal cortex, including rACC, and also that it caused activity decreases during pain in a subset of regions traditionally associated with pain processing, including the dorsal anterior cingulate cortex (dACC), insula, and thalamus. Later studies measured fMRI responses in the spinal cord and found that spinal responses to pain are reduced with placebo (Eippert et al. 2009b) and increased with nocebo (a "negative placebo" associated with expectations of increased symptoms; Geuter and Buchel 2013). Spinal changes provide evidence for placebo effects on ascending nociceptive signals, before cortical processing. Together, these neuroimaging studies of placebo provide evidence that not only does placebo cause real biological changes, but that placebos actually change responses to noxious stimuli in the central nervous system in ways that are relevant to pain experience.

2 Advantages of Meta-analyses of Expectancy-Based Pain Modulation

Clearly, our ability to observe the neural processes associated with placebo analgesia provides a new, and potentially strong, test of whether placebo effects cause "real" changes. Many cortical and subcortical brain regions have been implicated in individual studies of placebo analgesia and other forms of expectancy-based pain modulation. But findings from a given study can reflect either (a) fundamental mechanisms that support all instances of placebo analgesia or (b) idiosyncratic effects of that study's unique context and design. The best way to differentiate the former from the latter is to collapse across individual studies and identify commonalities using meta-analysis.

To elaborate, while placebo paradigms generally involve similar experimental paradigms (see Fig. 1), individual studies also vary substantially, not only as a function of technical details (e.g., sample size, fMRI scanner strength, acquisition parameters) but also in important experimental features. Studies vary in the type of pain they induce: Many apply noxious heat (Eippert et al. 2009a; Kong et al. 2009b; Wager et al. 2004, 2007b), some use lasers (Bingel et al. 2004; Lui et al. 2010), and others measure pain in patient populations (Harris et al. 2009; Lieberman et al. 2004; Price et al. 2007). Different pain modalities are associated with different effects in the brain (Baumgartner et al. 2010; Friebel et al. 2011), and different modalities may show different activity patterns and placebo responses (Liberman 1964). Likewise, studies differ in the type of pain they measure: Some ask participants to judge pain intensity (Keltner et al. 2006), while others also measure pain unpleasantness (Zubieta et al. 2005). Studies test different types of placebos, including topical ointments (Eippert et al. 2009a; Geuter et al. 2012; Wager et al. 2004), sham electrical stimulation (Lui et al. 2010), and sham acupuncture

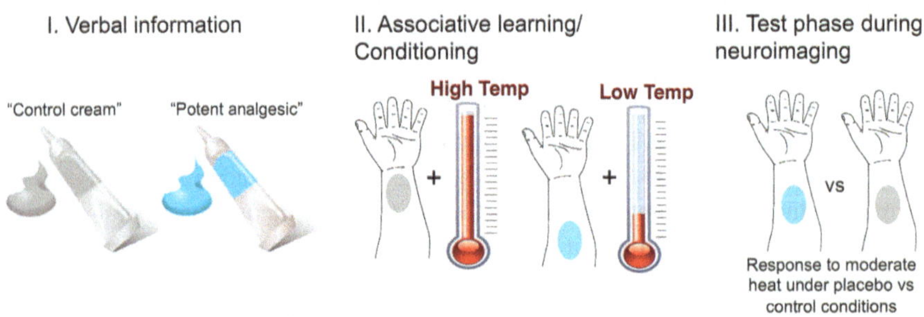

Fig. 1 Typical neuroimaging placebo paradigm. In a typical placebo study, participants are given an inert treatment (e.g., a topical cream) along with verbal instructions (e.g., "This is a potent analgesic") that induce expectations for pain relief. This is compared to a control condition—the same inert substance without expected pain relief. To reinforce verbal instructions, the placebo is paired with reduced stimulus intensity during an associative learning, or conditioning, phase. Finally, participants go through neuroimaging testing during a test phase during which the same stimuli are administered under both control and placebo conditions and experimenters test whether pain reports and brain responses are modulated by beliefs about treatment

(Kong et al. 2006). Individuals hold different beliefs about the efficacy of various treatments as a result of cultural influences and previous experiences (Barrett et al. 2006), and different placebos can induce slightly different effects (de Craen et al. 2000; Kaptchuk et al. 2000); therefore, each type of placebo might even be linked to unique mechanisms. Many other experimental features vary across experiments: whether a study combines verbal suggestions and conditioning to induce expectations about the placebo treatment (Lee et al. 2012; Wager et al. 2004) or uses verbal suggestion alone (Price et al. 2007), whether noxious stimuli vary in intensity (Atlas et al. 2010, 2012; Study 1 in Wager et al. 2004) or remain constant during a test phase (Lui et al. 2010; Study 2 in Wager et al. 2004; Wiech et al. 2010), and whether test stimuli are cued (Lui et al. 2010; Wager et al. 2004) or uncued (Kong et al. 2006). These experimental differences are clearly substantial, and therefore we need a way to identify brain responses that are consistent across these different experimental choices. Ultimately, with more and larger studies, we will understand more about the impact of these choices on brain placebo responses; for now, however, we focus on commonalities across studies and one distinction that is particularly highly powered—manipulations of treatment expectancies vs. stimulus expectancies—because there are a number of studies of each type.

Meta-analysis provides a way to combine individual experiments and determine which brain responses are consistently implicated across studies. Voxel-wise coordinates of individual contrasts from individual studies are added together and compared to random permutations to identify the regions that are consistently activated by a given psychological process (see Kober and Wager 2010 or Wager et al. 2007a for review). To our knowledge, three published meta-analyses of placebo analgesia exist to date. Amanzio et al. (2011) used Activation Likelihood Estimation (ALE; Eickhoff et al. 2009) to conduct a formal meta-analysis of 11 placebo studies (9 fMRI, 2 PET) and separately analyzed brain responses during pain anticipation and during noxious stimulation. We chose to combine placebo-induced increases in activation during pain anticipation with increases during noxious stimulation, as both reflect modulatory mechanisms. Our study also expands on this work by incorporating three more years of prolific neuroimaging research on placebo and expectancy—increasing the number of relevant studies from 11 to 25—and by using multilevel kernel density analysis (MKDA) instead of ALE. In brief, though ALE and MKDA now produce similar results, MKDA focuses on the distribution of statistical contrast maps rather than the distribution of peak coordinates alone, which ensures that studies that report many peaks in a location are not overrepresented (see Wager et al. (2007a) for a more thorough discussion of the relationship between MKDA and other meta-analytic approaches, including ALE). Two other more recent publications have reported *qualitative* meta-analyses, focusing on regions that are activated by three or more studies (Meissner et al. 2011; Wager and Fields 2013). These reports included both contrasts between placebo and control as well as brain–behavior correlations. Correlations are extremely useful in establishing links between brain activity and pain, but they do not isolate causal effects of placebo on brain responses. The meta-analysis presented here extends this work by applying *quantitative* meta-analysis, by focusing only on contrasts rather than correlations with behavior, and by acknowledging different forms of expectancy-based pain modulation.

We performed meta-analyses on 25 neuroimaging studies that manipulated and measured placebo analgesia and expectancy-based pain modulation during brain imaging with fMRI or PET (see Table 1). The studies varied in the experimental dimensions listed above, but all studies compared one condition that induced expectations for pain relief (e.g., placebo administration or a cue predictive of low intensity) with a second control condition, in which the physical treatment or stimulus was identical but there was no expectation for relief. We focused on contrasts between these conditions, rather than correlations with behavior or analyses of responders vs. nonresponders, as contrasts allow for stronger inferences on causal effects of placebo administration.

In our first meta-analysis, we combined (1) studies that manipulated expectations about treatments and tested responses to inert treatments (placebo and nocebo studies), (2) studies that manipulated expectations about treatments and tested responses during actual treatments (open-hidden paradigms), and (3) studies that manipulated expectations about noxious stimulus intensity (cue-based expectancy studies). All of these are types of placebo manipulations in that they manipulate the psychological context—usually a combination of instructions and prior experiences—and all elicit more positive expectations in the placebo condition than a matched control condition with the same physical testing conditions. This primary analysis isolates the brain mechanisms that show consistent increases with expectations about pain, and the brain regions whose pain-evoked activation is influenced by expectations. As a secondary analysis, we separated studies that manipulated expectations about treatments from those that measured expectations about stimuli. This analysis summarizes whether different types of expectations about pain rely on similar or different mechanisms.

3 Methods

3.1 Study Selection and Coordinate Identification

Forty neuroimaging studies of expectancy-based pain modulation were identified using literature searches in PubMed and Google Scholar, the authors' personal libraries, and examining references of relevant papers. We included only studies that (1) used experimental manipulations to induce pain relief; (2) reported formal comparisons (i.e., subtraction-based contrasts) between experimental and control conditions; and (3) reported voxel-wise results in either Montreal Neurological Institute (MNI) or Talairach/Tournoux coordinates. Brain–behavior correlations and ROI-wise analyses were not included in the meta-analysis in order to isolate the direct effects of experimental manipulations on brain responses.

Of the 40 studies originally identified, 25 studies met our criteria (see Table 1). Seventeen of these studies manipulated expectations using placebo manipulations,

Table 1 Experiments included in the meta-analyses

Study	Category	Imaging modality	Type of pain	Sample size	Contrasts associated with reduced activation	Contrasts associated with increased activation
Petrovic et al. (2002)	Placebo analgesia	PET rCBF	Thermal	9	N/A	1. (Pain + placebo)-(pain alone); 2. [(pain + placebo)-(pain alone)]-[(warm + placebo)-(warm alone)]
Wager et al. (2004): Study 1	Placebo analgesia	fMRI	Shock	24	1. (Control-placebo)-(intense-mild shock)	N/A
Wager et al. (2004): Study 2	Placebo analgesia	fMRI	Thermal	23	1. (Control-placebo) during early pain; 2. (Control-placebo) during late pain	1. (Placebo-control) during anticipation
Zubieta et al. (2005)	Placebo analgesia	PET μOR binding	Muscle (sustained)	14	N/A	1. (Placebo-control) opioid increase; carfentanil binding decrease
Koyama et al. (2005)	Stimulus expectancy	fMRI	Thermal	10	1. Decreased pain-induced activation evoked by decreased expectations (expect 48C, actual 50C versus expect 50C, actual 50C)	N/A
Bingel et al. (2006)	Placebo analgesia	fMRI	Laser	19	N/A	1. (Placebo-control) during laser stimulation
Kong et al. (2006)	Placebo analgesia	fMRI	Thermal	24	N/A	1. (After-before placebo treatment)-(placebo-control site)

(continued)

Table 1 (continued)

Study	Category	Imaging modality	Type of pain	Sample size	Contrasts associated with reduced activation	Contrasts associated with increased activation
Keltner et al. (2006)	Stimulus expectancy	fMRI	Thermal	27	1. High expectancy, high temp-low expectancy, high temp	N/A
Price et al. (2007)	Placebo analgesia	fMRI	Visceral (rectal) in patients with irritable bowel syndrome	9	1. Pre-placebo (B1)-placebo during pain AND post-placebo (B2)-placebo; ROIs	N/A
Wager et al. (2007b)	Placebo analgesia	PET μOR binding	Thermal	15	N/A	1. (Placebo-control) opioid increase, (painful-nonpainful heat)
Craggs et al. (2008)	Placebo analgesia	fMRI	Visceral (rectal) in patients with irritable bowel syndrome	9	N/A	1. Placebo-baseline during pain
Eippert et al. (2009a)	Placebo analgesia	fMRI	Thermal	48	1. (Control-placebo) during early pain; 2. (Control-placebo) during late pain	1. (Placebo-control) early pain; 2. (placebo-control) during pain, brainstem-specific
Lu et al. (2009)	Placebo analgesia	fMRI	Visceral (esophageal)	14	1. (Control-placebo) during pain	N/A
Watson et al. (2009)	Placebo analgesia	fMRI	Laser	11	N/A	1. (Placebo-control) during anticipation
Kong et al. (2009b)	Placebo analgesia	fMRI	Thermal	12	1. Control > high expectancy in placebo group	1. Placebo acupuncture group ($N = 12$), (pre-post-placebo) × (placebo-control) site
Kong et al. (2009a)	Expectancy effects during treatment	fMRI	Thermal	Different size for each contrast	1. Verum acupuncture (high-low expectancy) groups, (Pre-Post-treatment) $n = 24$; 2. ME of expectancy on high-expectancy site, (pre-post treatment) $n = 48$	1. Verum acup high-expectancy group, (pre-post-treatment)-(expected-Control site) $n = 12$

Harris et al. (2009)	Placebo analgesia	PET µOR binding	Chronic (fibromyalgia)	10	N/A	1. Opioid increase (binding decrease) After-before sham acupuncture
Lui et al. (2010)	Placebo analgesia	fMRI	Laser	31	N/A	1. Placebo/low exp vs. control/high exp during anticipation; 2. placebo/low exp vs. Control/high exp during pain
Atlas et al. (2010)	Stimulus expectancy	fMRI	Thermal	18	1. Assimilation/expectancy-induced reductions	1. Contrast/expectancy-induced incr
Wiech et al. (2010)	Stimulus expectancy	fMRI	Laser	16	1. High > low threat, across pain and non; 2. Intxn: [pain, high > low threat] > [no pain, high > low threat]	1. Low > high threat, across pain and non
Bingel et al. (2011)	Expectancy effects during treatment	fMRI	Thermal	22	1. Placebo-nocebo: decreases	1. Placebo–nocebo: increases
Lee et al. (2012)	Placebo analgesia	fMRI	Visceral (rectal) in patients with irritable bowel syndrome and healthy controls	17	1. Placebo-control during rectal pain: healthy volunteers (decreases); 2. placebo-control during rectal pain: IBS pts (decreases)	1. Placebo-control during rectal pain: healthy volunteers (increases); 2. placebo-control during rectal pain: IBS pts (increases)
Geuter et al. (2012)	Placebo analgesia	fMRI	Thermal	40	1. Antic: (control > strong placebo) > (control > weak placebo); 2. early pain: control > placebo; 3. late pain: control > placebo; 4. late pain: control > strong placebo; 5. late pain: control > weak placebo; 6. late pain:	1. Antic: placebo > control; 2. antic: strong placebo > control; 3. antic: Weak placebo > control; 4. early pain: placebo > control; 5. early pain: strong placebo > control; 6. early pain: weak placebo > control; 7. early

(continued)

Table 1 (continued)

Study	Category	Imaging modality	Type of pain	Sample size	Contrasts associated with reduced activation	Contrasts associated with increased activation
					(control > strong placebo) > (control > weak placebo)	pain: (strong placebo > control) > (weak placebo > control); 8. late pain: weak placebo > control; 9. late pain: (strong placebo > control) > (weak placebo > control)
Atlas et al. (2012)	Expectancy effects during treatment	fMRI	Thermal	21	1. Instruction-related decreases; 2. open-hidden: decreases	1. Instruction-related increases; 2. open-hidden: Increases
Kong et al. (2013)	Stimulus expectancy	fMRI	Thermal	46	1. For subjects with large exp fx, HC > LC	1. Low cue > high cue, pain; 2. low cue > high cue, antic

five manipulated expectations using cue-based information about stimulus intensity, and three measured expectancy effects during drug treatment using open-hidden administration paradigms. We divided the analyses and results into expectancy-related reductions (e.g., reduced activation with placebo vs. control during pain) or expectancy-related increases (e.g., increased activation with placebo vs. control during anticipation or pain). Our analysis of expectancy-related reductions included only contrasts that focused on brain responses during noxious stimulation, as this identifies regions in which pain processing is modulated by expectancy. However, because pain is thought to be influenced by both preparatory processes and modulation during stimulation, our analysis of expectancy-related increases includes contrasts of activation during both anticipation and stimulation periods. Our meta-analysis of expectancy-related increases also included PET studies that reported reductions in μ-opioid receptor (MOR) tracer binding (consistent with increases in endogenous MOR binding), as MORs comprise one well-supported mechanism of expectancy-based pain modulation.

We extracted peak voxel coordinates from relevant contrasts, and used the Tal2MNI algorithm (Matthew Brett; http://imaging.mrc-cbu.cam.ac.uk/imaging/MniTalairach) to convert Talaraich coordinates to MNI space. We identified 358 peaks from 61 contrasts in 25 studies (see Fig. 2). Some studies reported

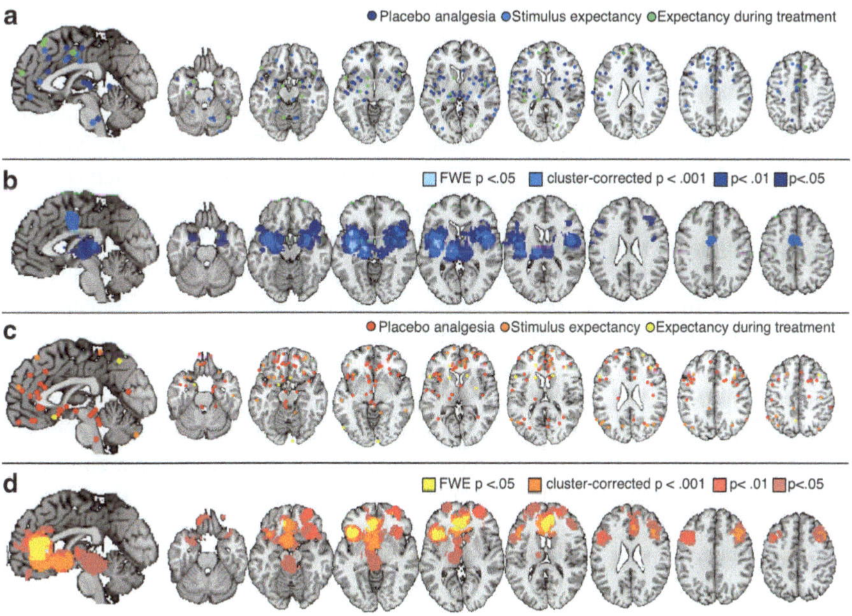

Fig. 2 Meta-analysis 1: expectancy-based pain modulation. (**a**) Peaks included in a meta-analysis of expectancy-based reductions during pain. (**b**) Brain regions that showed reliable reductions during placebo administration and expectations for reduced pain (see Table 3). (**c**) Peaks included in meta-analysis of modulatory increases during pain. (**d**) Regions that showed consistent increases during anticipation or pain stimulation with expectations for reduced pain (see Table 4)

Table 2 Meta-analysis details

Meta-analysis	Number of individual studies included	Number of contrasts	Number of peak coordinates
Decreases during pain: all paradigms	16	27	171
Decreases during pain: placebo studies	8	16	83
Decreases during pain: stimulus expectancy studies	5	6	56
Increases: all paradigms	19	34	187
Increases: placebo studies	13	26	122
Increases: stimulus expectancy studies	3	4	43

both voxel-wise reductions and increases, while others reported effects in only one direction (see Table 1). Table 2 provides detail on the number of peaks, contrasts, and studies included in each meta-analysis.

3.2 Analysis

We performed meta-analyses with MKDA, which summarizes the number of contrasts that activated in the local vicinity (here, within 15 mm) of each voxel in the brain, and uses Monte Carlo simulations to identify regions that are activated more frequently than would be expected by chance. The MKDA approach is described in detail in Wager et al. (2007a, 2009). Peak activations for each contrast were convolved with a spherical smoothing kernel with a 15-mm radius and a weighted average (with weights based on the square root of the sample size) of these was used to generate an activation frequency map. The map was thresholded at $p < 0.05$ family-wise error rate corrected across the whole brain using the maximum null hypothesis activation frequency from each of 5,000 Monte Carlo simulations. The null hypothesis was a random uniform distribution of activation peaks throughout gray matter, which was simulated by permuting the peak locations for each contrast and recalculating the activation frequency map for each iteration. Voxels that survived correction are reported below. We conducted six separate meta-analyses: (1) expectancy-induced reductions in activity (combined across stimulus and treatment expectancy); (2) Expectancy-induced increases; (3) treatment expectancy-induced reductions (placebo analgesia); (4) treatment expectancy-induced increases; (5) stimulus expectancy-induced reductions (cue-based manipulations that show assimilation toward expectations); and (6) stimulus expectancy-induced increases (cue-based manipulations that show contrast away from expectations, e.g., increased activation with expectation for reduced pain).

4 Results

4.1 Meta-analyses 1 and 2: Expectancy-Based Pain Modulation

This analysis combines across three standard types of experimental manipulations that all induce expectations about pain, either through placebo manipulations, pain-predictive cues, or open information about drug delivery. Included studies and contrasts are listed in Tables 1 and 2. The results reported below and in Tables 3 and 4 incorporate both height-corrected results (FWE-corrected $p < 0.05$) and spatial extent-corrected results (cluster-corrected $p < 0.001$).

4.1.1 Expectancy-Induced Reductions During Noxious Stimulation

As shown in Fig. 2b, experimentally manipulated expectations for reduced pain were associated with consistent decreases in activation during noxious stimulation in bilateral anterior insula, bilateral middle insula, left posterior insula, bilateral thalamus, bilateral amygdala, dorsal anterior cingulate (dACC), and bilateral lateral prefrontal cortex (see Table 3).

4.1.2 Expectancy-Induced Increases During Noxious Stimulation

Experimentally manipulated expectations for increased pain were associated with modulatory increases in medial and lateral orbitofrontal cortex (OFC), right anterior prefrontal cortex/superior orbital gyrus, pregenual/rostral ACC (pgACC), rostrodorsal ACC, left ventral striatum, left anterior insula, and midbrain surrounding the periacqueductal gray (PAG; see Fig. 2d and Table 4).

4.2 Meta-analyses 3 to 6: Stimulus Expectancies Versus Treatment Expectancies

We performed separate meta-analyses for studies that manipulated placebo-based expectations about treatments and those that manipulated expectations about stimuli on a trial-by-trial basis using conditioned cues (see Tables 1 and 2 for details). We then performed direct contrasts between these two forms of expectancy-based modulation to identify any regions that are differentially modulated by each form of expectancy. The results reported below and in Tables 5, 6, and 7 incorporate both height-corrected results (FWE-corrected $p < 0.05$) and spatial extent-corrected results (cluster-corrected $p < 0.001$).

4.2.1 Placebo-Induced Reductions During Noxious Stimulation

As shown in Fig. 3a and reported in Table 5, placebo-induced expectations for reduced pain were associated with consistent reductions during noxious stimulation in bilateral anterior insula, left middle insula, left posterior insula, dACC, bilateral medial thalamus, bilateral amygdala, and right lateral prefrontal cortex.

Table 3 Meta-analysis 1: reductions during pain[a]

	Name	x	y	z	Voxels	Studies active (%)
Height-corrected FWE $p < 0.05$	Left amygdala	−24	−4	−8	60	37.73
	Insula, L middle	−38	8	−2	63	37.25
	Insula, L posterior	−36	−10	0	233	44.42
		−38	−6	10	26	44.42
	Insula, L dorsal posterior/ rolandic operculum/OP4	−48	−10	12	5	25.64
	SII, L (rolandic operculum/ OP1)	−46	−22	14	4	27.28
	Insula, R middle	44	10	−2	3	25.61
		44	6	10	48	29.13
	Putamen, L	−30	−18	8	12	27.53
	Thalamus, L (premotor)	−14	−18	−2	109	40.1
Extent-corrected $p < 0.001$	SMA, L	−8	0	46	137	25.07
	Cingulate, L middle	0	−2	40	559	25.07
	Thalamus, L	−12	−22	12	346	32.44
	Rolandic operculum, L	−50	4	10	110	34.22
	Thalamus, L	−6	−10	6	41	27.35
	Thalamus, medial	0	−24	8	45	27.11
	Superior temporal gyrus, L (TE 1.0)	−42	−20	4	257	30.73
	Insula, L middle	−44	−4	−2	216	39.52
	Pallidum, L	−26	−12	0	176	50.18
	Insula, L	−36	−20	−4	96	35.35
	Hippocampus, L	−34	−8	−12	197	38.05
		−22	−14	−16	68	24.62
	Superior temporal gyrus, L	−42	2	−14	207	22.97
	Insula, R middle/rolandic operculum (OP4)	52	−6	12	88	21.14
		34	8	6	153	37.06
		52	0	4	209	24.59
		42	−6	6	234	24.59
	Putamen, R	30	0	−6	138	41.42
	Amygdala, R	28	−6	−14	91	21.72
		30	4	−16	120	19.6

[a]This table reports clusters and contiguous subclusters corresponding to Fig. 2b

4.2.2 Stimulus Expectancy-Induced Reductions During Noxious Stimulation

We found no regions that showed consistent increases in response to cue-based expectations for reduced pain. This is likely due to the small number of studies included in this analysis (5 contrasts from 3 studies; see Table 2).

Table 4 Meta-analysis 2: modulatory mechanisms/expectancy-induced increases[a]

	Name	x	y	z	Voxels	Studies active (%)
Height-corrected FWE $p < 0.05$	mOFC/sgACC	−8	38	−10	89	33.75
	pgACC, medial	−10	28	0	178	34.45
		4	40	0	96	49.56
		−2	36	10	524	52.58
	pgACC, R	−8	40	0	267	40.89
		6	44	14	362	39.19
	Insula, L anterior	−40	20	2	692	28.7
	Anterior PFC/superior orbital gyrus, R	28	54	−4	4	22.8
	IFG pars triangularis, L (latPFC)	−46	18	18	4	22.33
Extent-corrected $p < 0.001$	DLPFC, R (middle frontal gyrus)	42	20	36	409	27.07
		36	26	30	71	20.3
	rdACC, R	2	32	20	120	38.58
		12	40	22	48	28.02
		12	26	12	102	24.97
	rdACC, L	−10	34	16	138	25.12
		−2	24	14	102	28.43
	Insula, L anterior	−28	24	6	130	30.07
		−36	16	−8	233	25.62
		−34	12	8	52	26.35
	pgACC	−14	44	8	63	32.84
	Inferior frontal gyrus, pars triangularis (BA 45)	−50	18	6	69	26.35
		−42	32	2	125	26.35
	sgACC	8	28	0	79	36.14
		2	16	−6	160	22.79
	Caudate/ventral striatum, L	−10	10	−2	207	23.89
	Ventral striatum, R	2	0	−4	86	18.91
	Ventral striatum/globus pallidus, L	−8	−2	−4	157	24.85
	Ventral striatum, L	−14	10	−12	209	31.67
	Ventral striatum/sgACC/ olfactory cortex	−2	8	−12	336	26.39
	Inferior frontal gyrus, pars orbitalis	−32	28	−6	82	31.2
		−48	26	−6	50	28.7

[a]This table reports clusters and contiguous subclusters corresponding to Fig. 2d

4.2.3 Placebo-Induced Increases During Noxious Stimulation

Placebo-induced expectations for reduced pain were associated with consistent increases in activation during noxious stimulation in medial OFC, right lateral OFC, pgACC, right anterior prefrontal cortex, bilateral dorsolateral prefrontal

Table 5 Meta-analysis 3: placebo-induced reductions during pain[a]

Placebo-induced reductions	Name	*x*	*y*	*z*	Voxels	Studies active (%)
Height-corrected FWE $p < 0.05$	Insula, L anterior	−38	−2	−16	60	31.67
		−44	−4	−8	95	31.67
		−42	−18	2	65	32.62
	Insula, L middle	−36	−10	−4	153	46.35
		−34	4	−4	249	49.83
		−40	−6	6	310	49.83
Extent-corrected $p < 0.001$	dpIns, L (OP4)/Heschl's gyrus	−48	−16	10	141	36.1
	Insula, L anterior (BA44)	−42	10	2	82	36.92
	Putamen, L	−26	−12	2	66	53.69
	Putamen, L, contiguous with L anterior insula	−30	14	−2	241	33.44
	Amygdala, L	−26	−6	−10	159	39.01
	Amygdala, L, contiguous with L putamen	−26	4	−12	161	26.1
	Insula, L anterior	−38	8	−14	141	31.67

[a]This table reports clusters and contiguous subclusters corresponding to Fig. 3a

cortex, ventral striatum, left thalamus, and midbrain surrounding PAG (see Fig. 4a and Table 6).

4.2.4 Stimulus Expectancy-Induced Increases During Noxious Stimulation

Cue-based expectations for reduced pain were associated with consistent increases in left anterior prefrontal cortex and left superior parietal lobule/angular gyrus (see Fig. 4b and Table 7).

4.2.5 Treatment Expectancy Effects Versus Stimulus Expectancy Effects

A formal comparison of treatment expectancy studies and stimulus expectancies revealed that treatment expectancies were significantly more likely to reduce activation in left insula (anterior, middle, and posterior insula; see Fig. 3b and Table 8). There were no regions that showed reductions unique to stimulus expectancy, which is unsurprising given that the meta-analysis revealed no common reductions in the few studies included. We also examined differences in expectancy-related increases (regions that showed increases with placebo administration, showed greater activation following stimulus expectancy cues that predict low pain, or showed reductions with cues that predict high pain). We found that treatment expectancies were associated with larger increases in ventromedial prefrontal cortex (VMPFC), PAG, dorsomedial prefrontal cortex, left ventral striatum, pgACC, sgACC, left lateral prefrontal cortex, and left thalamus than stimulus

Table 6 Meta-analysis 4: placebo-induced increases[a]

	Name	x	y	z	Voxels	Studies active (%)
Height-corrected FWE $p < 0.05$	pgACC, L	−8	34	−6	399	43.47
		−2	40	0	415	56.31
		−12	28	4	108	37.84
		−4	32	10	340	50.09
		−4	42	12	223	50.68
	pgACC, R	4	38	18	76	40.54
	Ventral striatum, L	−6	6	−8	22	23.95
		−8	−2	−2	33	28.78
	Inferior frontal gyrus, pars triangularis, L	−46	24	0	151	25.46
	Insula, L anterior	−38	18	2	498	25.46
	rdACC, R	6	28	24	10	23.1
Extent-corrected $p < 0.001$	Midbrain surrounding the PAG	0	−32	−12	482	16.76
	Midbrain, contiguous with thalamus	−6	−20	−4	313	21.6
	mOFC (mid orbital gyrus)	−2	26	−14	147	26.2
	mOFC, L (rectal gyrus)	−6	36	−16	60	28.5
	sgACC	0	20	−6	202	26.02
		4	34	−8	193	36.76
	rACC	2	22	8	84	41.04
		12	24	12	38	16.3
		10	44	12	259	44.21
	VMPFC, L	−12	46	−10	14	16.42
	Insula, L anterior	−38	18	−10	92	18.21
		−40	10	−4	68	18.21
		−30	28	−2	99	25.46
		−40	32	2	103	25.46
	Inferior frontal gyrus, pars triangularis, LL	−40	24	12	364	32.87

[a]This table reports clusters and contiguous subclusters corresponding to Fig. 4a

Table 7 Meta-analysis 6: stimulus expectancy-induced increases[a]

Stimulus expectancy-induced increases	Name	*x*	*y*	*z*	Voxels	Studies active (%)
Height-corrected FWE $p < 0.05$	Angular gyrus, L/IPC (PFm)	−40	−60	46	319	81.66
	Angular gyrus, L/IPC (PGa)	−48	−60	36	158	81.66
	Superior frontal gyrus, L (aPFC)	−16	66	10	131	81.66
Extent-corrected $p < 0.001$	*None*					

[a]This table reports clusters and contiguous subclusters corresponding to Fig. 4b

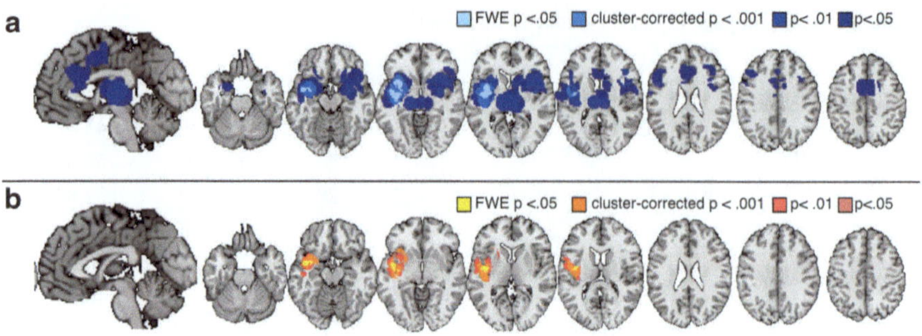

Fig. 3 Decreases: treatment expectancies vs. stimulus expectancies. (**a**) Brain regions that showed reliable reductions during placebo analgesia (see Table 5). (**b**) Differences between placebo analgesia and stimulus expectancy-induced reductions (placebo analgesia > stimulus expectancies; see Table 8). Left anterior insula was significantly more likely to show reductions with placebo analgesia than with stimulus expectancies

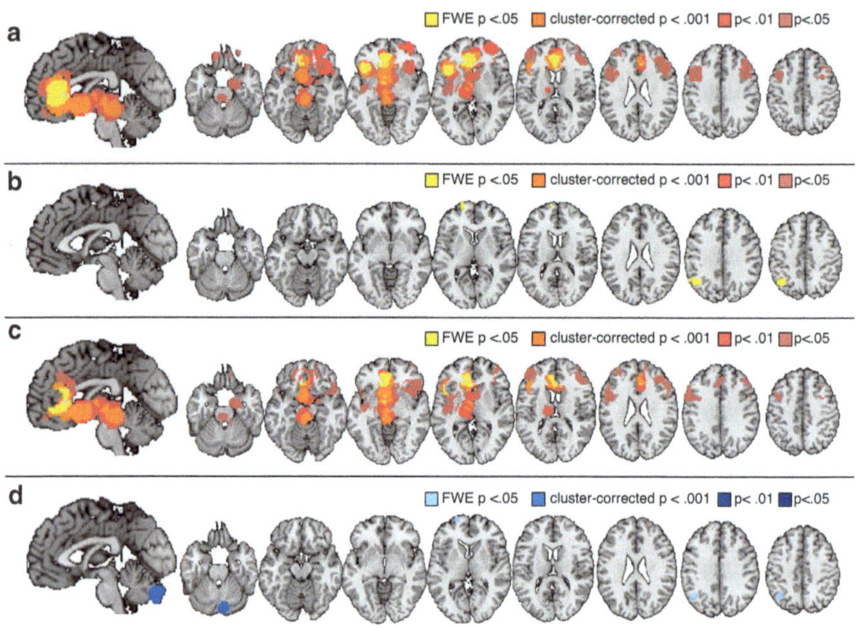

Fig. 4 Modulatory increases: treatment expectancies vs. stimulus expectancies. (**a**) Brain regions that showed reliable increases prior to or during placebo analgesia (see Table 6). (**b**) Brain regions that showed reliable increases as a function of stimulus expectancy (i.e., increased activity with expectation for reduced pain; see Table 7). (**c**) Differences between placebo analgesia and stimulus expectancy-induced increases (placebo analgesia > stimulus expectancies; see Table 8). (**d**) Regions that showed larger increases with stimulus expectancy than placebo analgesia (see Table 9)

Table 8 Contrast A: expectancy-induced reductions during pain[a]

Placebo-induced reductions > stimulus expectancy-induced reductions	Names	*x*	*y*	*z*	Voxels	Placebo studies active (%)	Stimulus expectancy studies active (%)
Height-corrected FWE $p < 0.05$	Insula, L anterior	−32	12	−6	7	33.44	0
	Insula, L posterior/ rolandic Operculum (OP3)	−42	−10	12	76	36.1	0
	Insula, L posterior/ superior temporal gyrus (TE 1.0)	−44	−18	0	38	32.62	0
	Insula, L middle	−42	−8	2	173	49.83	27.04
	Insula, L posterior/ superior temporal gyrus	−42	−2	−12	151	39.01	0
Extent-corrected $p < 0.001$	Rolandic operculum, L (OP4)	−52	−2	14	86	26.21	0
	Amygdala, L	−26	4	−10	142	26.1	0
Stimulus expectancy-induced reductions > placebo-induced reductions							
Height-corrected FWE $p < 0.05$	*None*						
Extent-corrected $p < 0.001$	*None*						

[a]This table reports clusters and contiguous subclusters corresponding to Fig. 3b

expectancies (see Fig. 4c and Table 9). Stimulus expectancies were more likely to activate left anterior prefrontal cortex, left superior parietal lobule, and the cerebellum (see Fig. 4d). We note that these differences should be considered tentative and exploratory, as the number of placebo studies far outweighed the number of stimulus expectancy studies included.

Table 9 Contrast B: expectancy-induced increases[a]

Placebo-induced increases > stimulus expectancy-induced increases	Name	*x*	*y*	*z*	Voxels	Placebo studies active (%)	Stimulus expectancy studies active (%)
Height-corrected FWE $p < 0.05$	DMPFC	4	42	24	1	23.17	0
	rdACC	6	28	24	10	23.1	0
	Inferior frontal gyrus, pars triangularis, L (BA44)	−46	16	6	88	25.46	0
	Ventral striatum, L	−8	−2	−2	33	28.78	0
	Ventral striatum, L	−6	6	−8	22	23.95	0
	pgACC, L	−10	42	10	141	43.56	0
		−4	32	10	360	54.35	0
		−12	26	2	153	37.84	0
		−6	38	−4	474	52.05	0
Extent-corrected $p < 0.001$	Midbrain surrounding the PAG	2	−32	−14	274	16.76	0
		−4	−24	−6	518	21.6	0
	mOFC (mid orbital gyrus, L)	0	26	−14	73	26.2	0
		0	46	−6	33	32.38	0
		−10	48	−10	9	16.42	0
	sgACC	4	34	−8	190	28.18	0
		0	20	−6	195	26.02	0
	WM_rACC	2	22	10	82	41.04	0
		12	24	12	27	16.3	0
	Inferior frontal gyrus, pars triangularis, L (BA 45)	−52	22	0	33	25.46	0
		−44	26	12	272	32.87	0

(continued)

Stimulus expectancy-induced increases > placebo-induced increases	Name	x	y	z	voxels	Placebo studies active (%)	Stimulus expectancy studies active (%)
Height-corrected FWE $p < 0.05$	Inferior parietal lobule, L (IPC (Pga))	−40	−58	54	50	0	81.66
	Angular gyrus, L (IPC (PFm))	−38	−60	44	192	0	81.66
	Angular gyrus, L (IPC (Pga))	−46	−62	34	84	0	81.66
	Superior frontal gyrus, L (aPFC)	−16	66	10	117	0	81.66
Extent-corrected $p < 0.001$	Cerebellum, R	4	−86	−28	198	0	62.2
		16	−82	−30	325	0	80.54
		8	−74	−32	386	0	80.54
		2	−82	−38	198	0	80.54
		12	−88	−40	158	0	62.2
		14	−76	−42	202	0	80.54

[a]This table reports clusters and contiguous subclusters corresponding to Fig. 4c, d

5 Discussion

We used formal meta-analysis to examine the brain mechanisms associated with placebo- and expectancy-based pain modulation, as identified by 25 neuroimaging studies published between 2002 and 2013. Relative to control conditions, expectations for pain reduction were associated with widespread reductions in brain responses during painful stimulation, with decreases in dACC, insula, thalamus, amygdala, striatum, and lateral prefrontal cortex. Expectations for reduced pain were also associated with increases in activation prior to and during noxious stimulation in a number of regions, including dorsolateral and ventromedial prefrontal cortex, rostral anterior cingulate cortex, the midbrain surrounding the PAG, left anterior insula, and the striatum. These regions reveal the most reliable neural mechanisms underlying placebo analgesia and expectancy-based pain modulation, and they can serve as regions of interest in future studies designed to directly isolate their specific contributions. In addition, we observed initial support for separate mechanisms underlying placebo-based treatment expectancy effects and stimulus expectancy effects, though results should be considered tentative. In this final section, we discuss these networks from the standpoint of cognitive and affective neuroscience, and we address outstanding questions that future studies can address.

5.1 Placebo Effects on Brain Regions Traditionally Associated with Pain Processing

We observed consistent reductions in dACC, bilateral insula, and thalamus, a subset of the regions that are most reliably activated by noxious stimulation (Duerden and Albanese 2011; Apkarian et al. 2005; Friebel et al. 2011; Peyron et al. 2000; Wager et al. 2013). Interestingly, secondary somatosensory cortex (SII) and dorsal posterior insula (dpIns) are conspicuously absent (although we saw left posterior insula when we collapsed across stimulus and treatment expectancy). SII and dpIns are most consistently and specifically activated by noxious stimuli in neuroimaging studies (Kross et al. 2011; Peyron et al. 2000), are thought to support pain's sensory components (Maihöfner et al. 2006), and were recently shown to be the only cortical regions in which intracranial stimulation can produce a sensation of pain (Mazzola et al. 2011).

Why might we see this distinction? Is this evidence that placebo analgesia does *not* alter the earliest levels of processing? We believe that the absence of SII and dpIns reductions in our analysis does not imply that early nociceptive processing is unaltered during placebo analgesia. First, we know from individual studies that placebo analgesia and nocebo hyperalgesia can influence spinal nociceptive reflexes and spinal responses to noxious stimuli (Eippert et al. 2009b; Goffaux et al. 2007; Matre et al. 2006; Geuter and Buchel 2013) which reveals that placebo can alter pain processing before ascending nociceptive signals even reach cortex. Second, behavioral investigations indicate that placebo alters both pain intensity and unpleasantness ratings (De Pascalis et al. 2002; Price et al. 1999), although early psychophysical investigations that used signal detection theoretic analyses suggested that placebo altered response biases without altering sensory discrimination (Clark 1969). Third, our meta-analysis collapsed across both pain modalities (see Table 1) and site of noxious stimulation, and SII and mid-to-posterior insula are sensitive to different types of pain (Baumgartner et al. 2010; Friebel et al. 2011) and contain a somatotopic map (Baumgartner et al. 2010), thereby representing different body sites in different precise locations. Finally, it is possible that the strength of nociceptive modulation varies across studies and contexts, and that effects in these regions exist only for the strongest contexts and the strongest placebo responders. Thus the fact that these regions did not emerge in our meta-analyses is not definitive evidence that sensory processing is unaffected during expectancy-based pain modulation.

Despite the fact that we cannot say with certainty that SII and dpIns are unaffected by placebo, we must also acknowledge that there is a large literature supporting a distinction between these two regions and those that were consistently modulated by placebo and expectancy and identified in the meta-analysis presented here. Medial thalamus, dACC, and insula are all targets of the spinothalamic tract (Dum et al. 2009). They are functionally connected when individuals rate noxious stimuli but not when individuals make non-nociceptive magnitude estimations (Baliki et al. 2009). They have been traditionally thought to support pain's affective (i.e., motivational and emotional) components (Peyron et al. 2000; Rainville 2002; Rainville et al. 1997, 1999; Tölle et al. 1999; Zubieta et al. 2001; Bushnell et al. 2013). The insula has also been implicated in interoception and thermosensory processing (Craig 2009; Craig et al. 2000).

If one interprets the results of the present meta-analysis and considers only the aforementioned pain literature, one might assume that our findings reveal that pain affect is reliably influenced by placebo administration. However, one glaring caveat must be acknowledged. The insula, thalamus, and ACC are the most widely activated regions across all task-based fMRI studies (Yarkoni et al. 2011). The dACC and insular cortices show intrinsic connectivity during resting state fMRI and have been referred to as a "salience network" (Seeley et al. 2007). They are implicated in many broad psychological processes, including simple maintenance of task sets (Dosenbach et al. 2006), interoception (Craig 2002; Critchley et al. 2004), conflict monitoring (Botvinick et al. 2004), and affective processing (Shackman et al. 2011). All of these psychological processes might be implicated in a standard placebo experiment, and so determining whether the presence of these brain regions reflects these nonspecific processes or reflects something unique about placebo requires more sophisticated analyses. In this regard, correlations between brain responses and the magnitude of placebo analgesia (or placebo "responder status," a dichotomous version) are informative, as they establish a relationship between brain activity and pain. Correlations between placebo analgesia magnitude and reductions in dACC, anterior insula, thalamus, and SII have been reported in multiple studies (reviewed in Koban et al. 2013). Nonetheless, such correlations do not provide strong evidence that the brain processes that are affected by placebo are directly associated with nociception or pain, and stronger tests are needed. As the question of specificity applies to all of the regions identified in the meta-analyses presented here, we discuss this issue in greater detail and propose several solutions—below (see Sect. 5.6).

With regard to the question of whether nociceptive processing is altered, we note that the vast majority of the studies included in the present analyses applied either a single level of noxious stimulation, or, if stimulus intensity varied, the paradigm included cues that gave information about upcoming stimulus intensity, thereby influencing stimulus expectancies (even in the context of placebo analgesia studies designed to test only treatment expectancy). A simple modification of the basic experimental paradigm depicted in Fig. 1 would greatly improve our ability to determine the extent to which pain-related processing is altered by placebo. If stimulus intensity *varies* during placebo and control conditions, researchers can examine intensity-related changes in the control condition to establish subject-specific regions of interest involved in pain processing, and then test for placebo effects on these responses. This would also allow for more sophisticated analyses, such as tests of placebo x intensity interactions (Wager et al. 2004). Another important direction is to test the effects of placebo on brain patterns that are validated to be sensitive and specific to pain across studies, such as that recently developed by Wager et al. (2013).

5.2 Reductions During Pain in the Amygdala

In addition to regions often implicated in pain processing, we observed placebo- and expectancy-induced reductions in bilateral amygdala. The amygdala does receive nociceptive input through the spinopontoamygdalar pathway (Bernard et al. 1992; Willis and Westlund 1997), and some fMRI studies have shown that it tracks changes in noxious input (Bornhovd et al. 2002) and is important for the modulation of nociception by behavioral context (Helmstetter 1992). Thus it is possible that amygdala modulation is consistent with a straightforward account expectancy-based reduction in nociception. However, the amygdala is also strongly implicated in cognitive and affective processes (Phelps 2006) such as vigilance (Davis and Whalen 2001), threat (LaBar et al. 1998; LeDoux 1995; Rogan et al. 1997), and uncertainty (Rosen and Donley 2006; Whalen 2007)—processes that often accompany or precede pain, but that also exist in the absence of noxious input, such as in response to salient cues (e.g., subliminal fear expressions; Whalen et al. 1998, 2004) that induce vigilance or change one's motivational and attentional state. As with the regions reviewed in the prior section, future experiments should directly test whether placebo-induced changes in amygdala responses relate more closely to pain processing or vigilance and uncertainty, as might be expected if the treatment context causes patients to feel calm and protected.

5.3 Modulatory Mechanisms

In addition to expectancy- and placebo-based reductions, we saw widespread increases in activation with expectation for decreased pain. Increases were apparent in the pgACC/rACC and the periacqueductal gray, two brain regions that have been linked with endogenous opioid release in animal models (Fields 2000, 2004) and in prior studies of placebo analgesia in humans (Levine et al. 1978) (Eippert et al. 2009a; Wager et al. 2007b; Zubieta et al. 2005). In addition, we observed consistent increases in the ventral striatum, a region that has been linked with dopamine binding and reward processing (Scott et al. 2007, 2008). The ventral striatum is also involved in learning about affective value, and works in concert with the VMPFC/OFC—another region that showed reliable increases with expectations for pain reduction—to track expected value and update expectations (Liljeholm and O'Doherty 2012; Murray et al. 2007; Schoenbaum et al. 2009). Relating placebo and expectancy effects with models of reinforcement learning is likely to be a fruitful direction for future research. We also note that VMPFC might have a more general role in ascribing value and meaning to stimuli (Roy et al. 2012), perhaps linking to meaning-based conceptions of placebo analgesia (Moerman and Jonas 2002). Finally, we also observed consistent placebo-based increases in the DLPFC, a region broadly implicated in executive function, including cognitive control, working memory, and rule maintenance (Miller 2000; Miller and Cohen 2001). For a more thorough discussion of these systems and their involvement in cognitive and affective functions, see Atlas and Wager (2013).

5.4 Treatment Expectancies Versus Stimulus Expectancies

Expectations about stimuli and expectations about treatment are often discussed interchangeably in the literature. Our first meta-analysis adopts this perspective to identify the brain mechanisms associated with expectancy-based pain modulation. However, we and others have argued that the two should be thought of as distinct processes (Atlas et al. 2010; Atlas and Wager 2012, 2013; Kirsch 1985, 1997). Stimulus expectancies are predictions about discrete events in the environment. Thus, stimulus expectancies are likely to rely upon transient processes that can change from moment to moment depending on the content of expectation, and may even recruit preparatory antinociceptive responses. We have hypothesized that cue-based stimulus expectancies about pain intensity are likely to depend on phasic responses in dopamine neurons and related systems involved in processing expected value and prediction error (Atlas et al. 2010). Indeed, quantitative modeling supports this account (Seymour et al. 2004, 2005), but the relationship between such signals and perceived pain has not been formally tested. Treatment expectancies, on the other hand, involve knowledge about one's overall context, and beliefs that one will be less affected by stimuli in the environment. Thus, we have hypothesized that treatment expectancies are likely to depend on sustained mechanisms, such as affective shifts and tonic opioid binding (Atlas et al. 2010; Atlas and Wager 2012).

The theoretical distinctions between these two types of expectancy motivated us to separate and formally compare them in our secondary set of meta-analyses. Placebo-based reductions and increases were nearly identical to our findings from the collapsed meta-analysis. This is because of the vast imbalance in the studies that were included: Our meta-analysis was heavily weighted toward studies of placebo-based treatment expectancy (17 treatment expectancy vs. 5 stimulus expectancy). The dearth of experiments relating stimulus expectancies with perceived pain is also likely responsible for the fact that our meta-analysis failed to identify any regions that showed reliable stimulus expectancy-induced reductions (assimilation with expectations) during pain. We did, however, observe consistent increases with expectations for low pain in the parietal cortex and anterior prefrontal cortex, which might be related to the frontoparietal network, a network involved in selective attention (Szczepanski et al. 2013). We note that these regions were *not* observed in our analysis of treatment expectancy-induced increases, providing at least some support for the notion of neural segregation, although we feel it would be premature to infer that attention networks are themselves altered by the two processes. We encourage researchers to consider these distinctions in the future, so we can better identify the similarities and differences between these two types of expectations.

5.5 Relationship to Prior Meta-analyses of Placebo Analgesia

It is important to consider the present findings in relation to a quantitative meta-analysis published previously (Amanzio et al. 2011). Both analyses examined placebo-induced reductions during noxious stimulation and found evidence for reductions in the cingulate cortex and insula/clustrum, although laterality and precise location differed [middle and posterior cingulate in Amanzio et al. (2011) vs. anterior and middle cingulate here; right posterior and left anterior insula in Amanzio et al. (2011) vs. bilateral anterior and left posterior insula here]. All other reductions reported in Amanzio et al. (2011) fell within the boundaries of the placebo-induced reductions reported here. We also observed reductions in bilateral amygdala and bilateral lateral prefrontal cortex, which were not observed in the previous meta-analysis.

Amanzio et al. (2011) separated placebo-induced increases during anticipation ("stage 1") and noxious stimulation ("stage 2"). The "stage 1" anticipatory increases identified by Amanzio et al. all overlap with or are directly adjacent to increases identified here. There were some slight differences when it came to "stage 2" increases. We did not observe consistent increases in the pons, the inferior parietal lobule, the postcentral gyrus, or the medial frontal gyrus, and the dorsolateral prefrontal cortex activations we observed were bilateral and posterior to the left DLPFC activation reported by Amanzio et al. (2011). We also found evidence of consistent placebo-induced increases in the ventral striatum and bilateral anterior insula/inferior frontal gyrus, which were not observed in the earlier meta-analysis.

Some of these differences are likely due to differences in power: The present analysis included twice as many studies than Amanzio et al. (2011), which points to a growing scientific interest in the brain basis of placebo analgesia as neuroimaging studies of placebo continue to be published. In addition, we used MKDE rather than ALE, which might have accounted for subtle differences in exact location of peaks. However, other differences are likely to stem from explicit decisions based on theoretical considerations. Amanzio et al. separated increases during anticipation and stimulation, whereas we collapsed across both phases in our analysis of modulatory increases. We decided to collapse across these periods since not all experiments are designed to separate anticipation and pain, and both periods involve modulatory mechanisms. In addition, we incorporated analyses of cue-based stimulus expectancies (though the comparisons discussed here refer specifically to our analysis of placebo-based treatment expectancy effects) to isolate expectancy-based changes and to determine whether there are reliable differences in treatment and stimulus expectancies.

5.6 Unanswered Questions: Extending the Meta-analysis

Throughout this discussion, we have deliberately avoided reverse inferences about the specific processes supported by the regions identified in our meta-analysis. As we have pointed out, the regions identified here are reliably influenced during placebo, but are also implicated in many other psychological processes. Meta-analyses tell us *where* placebo effects occur, but not *how* these brain regions—either individually or as a network—contribute to placebo effects on pain. How can future studies establish specificity?

One simple way to understand the contributions of these commonly activated brain regions in any particular study is to link placebo effects on the brain with placebo effects on behavior. This has been accomplished by (1) correlating individual differences in placebo effects on activation with effects on pain reports (Kong et al. 2006; Wager et al. 2004) [see (Koban et al. 2013) for review and meta-analysis]; (2) mediation analyses that identify regions that dynamically mediate the effects of experimental manipulations on measured behavior (Atlas et al. 2010); and (3) machine learning techniques that identify spatially distributed patterns of brain responses predictive of the magnitude of the placebo response across individuals (Wager et al. 2011). These brain-behavior approaches can help individual studies differentiate between regions that are simply activated by elements of the treatment context or experimental context and those that actually correlate with or even cause changes in subjective pain. Further specificity can be attained by differentiating between various aspects of the pain experience, e.g., pain intensity versus pain unpleasantness (De Pascalis et al. 2002; Kulkarni et al. 2005; Price et al. 1999). For example, one could use any of these methods to test formally the hypothesis that placebo effects on dACC reflect placebo-based reductions in pain unpleasantness.

Another way to specify the precise functional contributions of these regions is to design experiments that isolate the effects of different components that contribute to placebo responses. The standard placebo manipulation depicted in Fig. 1 (and employed in many of the experiments in the present meta-analysis) combines many features, only some of which have been directly investigated with neuroimaging tools. Participants generally receive both verbal suggestions and conditioning/paired association. A number of laboratory experiments have sought to separate the contributions of verbal suggestions and conditioning (Benedetti et al. 2003; Montgomery and Kirsch 1996, 1997), but such paradigms have yet to be extended to the neuroimaging domain. While one study examined the neural correlates of the conditioning phase as well as accompanying placebo analgesia during test (Lui et al. 2010), conditioning-based placebo was not directly compared to a placebo condition based on suggestion alone, nor was conditioning examined in the absence of explicit verbal instructions. Another key component of the placebo effect involves the psychosocial aspects of the patient–doctor relationship (Kaptchuk 2002). While few studies have formally investigated this interaction, some evidence comes from studies showing that patient responses are influenced by doctors' expectations (Gracely et al. 1985; Levine and Gordon 1984). One recent experiment attempted to identify neural mechanisms contributing to the patient–doctor relationship by focusing on brain responses in physicians as they administered treatment (Jensen et al. 2013). Future studies can elaborate on this work and directly address the interactive nature of this relationship by adapting designs from social neuroscience optimized to study interpersonal interactions.

Finally, placebo effects might induce changes in attention and/or induce positive affective shifts (Atlas and Wager 2013), and, as mentioned above, changes in regions like insula and dACC might reflect these nonspecific processes. Experiments that relate placebo effects, brain responses, and performance on well-validated attention and emotion experiments can evaluate the extent to which this is true. For example, one study (Scott et al. 2007) related placebo-induced changes in striatal dopamine binding with performance on a reward task (Knutson et al. 2001), suggesting that this region might play a role in the rewarding aspects of receiving treatment. We have interwoven attention probes with placebo administration (Atlas et al. 2014) and stimulus expectancy cues (Johnston et al. 2012) to determine whether expectancy-based pain modulation depends on changes in attention. Finally, studies have tested whether placebo involves changes in emotion processing by relating responses during placebo analgesia with placebo effects on responses to aversive images (Petrovic et al. 2010; Zhang and Luo 2009) and by testing whether responses to emotional stimuli are altered during placebo analgesia (Atlas et al. 2014).

In sum, we envision an iterative process that will ultimately lead to precision and specificity with regard to the contributions of the individual regions and networks identified here. The regions we have identified can serve as regions of interest for future studies designed to isolate specific elements of placebo analgesia and other forms of expectancy-based pain modulation. As studies on a specific subprocess (e.g., positive affect/reward processing) accumulate, future meta-analyses will determine which regions are reliably activated as a function of that process. This iterative science will ultimately provide us with a detailed picture of how distinct psychological and neural processes combine to influence pain under placebo, which can then be extended to develop targeted interventions at a clinical level.

Acknowledgments This work was supported in part by the Intramural Research program of the NIH's National Center for Complementary and Alternative Medicine.

References

Amanzio M, Benedetti F, Porro CA, Palermo S, Cauda F (2011) Activation likelihood estimation meta-analysis of brain correlates of placebo analgesia in human experimental pain. Hum Brain Mapp 34:738–752

Apkarian AV, Bushnell MC, Treede R-D, Zubieta J-K (2005) Human brain mechanisms of pain perception and regulation in health and disease. Eur J Pain (Lond) 9:463–484

Atlas LY, Wager TD (2012) How expectations shape pain. Neurosci Lett 520:140–148

Atlas LY, Wager TD (2013) Expectancies and beliefs: insights from cognitive neuroscience. In: Ochsner KN, Kosslyn SM (eds) Oxford handbook of cognitive neuroscience. Oxford University Press, Oxford, NY, pp 359–381

Atlas LY, Bolger N, Lindquist MA, Wager TD (2010) Brain mediators of predictive cue effects on perceived pain. J Neurosci 30:12964–12977

Atlas LY, Whittington RA, Lindquist MA, Wielgosz J, Sonty N, Wager TD (2012) Dissociable influences of opiates and expectations on pain. J Neurosci 32:8053–8064

Atlas LY, Wielgosz J, Whittington RA, Wager TD (2014) Specifying the non-specific factors underlying opioid analgesia: expectancy, attention, and affect. Psychopharmacology 231:813–823

Baliki MN, Geha PY, Apkarian AV (2009) Parsing pain perception between nociceptive representation and magnitude estimation. J Neurophysiol 101:875–887

Barrett B, Muller D, Rakel D, Rabago D, Marchand L, Scheder JC (2006) Placebo, meaning, and health. Perspect Biol Med 49:178–198

Baumgartner U, Iannetti GD, Zambreanu L, Stoeter P, Treede R-D, Tracey I (2010) Multiple somatotopic representations of heat and mechanical pain in the operculo-insular cortex: a high-resolution fMRI study. J Neurophysiol 104:2863–2872

Beecher HK (1955) The powerful placebo. J Am Med Assoc 159:1602–1606

Benedetti F, Pollo A, Lopiano L, Lanotte M, Vighetti S, Rainero I (2003) Conscious expectation and unconscious conditioning in analgesic, motor, and hormonal placebo/nocebo responses. J Neurosci 23:4315–4323

Bernard JF, Huang GF, Besson JM (1992) Nucleus centralis of the amygdala and the globus pallidus ventralis: electrophysiological evidence for an involvement in pain processes. J Neurophysiol 68:551–569

Bingel U, Lorenz J, Glauche V, Knab R, Glascher J, Weiller C, Buchel C (2004) Somatotopic organization of human somatosensory cortices for pain: a single trial fMRI study. NeuroImage 23:224–232

Bingel U, Lorenz J, Schoell E, Weiller C, Buchel C (2006) Mechanisms of placebo analgesia: rACC recruitment of a subcortical antinociceptive network. Pain 120:8–15

Bingel U, Wanigasekera V, Wiech K, Mhuircheartaigh RN, Lee MC, Ploner M, Tracey I (2011) The effect of treatment expectation on drug efficacy: imaging the analgesic benefit of the opioid remifentanil. Sci Transl Med 3:70ra14

Bornhovd K, Quante M, Glauche V, Bromm B, Weiller C, Buchel C (2002) Painful stimuli evoke different stimulus-response functions in the amygdala, prefrontal, insula and somatosensory cortex: a single-trial fMRI study. Brain 125:1326–1336

Botvinick MM, Cohen JD, Carter CS (2004) Conflict monitoring and anterior cingulate cortex: an update. Trends Cogn Sci 8:539–546

Bushnell MC, Ceko M, Low LA (2013) Cognitive and emotional control of pain and its disruption in chronic pain. Nat Rev Neurosci 14:502–511

Clark WC (1969) Sensory-decision theory analysis of the placebo effect on the criterion for pain and thermal sensitivity. J Abnorm Psychol 74:363–371

Craggs J, Price D, Perlstein W, Nicholas Verne G, Robinson M (2008) The dynamic mechanisms of placebo induced analgesia: evidence of sustained and transient regional involvement. Pain 139:660–669

Craig AD (2002) How do you feel? Interoception: the sense of the physiological condition of the body. Nat Rev Neurosci 3:655–666

Craig ADB (2009) How do you feel—now? The anterior insula and human awareness. Nat Rev Neurosci 10:12

Craig ADB, Chen K, Bandy D, Reiman EM (2000) Thermosensory activation of insular cortex. Nat Neurosci 3:184–190

Critchley HD, Wiens S, Rotshtein P, Ohman A, Dolan RJ (2004) Neural systems supporting interoceptive awareness. Nat Neurosci 7:189–195

Davis M, Whalen PJ (2001) The amygdala: vigilance and emotion. Mol Psychiatry 6:13–34

de Craen AJ, Tijssen JG, de Gans J, Kleijnen J (2000) Placebo effect in the acute treatment of migraine: subcutaneous placebos are better than oral placebos. J Neurol 247:183–188

De Pascalis V, Chiaradia C, Carotenuto E (2002) The contribution of suggestibility and expectation to placebo analgesia phenomenon in an experimental setting. Pain 96:393–402

Dosenbach NUF, Visscher KM, Palmer ED, Miezin FM, Wenger KK, Kang HC, Burgund ED, Grimes AL, Schlaggar BL, Petersen SE (2006) A core system for the implementation of task sets. Neuron 50:799–812

Duerden EG, Albanese M-C (2011) Localization of pain-related brain activation: a meta-analysis of neuroimaging data. Hum Brain Mapp 34:109–149

Dum RP, Levinthal DJ, Strick PL (2009) The spinothalamic system targets motor and sensory areas in the cerebral cortex of monkeys. J Neurosci 29:14223–14235

Eickhoff SB, Laird AR, Grefkes C, Wang LE, Zilles K, Fox PT (2009) Coordinate-based activation likelihood estimation meta-analysis of neuroimaging data: a random-effects approach based on empirical estimates of spatial uncertainty. Hum Brain Mapp 30:2907–2926

Eippert F, Bingel U, Schoell ED, Yacubian J, Klinger R, Lorenz J, Buchel C (2009a) Activation of the opioidergic descending pain control system underlies placebo analgesia. Neuron 63:533–543

Eippert F, Finsterbusch J, Bingel U, Buchel C (2009b) Direct evidence for spinal cord involvement in placebo analgesia. Science 326:404

Fields HL (2000) Pain modulation: expectation, opioid analgesia and virtual pain. Prog Brain Res 122:245–253

Fields HL (2004) State-dependent opioid control of pain. Nat Rev Neurosci 5:565–575

Friebel U, Eickhoff S, Lotze M (2011) Coordinate-based meta-analysis of experimentally induced and chronic persistent neuropathic pain. NeuroImage 1–11

Geuter S, Buchel C (2013) Facilitation of pain in the human spinal cord by nocebo treatment. J Neurosci 33:13784–13790

Geuter S, Eippert F, Attar CH, Büchel C (2012) Cortical and subcortical responses to high and low effective placebo treatments. Neuroimage:1–10

Goffaux P, Redmond WJ, Rainville P, Marchand S (2007) Descending analgesia–when the spine echoes what the brain expects. Pain 130:137–143

Gracely RH, Dubner R, Deeter WR, Wolskee PJ (1985) Clinicians' expectations influence placebo analgesia. Lancet 1:43

Harris RE, Zubieta J-K, Scott DJ, Napadow V, Gracely RH, Clauw DJ (2009) Traditional Chinese acupuncture and placebo (sham) acupuncture are differentiated by their effects on mu-opioid receptors (MORs). NeuroImage 47:1077–1085

Helmstetter FJ (1992) The amygdala is essential for the expression of conditional hypoalgesia. Behav Neurosci 106:518–528

Jensen KB, Petrovic P, Kerr CE, Kirsch I, Raicek J, Cheetham A, Spaeth R, Cook A, Gollub RL, Kong J, Kaptchuk TJ (2013) Sharing pain and relief: neural correlates of physicians during treatment of patients. Mol Psychiatry 19:392–398

Johnston NE, Atlas LY, Wager TD (2012) Opposing effects of expectancy and somatic focus on pain. PLoS ONE 7:e38854

Kaptchuk TJ (2002) The placebo effect in alternative medicine: can the performance of a healing ritual have clinical significance? Ann Intern Med 136:817–825

Kaptchuk TJ, Goldman P, Stone DA, Stason WB (2000) Do medical devices have enhanced placebo effects? J Clin Epidemiol 53:786–792

Keltner J, Furst A, Fan C, Redfern R, Inglis B et al (2006) Isolating the modulatory effect of expectation on pain transmission: a functional magnetic resonance imaging study. J Neurosci 26:4437–4443

Kirsch I (1985) Response expectancy as a determinant of experience and behavior. Am Psychol 40:1189–1202

Kirsch I (1997) Response expectancy theory and application: a decennial review. Appl Prev Psychol 6:69–79

Knutson B, Adams CM, Fong GW, Hommer D (2001) Anticipation of increasing monetary reward selectively recruits nucleus accumbens. J Neurosci 21:RC159

Koban L, Ruzic L, Wager TD (2013) Brain predictors of individual differences in placebo responding. In: Colloca L, Flaten MA, Meissner K (eds) Placebo and pain (89–101). Elsevier/Academic Press, Burlington, MA

Kober H, Wager TD (2010) Meta-analysis of neuroimaging data. WIREs Cogn Sci 1:293–300

Kong J, Gollub RL, Rosman IS, Webb JM, Vangel MG, Kirsch I, Kaptachuk TJ (2006) Brain activity associated with expectancy-enhanced placebo analgesia as measured by functional magnetic resonance imaging. J Neurosci 26:381–388

Kong J, Kaptachuk T, Polich G, Kirsch I, Angel M, Zyloney C, Rosen B, Gollub R (2009a) An fMRI study on the interaction and dissociation between expectation of pain relief and acupuncture treatment. NeuroImage 47:1066–1076
Kong J, Kaptachuk TJ, Polich G, Kirsch I, Vangel MG, Zyloney C, Rosen BR, Gollub RL (2009b) Expectancy and treatment interactions: a dissociation between acupuncture analgesia and expectancy evoked placebo analgesia. NeuroImage 45:940–949
Kong J, Jensen K, Loiotile R, Cheetham A, Wey H-Y, Tan Y, Rosen B, Smoller JW, Kaptchuk TJ, Gollub RL (2013) Functional connectivity of the frontoparietal network predicts cognitive modulation of pain. Pain 154:459–467
Koyama T, McHaffie JG, Laurienti P, Coghill RC (2005) The subjective experience of pain: where expectations become reality. Proc Natl Acad Sci USA 102:12950–12955
Kross E, Berman MG, Mischel W, Smith EE, Wager TD (2011) Social rejection shares somatosensory representations with physical pain. Proc Natl Acad Sci USA 108:6270–6275
Kulkarni B, Bentley DE, Elliott R, Youell P, Watson A, Derbyshire SW, Frackowiak RS, Friston KJ, Jones AK (2005) Attention to pain localization and unpleasantness discriminates the functions of the medial and lateral pain systems. Eur J Neurosci 21:3133–3142
LaBar KS, Gatenby JC, Gore JC, LeDoux JE, Phelps EA (1998) Human amygdala activation during conditioned fear acquisition and extinction: a mixed-trial fMRI study. Neuron 20:937–945
LeDoux JE (1995) Emotion: clues from the brain. Annu Rev Psychol 46:209–235
Lee H-F, Hsieh J-C, Lu C-L, Yeh T-C, Tu C-H, Cheng C-M, Niddam DM, Lin H-C, Lee F-Y, Chang F-Y (2012) Enhanced affect/cognition-related brain responses during visceral placebo analgesia in irritable bowel syndrome patients. Pain 153:1301–1310
Levine JD, Gordon NC (1984) Influence of the method of drug administration on analgesic response. Nature 312:755–756
Levine JD, Gordon NC, Fields HL (1978) The mechanism of placebo analgesia. Lancet 2:654–657
Liberman R (1964) An experimental study of the placebo response under three different situations of pain. J Psychiatr Res 33:233–246
Lieberman MD, Jarcho JM, Berman S, Naliboff BD, Suyenobu B, Mandelkern M, Mayer EA (2004) The neural correlates of placebo effects: a disruption account. NeuroImage 22:447–455
Liljeholm M, O'Doherty JP (2012) Contributions of the striatum to learning, motivation, and performance: an associative account. Trends Cogn Sci 16:467–475
Lu H-C, Hsieh J-C, Lu C L, Niddam DM, Wu Y-T, Yeh T-C, Cheng C-M, Chang F-Y, Lee S-D (2010) Neuronal correlates in the modulation of placebo analgesia in experimentally-induced esophageal pain: A 3 T-fMRI study. Pain 148:75–83
Lui F, Colloca L, Duzzi D, Anchisi D, Benedetti F, Porro CA (2010) Neural bases of conditioned placebo analgesia. Pain 151:816–824
Maihöfner C, Herzner B, Otto Handwerker H (2006) Secondary somatosensory cortex is important for the sensory-discriminative dimension of pain: a functional MRI study. Eur J Neurosci 23:1377–1383
Matre D, Casey KL, Knardahl S (2006) Placebo-induced changes in spinal cord pain processing. J Neurosci 26:559–563
Mazzola L, Isnard J, Peyron R, Mauguiere F (2011) Stimulation of the human cortex and the experience of pain: Wilder Penfield's observations revisited. Brain 135:631–640
Meissner K, Bingel U, Colloca L, Wager TD, Watson A, Flaten MA (2011) The placebo effect: advances from different methodological approaches. J Neurosci 31:16117–16124
Miller EK (2000) The prefrontal cortex and cognitive control. Nat Rev Neurosci 1:59–65
Miller EK, Cohen JD (2001) An integrative theory of prefrontal cortex function. Annu Rev Neurosci 24:167–202
Moerman DE, Jonas WB (2002) Deconstructing the placebo effect and finding the meaning response. Ann Intern Med 136:471–476
Montgomery G, Kirsch I (1996) Mechanisms of placebo pain reduction: an empirical investigation. Psychol Sci 7:174–176

Montgomery GH, Kirsch I (1997) Classical conditioning and the placebo effect. Pain 72:107–113
Murray EA, O'Doherty JP, Schoenbaum G (2007) What we know and do not know about the functions of the orbitofrontal cortex after 20 years of cross-species studies. J Neurosci 27:8166–8169
Petrovic P, Kalso E, Petersson KM, Ingvar M (2002) Placebo and opioid analgesia– imaging a shared neuronal network. Science 295:1737–1740
Petrovic P, Kalso E, Petersson KM, Andersson J, Fransson P, Ingvar M (2010) A prefrontal non-opioid mechanism in placebo analgesia. Pain 150:59–65
Peyron R, Laurent B, Garcia-Larrea L (2000) Functional imaging of brain responses to pain. A review and meta-analysis. Clin Neurophysiol 30:263–288
Phelps EA (2006) Emotion and cognition: insights from studies of the human amygdala. Annu Rev Psychol 57:27–53
Price DD, Milling L, Kirsch I, Duff A (1999) An analysis of factors that contribute to the magnitude of placebo analgesia in an experimental paradigm. Pain 83:147–156
Price DD, Craggs J, Verne G, Perlstein W, Robinson ME (2007) Placebo analgesia is accompanied by large reductions in pain-related brain activity in irritable bowel syndrome patients. Pain 127:63–72
Rainville P (2002) Brain mechanisms of pain affect and pain modulation. Curr Opin Neurobiol 12:195–204
Rainville P, Duncan GH, Price DD, Carrier B et al (1997) Pain affect encoded in human anterior cingulate but not somatosensory cortex. Science 277:968–971
Rainville P, Carrier B, Hofbauer RK, Bushnell MC, Duncan GH (1999) Dissociation of sensory and affective dimensions of pain using hypnotic modulation. Pain 82:159–171
Rogan MT, Stäubli UV, LeDoux JE (1997) Fear conditioning induces associative long-term potentiation in the amygdala. Nature 390:604–607
Rosen JB, Donley MP (2006) Animal studies of amygdala function in fear and uncertainty: relevance to human research. Biol Psychol 73:49–60
Roy M, Shohamy D, Wager T (2012) Ventromedial prefrontal-subcortical systems and the generation of affective meaning. Trends Cogn Sci 16:147–156
Schoenbaum G, Roesch MR, Stalnaker TA, Takahashi YK (2009) A new perspective on the role of the orbitofrontal cortex in adaptive behaviour. Nat Rev Neurosci 10:885–892
Scott DJ, Stohler CS, Egnatuk C, Wang H et al (2007) Individual differences in reward responding explain placebo-induced expectations and effects. Neuron 55:325–336
Scott DJ, Stohler C, Egnatuk C, Wang H, Koeppe R, Zubieta JK (2008) Placebo and nocebo effects are defined by opposite opioid and dopaminergic responses. Arch Gen Psychiatry 65:220–231
Seeley WW, Menon V, Schatzberg AF, Keller J, Glover GH, Kenna H, Reiss AL, Greicius MD (2007) Dissociable intrinsic connectivity networks for salience processing and executive control. J Neurosci 27:2349–2356
Seymour B, O'Doherty JP, Dayan P, Koltzenburg M et al (2004) Temporal difference models describe higher-order learning in humans. Nature 429:664–667
Seymour B, O'Doherty JP, Koltzenburg M, Wiech K et al (2005) Opponent appetitive-aversive neural processes underlie predictive learning of pain relief. Nat Neurosci 8:1234–1240
Shackman AJ, Salomons TV, Slagter HA, Fox AS, Winter JJ, Davidson RJ (2011) The integration of negative affect, pain and cognitive control in the cingulate cortex. Nat Rev Neurosci 12:154–167
Szczepanski SM, Pinsk MA, Douglas MM, Kastner S, Saalmann YB (2013) Functional and structural architecture of the human dorsal frontoparietal attention network. Proc Natl Acad Sci USA 110:15806–15811
Tölle TR, Kaufmann T, Siessmeier T, Lautenbacher S, Berthele A, Munz F, Zieglgänsberger W, Willoch F, Schwaiger M, Conrad B, Bartenstein P (1999) Region-specific encoding of sensory and affective components of pain in the human brain: a positron emission tomography correlation analysis. Ann Neurol 45:40–47

Wager TD, Atlas LY, Lindquist MA, Roy M, Choong-Wan W, Kross E (2013) An fMRI-based neurologic signature of physical pain. N Engl J Med 368:1388–1397

Wager TD, Fields H (2013) Placebo analgesia. In: McMahon S, Koltzenburg M, Tracey I, and Turk DC (eds) Wall and Melzack's Textbook of pain, pp 362–373. Philadelphia: Elsevier/ Saunders

Wager TD, Rilling JK, Smith EE, Sokolik A, Casey KL, Davidson RJ, Kosslyn SM, Rose RM, Cohen JD (2004) Placebo-induced changes in FMRI in the anticipation and experience of pain. Science 303:1162–1167

Wager T, Lindquist M, Kaplan L (2007a) Meta-analysis of functional neuroimaging data: current and future directions. Soc Cogn Affect Neurosci 2:150–158

Wager TD, Scott DJ, Zubieta J-K (2007b) Placebo effects on human μ-opioid activity during pain. Proc Natl Acad Sci USA 104:11056–11061

Wager TD, Lindquist MA, Nichols TE, Kober H, Snellenberg JXV (2009) Evaluating the consistency and specificity of neuroimaging data using meta-analysis. NeuroImage 45:S210–S221

Wager TD, Atlas LY, Leotti LA, Rilling JK (2011) Predicting individual differences in placebo analgesia: contributions of brain activity during anticipation and pain experience. J Neurosci 31:439–452

Watson A, El-Deredy W, Iannetti G, Lloyd D, Tracey I, Vogt B, Nadeau V, Jones A (2009) Placebo conditioning and placebo analgesia modulate a common brain network during pain anticipation and perception. Pain 145:24–30

Whalen PJ (2007) The uncertainty of it all. Trends Cogn Sci 11:499–500

Whalen PJ, Rauch SL, Etcoff NL, McInerney SC, Lee MB, Jenike MA (1998) Masked presentations of emotional facial expressions modulate amygdala activity without explicit knowledge. J Neurosci 18:411–418

Whalen PJ, Kagan J, Cook RG, Davis FC, Kim H, Polis S, McLaren DG, Somerville LH, McLean AA, Maxwell JS, Johnstone T (2004) Human amygdala responsivity to masked fearful eye whites. Science 306:2061

Wiech K, C-s L, Brodersen KH, Bingel U, Ploner M, Tracey I (2010) Anterior insula integrates information about salience into perceptual decisions about pain. J Neurosci 30:16324–16331

Willis WD, Westlund KN (1997) Neuroanatomy of the pain system and of the pathways that modulate pain. J Clin Neurophysiol 14:2–31

Yarkoni T, Poldrack RA, Nichols TE, Van Essen DC, Wager TD (2011) Large-scale automated synthesis of human functional neuroimaging data. Nat Methods 8:665–670

Zhang W, Luo J (2009) The transferable placebo effect from pain to emotion: changes in behavior and EEG activity. Psychophysiology 46:626–634

Zubieta JK, Smith YR, Bueller JA, Xu Y, Kilbourn MR, Jewett DM, Meyer CR, Koeppe RA, Stohler CS (2001) Regional μ opioid receptor regulation of sensory and affective dimensions of pain. Science 293:311–315

Zubieta J-K, Bueller JA, Jackson L, Scott DJ et al (2005) Placebo effects mediated by endogenous opioid activity on μ-opioid receptors. J Neurosci 25:7754–7762

Placebo Analgesia: Cognition or Perception

Debora L. Morton, Wael El-Deredy, and Anthony K.P. Jones

Contents

1 Introduction 72
2 Prior Expectations 72
3 Signal Detection Errors and Cognitive Bias 73
4 Anticipatory Responses 74
5 Opioids in Placebo Analgesia and Distraction 75
6 Altered Sensory Processing 76
7 Loss of Prefrontal Regions 77
References 78

Abstract

Placebo analgesia has become a well-studied phenomenon that encompasses psychology, physiology and pharmacology. In this chapter we explore the complex interactions between these disciplines in order to argue that the placebo response is more than a simple change in perception but is a cognitive style driven by prior expectations. The expectation of treatment effect is shaped by prior information and prior experience which our brain uses to predict future events. In the case of placebo analgesia the prediction of pain relief overrules the actual feeling of pain leading to a decrease in pain sensation. This altered sensation can be attributed to personality traits, altered error monitoring processes, changes in anticipatory responses to pain and activation of the endogenous opioid system. In conclusion we discuss how altered sensory processing by descending pain modulation may play a part in placebo analgesia and how the loss of the brains prefrontal regions can make it impossible to have a placebo response.

D.L. Morton • W. El-Deredy • A.K.P. Jones (✉)
Human Pain Research Group, Institute of Brain, Behaviour and Mental Health, University of Manchester, Salford, UK
e-mail: anthony.jones@manchester.ac.uk

F. Benedetti et al. (eds.), *Placebo*, Handbook of Experimental Pharmacology 225,
DOI 10.1007/978-3-662-44519-8_4

Keywords
Placebo • Analgesia • Expectation • Cortex • Conditioning • Opioids • EEG • ERP

1 Introduction

The placebo response, once considered a nuisance in clinical trials, is now being investigated in its own right as a way to enhance treatment effects endogenously. Conditions such as pain and depression, where the outcome measures are continuous, subjective and are based on self-reports, are most likely to be subject to manipulation by placebo (Hrobjartsson and Gotzsche 2010), but the placebo response has also been noted in less subjective disorders such as Parkinson's disease (Colloca et al. 2004; De La Fuente-Fernandez et al. 2001) and asthma (Kaptchuk et al. 2008; Kemeny et al. 2007). The most studied of these conditions to manipulations by placebo is pain. Studies of placebo analgesia give us great insight into how psychological manipulations can cause physical changes in perception.

Placebo response rates are highly variable, ranging from no response to a full response. A placebo treatment will work if it has "meaning" to the individual receiving it and it is this "meaning" that is thought to cause the variance seen in placebo response rates (Moerman 2002).

Placebo response is highly dependent on prior expectation. Keeping expectations the same leads to a reproducible placebo response (Morton et al. 2009). Conversely, varying expectations, such as altering the name of the placebo, causes the response to become irreproducible (Whalley et al. 2008). Placebo response rates also vary inter-individually when the mode of treatment changes. For example, no relationship was found between subjects' responses to placebo pills and sham acupuncture (Kong et al. 2013).

At their simplest, the expectations generated by a treatment cause a change in the interpretation of the sensory information which is used to represent and understand the environment. In the case of placebo analgesia this leads to the individual experiencing a decrease in pain perception. However, the variability of the placebo response suggests that the mechanisms behind these changes in perception are much more complex. Here we look at evidence demonstrating that the placebo response is brought about by an enduring cognitive change in information processing.

2 Prior Expectations

If perception is receiving information about your environment, cognition can be viewed as learning or knowing about your environment. Learning is fundamental to placebo responsiveness. For instance, regular use of paracetamol leads to associations with the tablets' size, shape, colour, packaging and taste with pain relief. In order to learn, one must first receive information and it is this prior information that allows us to generate cognitive factors such as expectations and beliefs regarding future events. Expectations of analgesia are known to modify responses to analgesic treatment, a phenomena that is illustrated by Colloca et al 2004. In this study, patients unaware that they were receiving morphine via a computer-controlled infusion (hidden administration) experienced a significant reduction in analgesia compared to patients explicitly told they would be receiving morphine to help with their pain (open administration) (Colloca et al. 2004). The placebo component of the treatment is thought to be the difference between the open administration of the treatment and the hidden administration of treatment. The strength of these treatment expectations comprises an important component of the placebo response. Parkinson's patients given varying expectations of receiving active medication when given a placebo only experienced a significant release of dopamine when they were informed that they had a 75 % probability of receiving active medication (Lidstone et al. 2005).

Learning about treatments comes not only through our own experience, but also from knowledge we have gained from others. Gaining information from observing other people can in itself generate expectations of treatment outcome possibly by establishing "a self-projection into the future outcome (pp 33)" (Colloca and Benedetti 2009).

How do we get from an expectation of treatment to an actual placebo response? In the case of experimental placebo analgesia, one would expect that once the subject is exposed to pain after the placebo administration, they would realise that their expectation of pain relief was incorrect and would not experience an analgesic effect. Of course in some instances this is true and is a reason for the variability in magnitude of placebo response. However, in placebo responders this doesn't happen and may be explained by how our brains process sensory inputs.

3 Signal Detection Errors and Cognitive Bias

In order to quickly interpret the environment our brain constantly generates predictions about what our senses are telling us (Kveraga et al. 2007). These predictions use our past experiences and any prior information of the situation to create a picture of what is actually happening. Changes in our environment produce sensory information that can be incompatible with the model of the environment that has been generated in the brain (Yu and Dayan 2005). If the brain's predictions (top-down) and the sensory representations (bottom-up) don't match up, the two sets of information are thought to integrate through an error minimization pathway. A large error signal is then projected to a higher neural region where a new prediction refined by the error signal is generated (Friston 2005). Representing this as a computational model has shown how top-down inputs reduce the uncertainty of the stimulus representation when compared with bottom-up processing alone and leads to faster processing speeds (Siegel et al. 2000). In the placebo response, the expectation of treatment is thought to create uncertainty about incoming sensory information. Siegel's model shows how top-down/bottom-up synchrony can lead to the biased processing of top-down information.

As individuals our level of cognitive flexibility to error varies (Allan and Siegel 2002). Therefore what is immediately noticeable to one person as violating their prior expectations may be totally overlooked by another. Because there is such variability in placebo response both intra- and inter-individually, researchers are interested in being able to predict placebo responses even before placebo is administered. Studies comparing personality and placebo response indicate that suggestibility (De Pascalis et al. 2002; Morton et al. 2010a, b) and optimism (Geers et al. 2010; Morton et al. 2009) may be important correlates of placebo magnitude. How placebo responders weigh perceptual information against prior expectations has been previously tested. Screening of subjects in a visual perceptual task resulted in an experimental population of which half had a tendency to rely heavily on prior expectations, and half who tended to rely on the current perceptual information (Morton et al. 2011). Individuals who used prior expectations when making perceptual decisions in both the perceptual task and the placebo manipulation were found to have greater magnitude of placebo response (Morton et al. 2010b). These results suggest that placebo responders "ignore" the incoming sensory information to base their decisions on their prior expectations, which creates a conflict between the incoming pain signals and cognitive control. In this scenario, the placebo response should have a direct influence on electrophysiological markers of error processing. This has been shown by Koban et al. (2012) when they hypothesised that placebo analgesia "may induce a transient change in the reactivity of cognitive control networks in order to adjust for the mismatch between predicted and experienced pain" (pp 7) . The authors found that placebo analgesia was related to altered error monitoring processes in a go/nogo task. The go/nogo task was specifically designed to cause a large number of response errors and therefore a large event-related potential (ERP) on the EEG that corresponded to error processing and adjustments in behavioural control and error awareness. The error processing potential amplitude was significantly increased for placebo responders in the placebo condition compared to controls. Source reconstructions of the EEG recordings showed that this effect was probably caused by increased activation of specific medial frontal and lateral prefrontal regions, regions previously demonstrated to be vital in placebo analgesia (Krummenacher et al. 2010; Wager et al. 2004). Importantly these areas are also associated with adaptive control brain mechanisms (Botvinick et al. 2001; Ridderinkhof et al. 2004) and adjustments to expectations (Koban et al. 2012; Montague and Lohrenz 2007).

4 Anticipatory Responses

The anticipation of less pain during a placebo treatment has been suggested as an important component of placebo analgesia. Imaging a placebo conditioning procedure using fMRI showed activation in the left dorsolateral prefrontal cortex, medial frontal cortex and the anterior mid-cingulate cortex. These same areas were also found to be modulated during the anticipation of placebo analgesia (Watson et al. 2009). Learnt analgesia can have a significant effect on future anticipatory responses to pain. In a repeated placebo paradigm, participants in the placebo group not only anticipated less pain than controls after the administration of placebo but also demonstrated lowered anticipatory responses to pain before placebo administration when the treatment was repeated 2–6 weeks later (Morton et al. 2010a). Using a penalised regression procedure (LASSO-PCR) to create a model of re-analysed data from an earlier experiment (Wager et al. 2004), Wager et al. (2011) were able to predict 12 % of the variance found in the magnitude of placebo analgesia. Large magnitude placebo analgesia was related to increases in anticipatory responses in the prefrontal cortex and correlated with prior expectations of analgesia, and reduced anticipatory responses in somatosensory area 2/temporal regions. The latter probably reflects the shifting of attention away from the painful stimulus (Coghill et al. 1999). Together, these results suggest that an enduring cognitive change in anticipatory pain processing can be produced by placebo analgesia, and the engagement of emotional appraisal pathways is responsible for some of the variation in placebo analgesia.

5 Opioids in Placebo Analgesia and Distraction

Many studies have associated placebo analgesia with the activation of the endogenous opioid system and with brain areas that include the prefrontal, limbic and brainstem regions (Wager et al. 2007; Zubieta et al. 2005). Changes in activity of these brain regions are related to reductions in the physical and emotional aspects of pain experience. Placebo response is most likely initiated in the dorsolateral prefrontal cortex which is regarded as a cognitive-evaluative area. The placebo analgesic effect relies on enhanced functional coupling of the rostral anterior cingulate cortex with the hypothalamus, and brainstem areas such as the opioid receptor-rich periaqueductal grey and rostral ventral medulla (Amanzio and Benedetti 1999; Eippert et al. 2009; Wager et al. 2004, 2007), areas which have consistently shown expectancy-induced increases in relation to placebo analgesia (Atlas et al. 2010; Craggs et al. 2007; Eippert et al. 2009). The activity seen during placebo analgesia within all key regions of the descending pain modulatory system is significantly decreased with naloxone, an opioid antagonist (Amanzio and Benedetti 1999; Eippert et al. 2009; Levine et al. 1979; Zubieta et al. 2005). The placebo-dependent reduction of BOLD responses in fMRI and its reversal by naloxone is most evident in the dorsal anterior cingulate cortex (Eippert et al. 2009). Modulation of this region has been previously demonstrated in expectation manipulations (Keltner et al. 2006). During an fMRI study to image the spinal cord during pain, participants were required to do a continuous performance task (the N-back test) in order to distract them from the painful stimulus. The distraction task significantly reduced spinal responses to painful events whilst administration of naloxone during the task selectively blocked the distraction-induced reductions on reported pain (Sprenger et al. 2012). This indicates that opioids are at least partially required for both placebo responses and distraction effects. However, evidence shows placebo analgesia is not always mediated by opioids with some placebo responses being naloxone insensitive (Amanzio and Benedetti 1999; Vase et al. 2005). For example, Eippert et al. (2009) produced a blockade of placebo-induced decreases in BOLD responses, with naloxone, in regions associated with pain. However, the behavioural response was not completely blocked by naloxone as there was no significant increase in pain ratings after its administration. This suggests that pain self-reports due to placebo can be distinct from the physiological process of nociception which firmly implicates an additional non-opioidergic component to placebo analgesia.

During opioid analgesia and placebo analgesia there is consistent overlapping of brain regions involved in pain. In opioid analgesia there is more activation in the rostral anterior cingulate cortex and the anterior insula, whilst placebo analgesia generates greater responses in the lateral orbitofrontal cortex and ventrolateral prefrontal cortex. It is thought that this difference can be accounted for by the error signal generated by the discrepancy between actual pain and expectations of pain relief in placebo analgesia that is not present in opioid analgesia (Petrovic et al. 2010; Wager and Roy 2010). Colloca et al. (2004) open/hidden paradigm discussed earlier demonstrated that expectations of pain relief influence the magnitude of analgesia. Using this same paradigm to test the relationship between expectations and the opiate remifentanil, Atlas et al. (2012) showed that a hidden administration of remifentanil (no expectation of analgesia) influenced different brain regions when compared to an open administration of remifentanil (expectation of analgesia). Expectation of analgesia activated lateral and ventromedial prefrontal cortices and caused reduced responses in amygdala and pain-processing thalamic and somatosensory regions whereas analgesia caused by remifentanil without expectation of analgesia produced strong decreases in the anterior cingulate cortex and the weakest effects on somatosensory areas (S2/dorsal posterior insula). This suggests expectation operated independently but alongside remifentanil to reduce pain sensation. What these studies show us is that opioids, distraction, and placebo may have a common effect on pain, but they involve dissociable brain regions.

6 Altered Sensory Processing

As discussed in the previous section, the periaqueductal grey and the rostral ventral medulla are important in the production of the opioid-mediated placebo response. These same areas are also involved in the descending inhibition of pain by diffuse noxious inhibitory control (DNIC). DNIC was first described by LeBars et al. (1979) and is an endogenous pain-modulating system which includes descending inhibitory projections coordinated in the rostral ventral medulla. DNIC is a mechanism by which the response to painful stimulation by dorsal horn wide dynamic range neurons is inhibited by a second painful stimulus (counter-irritation). This response has been previously shown to reduce the amplitude of a spinal/nociceptive flexion reflex (RIII) (Willer et al. 1989, 1990). It has been suggested that the opioid-dependent placebo response may be attributed to, or work in parallel with, the inhibition of nociceptive processing in the dorsal horn of the spinal cord. Experimentally, expectations of hyperalgesia (nocebo) have been shown to block the normal decrease in both pain perception and the nociceptive reflex activity that is usually seen during counter-irritation (Goffaux et al. 2007). In contrast, fMRI imaging of the cervical spinal cord during painful heat together with placebo treatment significantly reduced spinal activity in response to heat compared to no treatment (Eippert et al. 2009). These findings suggest that the modulation of pain by placebo affects nociceptive signal processing at the earliest stage of the central nervous system.

7 Loss of Prefrontal Regions

The activation of opioid transmission has been seen in prefrontal brain areas (Eippert et al. 2009; Zubieta et al. 2005). In neurodegenerative conditions such as Alzheimer's disease, loss of prefrontal lobes can have severe implications for treatment effects. Benedetti et al. (2006) applied a local anaesthetic either openly or covertly to the skin of Alzheimer's patients to reduce burning pain after venipuncture. In this paradigm, as in Colloca et al. (2004), the placebo component of the treatment was shown by the difference in analgesia after expected and unexpected application of the anaesthetic. Frontal lobe damage often seen in Alzheimer's can be assessed using the frontal assessment battery, a series of simple tests which identifies impairments in cognition and motor behaviour. Patients with reduced frontal assessment battery scores showed a reduced placebo component of treatment and the reduction in placebo response was correlated with reduced cognitive status and the reduced functional connectivity of the frontal lobes to the rest of the brain. Losing the placebo component reduced the effectiveness of the treatment so much that a dose increase was needed to ensure sufficient analgesia.

Of particular interest in placebo analgesia is the involvement of the dorsolateral prefrontal cortex, an area known for cognitive and attention-related pain regulation (Lorenz et al. 2003; Miller and Cohen 2001; Peyron et al. 2000) that has been repeatedly identified in expectation-related placebo analgesia (Wager et al. 2004; Zubieta et al. 2005). Disruption of the dorsolateral prefrontal cortex has been shown to interfere with placebo analgesia. Krummenacher et al. (2010) used sham repetitive transcranial magnetic stimulation (rTMS) and an expectation of pain relief to induce an increase in pain threshold and pain tolerance indicative of a placebo response. Then using low-frequency rTMS, they artificially inhibited the function of the dorsolateral prefrontal cortex, which disrupted the placebo response and decreased pain threshold and pain tolerance. Previously, the dorsolateral prefrontal cortex has been related to the generation, maintenance and manipulation of cognitive representations (Miller and Cohen 2001; Pacheco-Lopez et al. 2006) and it has also been implicated in general attentional processes (Miller and Cohen 2001). The

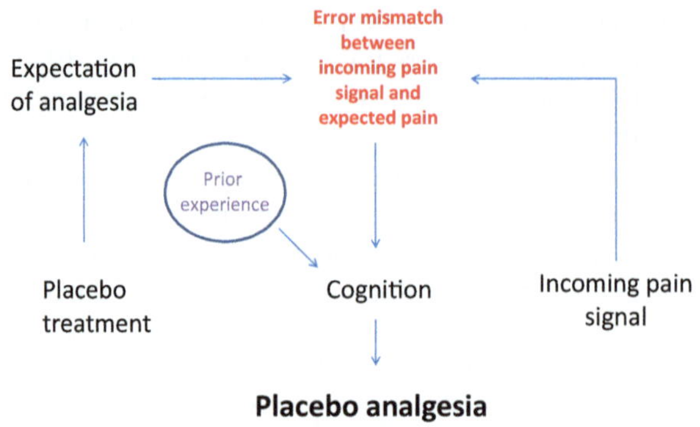

Fig. 1 The effect of cognition on placebo response using pain as an example. The placebo treatment causes an expectation of pain relief which is out of line with the incoming pain signal. Cognition uses prior experience (learnt information) to process the error signal leading to placebo analgesia

authors suggest that the loss of placebo analgesia after rTMS can be explained by the effects of disrupting the cognitive representation of analgesia and the directing of attention towards the painful stimulus.

Conclusion

If perception is the information we receive about a stimulus, cognition is how we have learnt to deal with that information. In the context of a placebo response, the stimulus information we receive is not variable but how we have learnt to deal with it using the expectations we have formed from our prior experiences is. To suggest that placebo response is due to a simple change in perception is to suggest that the placebo response is formed by a simple mechanism. Instead we see that a network of brain areas is responsible for the formation of a response, and that the frontal cortex, particularly the dorsolateral prefrontal cortex, is the core area for the cognitive modulation of pain. With no prefrontal cortex there can be no cognitive input, and with no cognitive input there can be no placebo response (Fig. 1).

References

Allan LG, Siegel S (2002) A signal detection theory analysis of the placebo effect. Eval Health Prof 25(4):410–420

Amanzio M, Benedetti F (1999) Neuropharmacological dissection of placebo analgesia: expectation-activated opioid systems versus conditioning-activated specific subsystems. J Neurosci 19(1):484–494

Atlas LY, Bolger N, Lindquist MA, Wager TD (2010) Brain mediators of predictive cue effects on perceived pain. J Neurosci 30:12964–12977

Atlas LY, Whittington R, Lindquist MA, Wielgosz J, Sonty N, Wager TD (2012) Dissociable influences of opiates and expectations on pain. J Neurosci 32(23):8053–8064

Benedetti F, Arduino C, Costa S, Vighetti S, Tarenzi L, Rainero I, Asteggiano G (2006) Loss of expectation-related mechanisms in Alzheimer's disease makes analgesic therapies less effective. Pain 121(1–2):133–144
Botvinick MM, Braver TS, Barch DM, Carter CS, Cohen JD (2001) Conflict monitoring and cognitive control. Psychol Rev 108:624–652
Coghill RC, Sang CN, Maisog JM, Iadarola MJ (1999) Pain intensity processing within the human brain: a bilateral, distributed mechanism. J Neurophysiol 82(4):1934–1943
Colloca L, Benedetti F (2009) Placebo analgesia induced by social observational learning. Pain 144(1–2):28–34
Colloca L, Lopiano L, Lanotte M, Benedetti F (2004) Overt versus covert treatment for pain, anxiety, and Parkinson's disease. Lancet Neurol 3(11):679–684
Craggs JG, Price DD, Verne GN, Perlstein WM, Robinson MM (2007) Functional brain interactions that serve cognitive-affective processing during pain and placebo analgesia. NeuroImage 38(4):720–729
De La Fuente-Fernandez R, Ruth TJ, Sossi V, Schulzer M, Calne DB, Stoessl AJ (2001) Expectation and dopamine release: mechanism of the placebo effect in Parkinson's disease. Science 293(5532):1164–1166
De Pascalis V, Chiaradia C, Carotenuto E (2002) The contribution of suggestibility and expectation to placebo analgesia phenomenon in an experimental setting. Pain 96(3):393–402
Eippert F, Bingel U, Schoell ED, Yacubian J, Klinger R, Lorenz J, Buchel C (2009) Activation of the opioidergic descending pain control system underlies placebo analgesia. Neuron 63 (533):543
Friston KJ (2005) A theory of cortical responses. Philos Trans R Soc B 360:815–836
Geers AL, Wellman JA, Fowler SL, Helfer SG, France CR (2010) Dispositional optimism predicts placebo analgesia. J Pain 11(11):1165–1171
Goffaux P, Redmond WJ, Rainville P, Marchand S (2007) Descending analgesia - when the spine echoes what the brain expects. Pain 130:137–143
Hrobjarrtsson A, Gotzsche PC (2010) Placebo interventions for all clinical conditions. Cochrane Database Syst Rev 1:CD003974. doi:10.1002/14651858.CD003974.pub3
Kaptchuk TJ, Kelley JM, Deykin A, Wayne PM, Lasagna LC, Epstein IO, Kirsch I, Wechsler ME (2008) Do "placebo responders" exist? Contemp Clin Trials 29(4):587–595
Keltner JR, Furst A, Fan C, Redfern R, Inglis B, Fields HL (2006) Isolating the modulatory effect of expectation on pain transmission: a functional magnetic resonance imaging study. J Neurosci 26(16):4437–4443
Kemeny ME, Rosenwasser LJ, Panetierri RA, Rose RM, Berg-Smith SM, Kline JN (2007) Placebo response in asthma: a robust and objective phenomenom. J Allergy Clin Immunol 199 (6):1375–1381
Koban L, Brass M, Lynn MT, Pourtois G (2012) Placebo analgesia affects brain correlates of error processing. PLoS ONE 7(11):e49784
Kong J, Spaeth R, Cook A, Kirsch I, Claggett B, Vangel M, Gollub RL, Smoller JW, Kaptchuk TJ (2013) Are all placebo effects equal? Placebo pills, sham accupuncture, cue condtioning and their association. PLoS ONE 8(7):e67485
Krummenacher P, Candia V, Folkers G, Schedlowski M, Schönbächler G (2010) Prefrontal cortex modulates placebo analgesia. Pain 148(3):368–374
Kveraga K, Boshyan B, Bar B (2007) Top-down predicitons in the cognitive brain. Brain Cogn 65:145–168
LeBars D, Dickenson AH, Besson JM (1979) Diffuse noxious inhibitory controls (DNIC). I. Effects on dorsal horn convergent neurones in the rat. Pain 6(3):283–304
Levine JD, Gordon NC, Fields HL (1979) Naloxone dose dependently produces analgesia and hyperalgesia in postoperative pain. Nature 278(5706):740–741
Lidstone SC, De La Fuente-Fernandez R, Stoessl AJ (2005) The placebo response as a reward mechanism. Sem Pain Med 3(1):37–42

Lorenz J, Minoshima S, Casey KL (2003) Keeping pain out of mind: the role of the dorsolateral prefrontal cortex in pain modulation. Brain 126(5):1079–1091

Miller EK, Cohen JD (2001) An integrative theory of prefrontal cortex function. Annu Rev Neurosci 24:167–202

Moerman D (2002) Meaning, medicine, and the "placebo effect". Cambridge University Press, Cambridge

Montague PR, Lohrenz T (2007) To detect and correct: norm violations and their enforcement. Neuron 56:14–18

Morton DL, Watson A, El-Deredy W, Jones AKP (2009) Reproducibility of placebo analgesia: effect of dispositional optimism. Pain 146(1):194–198

Morton DL, Brown CA, Watson A, El-Deredy W, Jones AKP (2010a) Cognitive changes as a result of a single exposure to placebo. Neuropsychologia 48(7):1958–1964

Morton DL, El-Deredy W, Watson A, Jones AKP (2010b) Placebo analgesia as a case of a cognitive style driven by prior expectation. Brain Res 1359:137–141

Morton DL, El-Deredy W, Morton AS, Elliott R, Jones AKP (2011) Optimism facilitates the utilisation of prior cues. Eur J Pers 25(6):424–430

Pacheco-Lopez G, Engler H, Niemi MB, Schedlowski M (2006) Expectations and associations that heal: immunomodulatory placebo effects and its neurobiology. Brain Behav Immun 20 (5):430–446

Petrovic P, Kalso E, Petersson KM, Andersson J, Fransson P, Ingvar M (2010) A prefrontal non-opioid mechanism in placebo analgesia. Pain 150(1):59–65

Peyron R, Laurent B, Garcia-Larrea L (2000) Functional imaging of brain responses to pain. A review and meta-analysis. Clin Neurophysiol 30(5):263–288

Ridderinkhof KR, Ullsperger M, Crone EA, Nieuenhuis S (2004) The role of the medial frontal cortex in cognitive control. Science 306:443–447

Siegel M, Kording KP, Konig P (2000) Integrating top-down and bottom-up sensory processing by somato-dendritic interactions. J Comput Neurosci 8:161–173

Sprenger C, Eippert F, Finsterbusch J, Bingel U, Rose M, Bnchel C (2012) Attention modulates spinal cord responses to pain. Curr Biol 22(11):1019–1022

Vase L, Robinson ME, Verne GN, Price DD (2005) Increased placebo analgesia over time in irritable bowel syndrome (IBS) patients is associated with desire and expectation but not endogenous opioid mechanisms. Pain 115(3):338–347

Wager TD, Roy M (2010) Separate mechanisms for placebo and opiate analgesia? Pain 150(1):8–9

Wager TD, Rilling JK, Smith EE, Sokolik A, Casey KL, Davidson RJ, Kosslyn SM, Rose RM, Cohen JD (2004) Placebo-induced changes in fMRI in the anticipation and experience of pain. Science 303(5661):1162–1167

Wager TD, Scott DJ, Zubieta JK (2007) Placebo effects on human {micro}-opioid activity during pain. Proc Natl Acad Sci USA 104:11056–11061

Wager TD, Atlas LY, Leotti LA, Rilling JK (2011) Predicting individual differences in placebo analgesia: contributions of brain activity during pain anticipation and pain experience. J Neurosci 31(2):439–452

Watson A, El-Deredy W, Iannetti GD, Lloyd D, Tracey I (2009) Placebo conditioning and placebo analgesia modulate a common brain network during pain anticipation and perception. Pain 145:24–30

Whalley B, Hyland ME, Kirsch I (2008) Consistency of the placebo effect. J Psychosom Res 64:537–541

Willer JC, De Broucker T, Le Bars D (1989) Encoding of nociceptive thermal stimuli by diffuse noxious inhibitory controls in humans. J Neurophysiol 62(5):1028–1038

Willer JC, Le Bars D, De Broucker T (1990) Diffuse noxious inhibitory controls in man: Involvement of an opioidergic link. Eur J Pharmacol 182(2):347–355

Yu AJ, Dayan P (2005) Uncertainty, neuromodulation, and attention. Neuron 46(4):681–692

Zubieta JK, Bueller JA, Jackson LR, Scott DJ, Xu Y, Koeppe RA, Nichols TE, Stohler CS (2005) Placebo effects mediated by endogenous opioid activity on {micro}-opioid receptors. J Neurosci 25(34):7754–7762

Pain-Related Negative Emotions and Placebo Analgesia

Magne Arve Flaten

Contents

1 Active Placebo 85
2 Is There One or More Placebo Effects? 87
3 Pain Is Modulated by Emotional Valence and Arousal 88
4 Emotions and the Placebo Response 88
5 Individual Differences in Emotions and Their Relation to Placebo Analgesia 91
6 Emotions and Homeostasis 91
7 Compensatory Reactions 93
8 Implications for Pharmacology 93
References 94

Abstract

Individuals undergoing treatment for a symptom like pain expect that the treatment will reduce the pain. Many studies show that healthy volunteers or patients in pain report less pain after inactive treatment, if they believe that active medication has been administrated. The reduction of pain can be partly blocked by systemic administration of naloxone, an opioid antagonist. There is reduced central nervous system activation to painful stimuli in individuals who have been given a placebo and told it is a painkiller. These findings suggest that the expectation of pain relief generates central nervous system opioid activity that inhibits pain transmission to the cerebral cortex. Expectations may thus lead to changes in central nervous system activity that reduces pain. It is proposed that expectations activate a homeostatic system that corrects perturbations to the system via negative feedback. The nocebo effect is the opposite of the placebo effect, and is due to induction of negative emotions. Part of the treatment of

M.A. Flaten (✉)
Department of Psychology, Norwegian University of Science and Technology, N-7491, Dragvoll, Trondheim, Norway
e-mail: magne.a.flaten@svt.ntnu.no

F. Benedetti et al. (eds.), *Placebo*, Handbook of Experimental Pharmacology 225,
DOI 10.1007/978-3-662-44519-8_5

many symptoms and diseases is due to autonomic adjustments controlled by the central nervous system. The involvement of emotional processes in placebo effects could have important consequences for interpretation of data from randomized controlled trials.

Keywords
Pain • Placebo analgesia • Placebo effect • Placebo response • Nocebo • Emotion

A treatment of symptoms or diseases like pain has at least three elements. One element is the cure itself, i.e., the drug, or other procedures that reduce the symptom. Secondly, the natural history of the disease also plays a role, as many diseases or symptoms vary in intensity or severity across time. A period of intense pain may be followed by a period with less pain. The third factor, the topic of this chapter, is regarding the expectations the patient has about the treatment.

After the ingestion of a painkiller, e.g., one expects that pain will be reduced. This expectation of improvement is the reason why people seek treatment. Has this expectation any consequences for the symptom? This is the question dealt with in this chapter.

Expectations are beliefs that some event will occur in the future. Response expectations (Kirsch 1999) are particularly important. Those are expectations of how one automatically will react to certain events, like the intake of a painkiller. It is hypothesized that response expectations can generate autonomic reactions. There are at least two important dimensions in expectations: one relates to the confidence that the response will occur, i.e., how sure you are that the painkiller will reduce pain. The other dimension is the magnitude of the response, i.e., how much the painkiller will reduce pain (Flaten 2010).

Expectations are at the top of a top-down system, where cognitions and emotions modulate sensory input to the brain, or modulate the representation of sensory input in the brain. Expectations, attitudes, and schemas govern how we react to events. Stress is one way of reacting to situations perceived as emotionally negative and uncontrollable, and stress has been linked to a number of symptoms. In short, cognitions and emotions may affect our health by changing our perceptions and, thereby, modulating stress. However, data said to support such positions are often difficult to interpret (Kemeny and Schedlowski 2007). A problem with data that link cognitions to stress and disease is the direction of causality: do cognitions affect stress and thereby modulate symptoms, or does stress modulate cognitions? Pain can increase stress, anxiety, and fear, but can negative emotions also affect pain and other symptoms?

Levine and Gordon (1984) is a classic study that makes several important points. They studied postoperative pain in patients that had third molars extracted under ordinary anesthesia. Without further treatment, postoperative pain increases linearly for at least 4 h after the operation (Levine et al. 1978). In Levine and Gordon (1984), the natural history control group received intravenous infusions of saline that was controlled by a pump that gave no cues to the patient that the infusion was performed. This procedure ensured that the natural history group received the same amount of saline as the "open infusion" group. In this group, administration of saline was performed by a person sitting at the bedside of the patient. Thus, the only difference between the natural history group and the open infusion group was that the latter group knew it was getting an infusion that could reduce the pain, whereas the natural history group did not have this knowledge. The results showed a decrease in pain report in the open infusion group compared to the natural history group of about 1.5 cm on a 10 cm scale. Since the only difference between the groups was their knowledge about the pain treatment, the difference in pain report was interpreted as a placebo analgesic response.

Some studies have shown that placebo analgesia is unaffected by naloxone. Vase et al. (2005) showed large placebo analgesic responses to experimentally induced pain in irritable bowel syndrome patients, but infusions of naloxone did not affect the placebo responses. Amanzio and Benedetti (1999) showed that the method of induction of expectations controlled whether placebo analgesia was naloxone reversible or not. They found that placebo analgesia induced by verbal information was completely antagonized by naloxone. Placebo analgesia induced via personal experience was naloxone reversible if the subjects had learned that morphine was an effective painkiller. However, if the subjects had been exposed to a non-opioid painkiller, subsequent placebo analgesic responses were not affected by naloxone.

Levine and Gordon (1984) gave placebo analgesia a physiological basis and started the neurobiological study of placebo effects. The finding that naloxone reduces placebo analgesic responses has been replicated several times (Carlino et al. 2011; Meissner et al. 2011). Administration of naloxone is indirect evidence that endorphins are involved in placebo analgesia. Lipman et al. (1990) showed increased β-endorphin sampled from cerebrospinal fluid in chronic pain patients after intrathecal saline, but only in the patients showing a placebo analgesic response. In patients not showing a placebo response, no change in β-endorphin was observed.

One central issue has been the role of endorphins and their antagonism by naloxone in experimental and clinical pain. As reviewed in ter Riet et al. (1998) most studies on experimental pain have shown that naloxone does not affect pain ratings, showing that administration of a painful stimulus does not elicit endorphin release. This is a critical point when it comes to interpretation of the finding that naloxone reverses placebo analgesic responses, as the observation that naloxone increases pain compared to a group receiving placebo could be due to antagonism of pain-elicited (and not placebo-elicited) endorphin release. Thus, it is crucial that naloxone can be shown to not affect pain in this type of experiment.

Studies on clinical pain, on the other hand, have shown that naloxone may increase pain levels. This complicates the interpretation of naloxone-induced reductions in placebo analgesia. Therefore, other methods have been used to assess the neurobiological mechanisms underlying placebo analgesia.

Benedetti et al. (1998) recorded respiratory parameters to a placebo, since one pronounced effect of opioids is respiratory depression. Thus, it was expected that a placebo respiratory depressant response should be observed together with a placebo analgesic response. A placebo depressant response was observed, which is indirect evidence of involvement of endorphins.

One objective way of assessing placebo analgesia is by electroencephalography. Experimentally induced pain stimuli with abrupt onset generate event-related or evoked potentials that can be detected in the electroencephalogram (Apkarian et al. 2005; Granovsky et al. 2008). The potentials reflect cortical activity to pain stimulation. The potentials are highly correlated with pain report (Granovsky et al. 2008). As placebo analgesic responses are hypothesized to reflect reduced pain experience, placebo analgesia should be associated with reduced evoked potentials to pain stimuli. Such a finding would support the idea that placebo analgesia is due to endogenous opioid descending mechanisms that reduce pain signals to the brain.

Watson et al. (2007) found that experimental induction of pain to the arm elicited cortical electrical activity that was reduced by application of a placebo cream. A placebo effect was also seen in the behavioral data. Wager et al. (2006) found smaller pain-elicited potentials in the placebo condition compared to the natural history condition, but only for the first half of the stimulations. In the second half of the experiment, however, no difference was found between the conditions. Aslaksen et al. (2011) also found evidence of reduced event-related potentials to painful stimulation under the influence of a placebo. Together, these studies provide indirect evidence that placebo analgesia is due to reduced pain signals to the brain.

Goffaux et al. (2007) took advantage of the fact that application of a second painful stimulus often reduces the pain elicited by a first stimulus. Goffaux et al. (2007) told one group of subjects that the application of the second stimulus would decrease their pain, while another group was told that the second painful stimulus would increase their pain. Pain ratings were clearly affected in the direction of the information, and pain-elicited reflexes, measured by electromyography from the stimulated leg, were also affected by the information. Interestingly, event-related potentials recorded at about 200 ms and later showed large differences between the groups, with smaller potentials in the group that expected the second painful stimulus to reduce the pain. This study has since been replicated (Bjørkedal and Flaten 2012) and provides evidence that expectations can reduce pain signals to the brain, since both the pain-reflex and the event-related potentials were reduced by information that a stimulus would reduce pain.

There is more detailed evidence that the placebo analgesic effect is partly due to activation of a descending pain-modulatory pathway. Eippert et al. (2009a) showed that pain-related activity in the spinal cord, in the segment in which the relevant pain pathway synapsed with second-order neurons, was reduced under placebo. This finding fits well with another finding by the same group (Eippert et al. 2009b) where administration of naloxone reduced reported placebo analgesia, and neural placebo responses in pain-related areas of the brain. These areas include the insula, the thalamus, and the rostral anterior cingulate cortex. Naloxone modulation of placebo-induced responses in important structures of the descending pain control system that involves the periaqueductal gray and the rostral ventromedial medulla was also observed. Furthermore, naloxone abolished placebo-induced coupling between rostral anterior cingulate cortex and the periaqueductal gray, which predicted neural and behavioral placebo effects, and activation of the rostral ventral medulla. Wager et al. (2004) reported that the prefrontal cortex was activated after administration of the placebo, but before administration of the painful stimulus. This activation was taken as evidence that expectations, associated with cortical activity, in ways still unknown controlled activity in the pain-modulatory network that involved the periaqueductal gray. Taken together, these findings support the notion that placebo analgesia is due to activity in a top-down system, where prefrontal cortical areas control a pain-modulating system that involves opioid mechanisms. However, as studies have shown that placebo analgesia may be opioid independent, this is just one possible mechanism underlying placebo analgesia.

1 Active Placebo

It has been proposed that expectations should be strengthened by administration of an actual drug instead of a placebo, with a consequent increase in the placebo response. Subjective effects of the drug should assure the individual that an active drug had been administrated, thus strengthening expectations of drug effects. Lyerly et al. (1964) reported that placebo effects tended to be stronger after administration of active drug. The fact that some drugs give rise to subjective side effects is a problem in testing of novel drugs, as it is difficult to do a double-blind comparison with placebo when the active drug has subjective effects. The participant in the active drug group then knows which group he/she belongs to. To solve this problem, an increasing number of clinical drug trials use active placebos instead of the ordinary inactive placebo. An active placebo is a drug that has similar subjective effects to the drug being tested, but is without specific effect on the relevant symptom.

The placebo response to active placebos has been extensively investigated in only a few studies. Lyerly et al. (1964) reported that drug-related information produced strongest effects when coupled with the administration of an actual drug. However, no statistical analysis was performed that confirmed this. Brodeur (1965) obtained similar findings. Flaten et al. (1999) [see also Flaten et al. (2004)] reported that information that a stimulant had been administrated increased reported tension compared to a group that did not receive any information, which is an example of a placebo response. When identical information was administrated together with administration of carisoprodol, a muscle relaxant, tension increased even more. Thus, when there was a conflict between the information and the drug response, the individual behaved in accordance with the information. Furthermore, the effect of the drug potentiated the placebo response to the information (Fig. 1). These results indicate that placebo responses can be strengthened by drug effects.

Bjørkedal and Flaten (2011) used caffeine as an active placebo, and tested the hypothesis that side effects of drugs can enhance expectancies and placebo responses. The logic underlying randomized placebo-controlled clinical trials

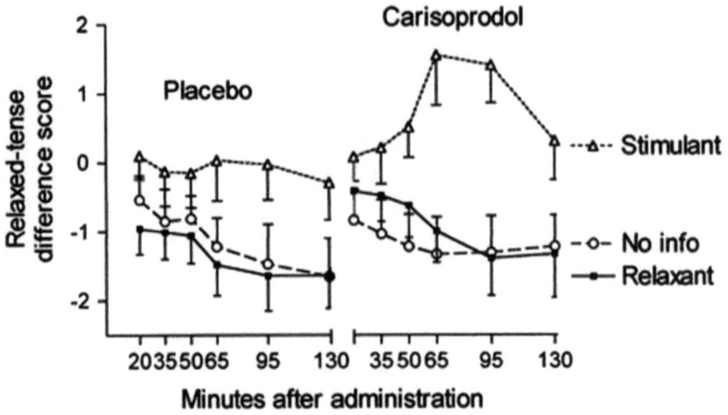

Fig. 1 Reported tension across time for groups receiving information that they received a relaxant, a stimulant, or received no information about the content of the capsules. The *left part* displays responses to information about the drug effect. All participants received placebo. The *right part* displays responses to the same information, but these groups also received the muscle relaxant carisoprodol. Note the increase in reported tension as increasing amounts of the drug were absorbed. The data are expressed as difference from pretest, performed before administration of the capsules. Thus, scores above zero indicate reports of increased tension [Reprinted from Flaten et al. (1999) with kind permission from the publisher]

(RCT) is that the psychosocial effects associated with receiving the treatment, regression to the mean and spontaneous remission should be identical in the drug and placebo groups. Hence, any differences between the groups should be ascribed to the pharmacological effects, according to the logic of the RCT. However, if the placebo response is stronger in one of the conditions, the logic underlying the RCT would not hold. Pain was induced by laser stimuli to healthy subjects before and after administration of a drink with 0 or 4 mg/kg body weight caffeine. The drink was administered with information that it contained a painkiller or that it was a placebo. Caffeine reduced pain, and information that a painkiller was administered increased the analgesic effect of caffeine, compared to caffeine administered with no information. Expectations mediated this effect. Information and expectancies had no effect on pain intensity when 0 mg/kg was administered.

Thus, the analgesic effect of caffeine was increased by information that a painkiller was administered. This was most likely due to the interaction of the pharmacological action of caffeine and expectancies. Thus, psychosocial effects accompanying were stronger when an active drug was administered compared to a placebo.

Taken together, there is some evidence that subjective feelings induced by a drug may increase placebo effects. However, there are methodological problems with studying active placebos, especially in the field of pain, as the drug may have effects on the dependent variable.

2 Is There One or More Placebo Effects?

The placebo effect is due to expectations that one has received active and effective treatment, and this is a common factor for all types of treatment, whether it be treatment against pain, Parkinson's disease, heart failure, or other diseases. One question is whether different expectations generate different reactions in different diseases. Expectations of analgesia are different from expectations that treatment should reduce the consequences of Parkinson's disease. The question is whether these two expectations generate different physiological reactions.

There are different views on the placebo response. One is the view that the mind controls the body in specific ways, and that specific expectations have specific effects. Thus, an expectation that a purported pain-relieving cream is applied to one hand should have an effect on that hand and not at other extremities. Exactly this result was obtained by Benedetti et al. (1999) and Montgomery and Kirsch (1997). A placebo analgesic response that was specific to one part of the body and not to other parts could be viewed as supporting the hypothesis that expectations have specific effects on some organs or response systems, and not others.

Watson et al. (2006) used a similar procedure, with a placebo cream applied to one arm and not the other, and with pain stimulation to both arms. It was found that one-third of the participants responded with a specific placebo response, i.e., placebo analgesia was observed in the arm where the placebo cream was applied but not in the arm without the placebo cream. However, one-third of the participants displayed placebo responses in both arms. Watson et al. (2006), in a similar procedure, found placebo effects in both arms even if a placebo cream was applied to one arm only. These findings are in line with a conception of the placebo response as a general psychophysiological reaction that affects multiple response systems.

The view presented here is that the placebo response, or part of it, is a general psychophysiological mechanism involved in stress and homeostatic regulation. From this viewpoint, the placebo response can be understood as regulation of psychophysiological processes. This may be illustrated by the general bodily processes occurring after the administration of treatment. As shown by several authors (Aslaksen and Flaten 2008; Aslaksen et al. 2011; Petersen et al. 2012; Scott et al. (2007)), administration of a painkiller reduces stress and negative emotions, which can improve several symptoms. Pain, e.g., is often increased by negative emotions like anxiety, and a reduction in anxiety reduces pain. Thus, changes in general psychophysiological processes like stress and anxiety may produce the results termed placebo effects. Several studies have shown that placebos can change the level of general arousal, supporting the idea that at least part of the placebo effect is a general process related to arousal and possibly homeostatic mechanisms (Flaten 1998; Flaten and Blumenthal 1999; Flaten et al. 2003) that can be assessed via psychophysiological methods (Flaten 1993).

3 Pain Is Modulated by Emotional Valence and Arousal

The unpleasantness of pain can be modulated by the context in which pain is experienced. Stimuli that induce negative emotions often increase pain, whereas stimuli that generate positive emotions often reduce pain. Thus, a placebo may reduce negative emotions or induce positive emotions, and thereby reduce pain via emotional mechanisms.

A number of experiments shown that negative emotional valence increases pain and positive emotions decrease pain (Rhudy et al. 2008). Emotions have been induced by photos from the International Affective Picture System (IAPS) that have been presented for several seconds, and pain has been induced after offset of the photos to control for the possibly confounding effect of attention to the slides. Rhudy et al. (2008) found that pain report, as well as the nociceptive flexion reflex and skin conductance responses, decreased linearly as a function of increasing positive emotional valence. The effect is not large, but is reliable.

The relationship between pain and emotional valence suggests that also arousal could be important, since highly positive and negative emotions are associated with larger arousal. At the high end of arousal and valence, as in severe stress, pain is reduced, called stress-induced analgesia. Herta Flor has shown that conditioned stress, induced by presenting a conditioned stimulus that has been paired with a difficult task and a distracter, reduces pain via opioid mechanisms (Flor and Grusser 1999). Thus, the relationship of emotional valence and/or arousal to pain only holds for positive emotions and moderately intense negative emotions.

4 Emotions and the Placebo Response

Does administration of treatment have an alleviating effect on negative emotions that should reduce pain and initiate a negative feedback loop (Flaten et al. 2006; Price et al. 1999). Aslaksen and Flaten (2008) (Fig. 2) induced experimental heat pain before and after administration of capsules containing corn starch, with information that they contained a powerful painkiller. A natural history control received only the painful stimulation. Placebo analgesia was observed as lower pain report in the group that received information that a painkiller had been administrated. Reported stress was also associated with the placebo analgesic response. Heart rate variability is an index of sympathetic and parasympathetic influences on heart rate, and this measure paralleled the stress data and showed lower sympathetic activation to painful stimulation after administration of a placebo. Reduced sympathetic response to painful stimulation after placebo administration has also been reported by Pollo et al. (2003). Thus, placebo analgesia was associated with reduced stress and negative emotions.

To determine whether there is a causal relation between emotions and placebo effects, emotions must be recorded in the absence of pain or other symptoms. A placebo analgesic response will be accompanied by reduced stress as pain levels are decreased. The reduced pain is most likely the reason for the reduced stress, and a

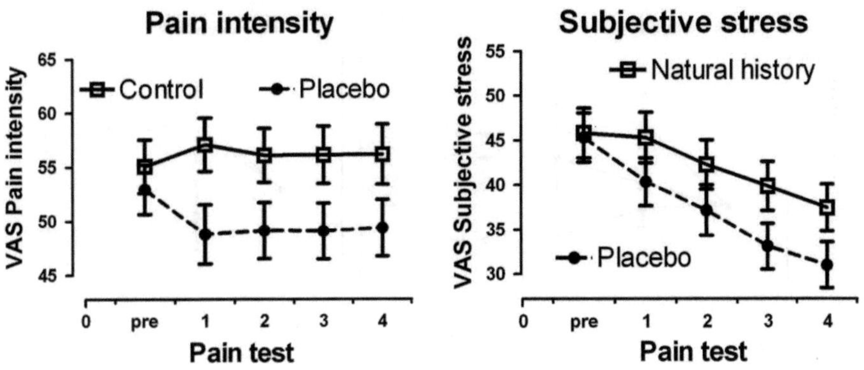

Fig. 2 *Left panel*: Reported pain intensity before and after administration of a placebo in the placebo condition. In the Control condition, pain was applied five times to control for the natural history of pain. The placebo analgesic response is the difference between the conditions in pain tests 1–4. *Right panel*: Reported stress levels before and after administration of a placebo in the placebo condition. In the natural history condition, pain was applied without any treatment or suggestion of treatment [Reprinted from Aslaksen and Flaten (2008) with kind permission from the publisher]

design where stress and pain or other symptoms are measured at different times is needed to allow conclusions about causality to be drawn.

Aslaksen et al. (2011) recorded stress and arousal in the absence of pain to observe whether information that a painkiller was administrated reduced reported stress. Heat pain was induced before and after administration of a placebo with information that it was a potent painkiller. Administration of the placebo reduced stress in two measurements after placebo, about 10 and 25 min after placebo administration, and reduced stress explained 17 and 26 % of the variance in the placebo analgesic response, respectively.

Scott et al. (2007) observed that administration of a placebo reduced pain, as well as negative affect and fear. The reduction in negative affect and fear was observed after placebo administration, but prior to pain administration, so the reduction in negative emotions was not confounded with the reduction in pain. Thus, the placebo reduced negative emotions prior to and independent of the subsequent reduction in pain, suggesting a causal link between the two. Another finding by Scott et al. (2007) was that positive emotions increased after administration of the placebo. These findings fit well with those of Vase et al. (2005) who also recorded emotions in the absence of pain. They first presented phasic painful stimuli for 20 min, and after this first phase of the study expectations of pain levels, anxiety, and desire for pain relief were recorded in the absence of painful stimuli. Thereafter, a second phase of phasic painful stimulation was in effect for about 20 min. They showed that expected pain levels, desire for pain relief, and anxiety accounted for 58 % of the placebo effect in the second phase of the experiment. However, change in expected pain was the only unique predictor of placebo analgesic responding. Similar findings were obtained by Petersen et al. (2012)

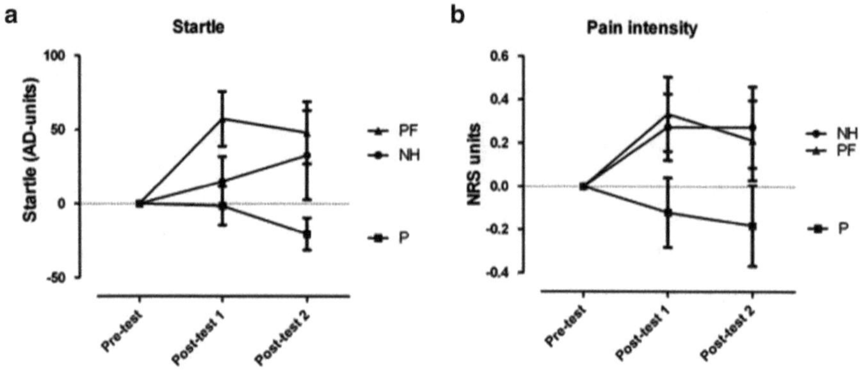

Fig. 3 Condition by test interactions in the startle reflex (**a**) and in reported pain intensity (**b**). *Vertical bars* denote ±1 SEM. *Negative numbers* indicate a reduction in response compared to the pretest, whereas *positive numbers* indicate an increase in response compared to the pretest. *NH* Natural History, *P* Placebo, *PF* Placebo + Fear. Startle reflex data show fear potentiated startle in the PF condition where the participants expected a shock to the fingers. The pain intensity data show that the induced fear abolished placebo analgesia [Reprinted from Lyby et al. (2012) with kind permission from the publisher]

who exposed patients with postoperative pain to a placebo manipulation where they received open or hidden administrations of lidocaine. The placebo reduced the area of pinprick hyperalgesia, and this placebo analgesic effect was associated with low levels of negative affect.

However, a definitive test of the hypothesis would be to induce stress or negative emotions, to investigate whether this reduces placebo analgesia. This study was performed by Peter Lyby et al. (2012). Negative emotions were induced by the anticipation of electric shock in order to investigate whether the negative emotions reduced the placebo analgesic effect. Startle eyeblink reflexes were recorded as an objective measure of negative emotions, as the startle reflex is increased by fear and other negative emotions. For pain intensity there was a trend towards a placebo effect that was abolished by induced fear, and was most pronounced in subjects who were highest in measures of fear of pain. Administration of the placebo caused a reduction in startle eyeblink reflex amplitude. However, this effect was canceled by the induced negative emotions, and was strongest among subjects high in fear of pain. Thus, induced fear abolished placebo analgesia, and mostly so in subjects who had high scores on measures of fear (Fig. 3).

In sum, these findings suggest that administration of a placebo decreases stress and negative emotions that mediate decreases in pain, i.e., placebo analgesia.

5 Individual Differences in Emotions and Their Relation to Placebo Analgesia

The observation that emotions modulate pain is related to the finding that there are individual differences in the placebo response. Individual differences in personality traits or other stable characteristics are problematic for methodological reasons. Causal inference cannot be made from observations of correlations between placebo analgesic responding and a score on measures of personality. However, an observation that a particular trait is associated with a modulation of placebo responding can be translated into an experimental test, as done by Lyby et al. (2012). Secondly, all subjects must be exposed to both the placebo and natural history conditions in order to compute a placebo response for each individual, to correlate the placebo response to other variables. A within-subject design must thus be employed. This design may induce variability due to the order of the conditions, as the subjects are more nervous at the beginning of the first session, and pain may be higher in that session. This can interfere with the placebo response. Thus, the placebo response can be underestimated in subject where the placebo session is run before the natural history session.

Only a few studies have looked at whether a trait measure of emotional responding can affect placebo analgesia. Lyby et al. (2010, 2011) showed that high fear of pain, a trait measure of how fearful individuals are for painful stimulation, reduced the placebo analgesic response. Additionally, individuals higher in fear of pain reacted with increased anticipatory stress when anticipating painful stimulation, and also reported increased pain. Taken together, these findings suggest that increased levels of stress or negative emotions reduce placebo analgesic responding. This is most pronounced in subjects high in fear of pain, as they react with increased fear and nervousness in the anticipation of pain.

6 Emotions and Homeostasis

The primary goal of autonomic function is to preserve homeostasis. Disease, symptoms, or normal stress reactions that we all experience during the day activate negative feedback mechanisms that correct or reduce the impact of stressors (Lovallo 2005). The idea proposed here is that placebo reactions are due to negative feedback mechanisms, activated by information that treatment has been administrated. Expectations have small effects if there is no deviation from homeostatic levels (Flaten et al. 2004). In such cases, compensatory reactions may result (Flaten et al. 1997; Siegel 1975, 1976).

A disease or symptom represents a deviation from the normal value or set point in a homeostatic system. A painful condition represents a deviation from a normal pain-free state, and autonomic and behavioral reactions are elicited to reduce the pain. The autonomic reactions involve activation of descending inhibitory pathways that reduce the pain signal to the brain. Thus, the pain signal activates a negative feedback loop that reduces the pain signal, via activation of endorphins, an example of a homeostatic mechanism. The contribution of expectation would be to increase the negative feedback and reduce response latency. Additionally, behavioral reactions, e.g., avoidance of movement, protection of injured body part, etc., avoid further increases in pain.

The research presented here indicates that expectations of pain treatment activate the same descending pathways that are activated by pain. The finding that naloxone reduces placebo analgesia, and that the event-related potential to painful stimulation is reduced by expectations, indicate that expectations activate opioid descending pathways that inhibit the pain signal to the brain. On the other hand, the pain-inhibitory system is also activated by high levels of stress and signals of impending stress. Thus, different stimuli may activate the pain-inhibitory system.

Studies on the neurobiology of placebo analgesia support the idea that expectations of treatment effects elicit activity in a homeostatic mechanism. A pain stimulus elicits activity in sensory fibers, Aδ-fibers that transmit the sharp and distinct first pain, and C-fibers that transmit the duller and longer lasting second pain. The pain signal is transmitted to thalamic nuclei and then the primary and secondary somatosensory cortices where the feeling of pain is generated. Pain stimuli also elicit activity in a number of other brain areas that are referred to as the pain matrix or pain network, one of these areas being the periaquaductal gray. The pain-modulatory system consists of cerebral nuclei, in the periaquaductal gray and the ventral medulla, that control activity in descending pathways (Fields et al. 2006). Stimulation of these nuclei by electrodes or by microinjections of morphine generates activity in the descending pathways, with a resultant inhibition of pain transmitting pathways. Thus, the pain signal elicits activity in descending pathways that reduce the pain signal to the brain, an example of negative feedback and homeostatic control.

Expectations seem to elicit activity in the same homeostatic system. As noted above, several studies have shown that administration of naloxone reverses or inhibits placebo analgesia. This finding is consistent with the central role for opioids in the pain-inhibitory system. Positron emission tomography studies have shown that the placebo analgesic response is correlated with activity in the periaquaductal gray (Wager et al. 2007; Zubieta et al. 2005) which is further evidence that placebo analgesia involves activity in a pain-inhibitory negative feedback loop.

In the absence of a deviation from homeostatic levels, expectations will have no or only weak effects. Several reports have investigated placebo responses in this paradigm. Flaten (1998) and Flaten et al. (1999) gave subjects information that they would receive a relaxant or stimulant drug, and found some evidence that the subjects reported being more tense compared to a control group that did not receive any information. If there is no deviation from normal levels, then homeostatic mechanisms are not brought into play and expectations have no or little effect.

7 Compensatory Reactions

The hypothesis that placebo effects represent activation of a homeostatic mechanism is congruent with findings of drug-compensatory reactions in healthy organisms (Eikelboom and Stewart 1982; Flaten 2009). Drugs administrated to healthy volunteers or animals have been shown to support drug-compensatory reactions. Flaten et al. (1997) administrated the muscle relaxant carisoprodol repeatedly to healthy volunteers. An unconditioned effect of carisoprodol is to reduce blink reflex magnitudes, and repeated administration of carisoprodol led to the development of tolerance, i.e., the drug's inhibitory effect on the blink reflex was gradually reduced. Flaten et al. (1997) showed that conditioned stimuli that signaled that carisoprodol was to be administrated gave rise to an increase in blink reflexes, i.e., a conditioned compensatory reaction. Thus, the conditioned compensatory reaction was associated with a reduced effect of the drug.

The findings of conditioned responses that compensate for the drug response suggest a homeostatic mechanism: For a healthy individual, the drug response represents a deviation from homeostasis, and physiological reactions that compensate for and reduce the drug response are elicited by the drug response. This is a form of tolerance. Siegel's (1975, 1976) work showed that tolerance could be elicited by presentations of signals of drug administration. This form of associative tolerance showed that learning could produce physiological changes that had great importance for adjustment. However, several studies have observed associative tolerance without any compensatory conditioned response.

Compensatory reactions may be seen as the opposite of placebo reactions. Nocebo reactions in the form of hyperalgesia (an increase in feelings of pain) result to a large part from anxiety (Benedetti et al. 2006). There is a large literature showing that negative emotions in many cases increase pain (Rhudy et al. 2008). In this perspective, nocebo reactions are not viewed as compensatory and homeostatic reactions exemplifying negative feedback mechanisms. Rather, nocebo reactions act as positive feedback loops, where increased anxiety leads to increased pain that in turn leads to increased anxiety, i.e., the opposite of homeostatic control and compensatory mechanisms.

8 Implications for Pharmacology

An assumption underlying the double-blind design is that the placebo effect is the same in both the active and placebo arms. However, there is some evidence that the placebo effect is stronger in the active arm compared to the placebo arm (Bjørkedal and Flaten 2011; Flaten et al. 1999). This could be due to the subjective effects of the drug that un-blinds the participants in the active arm and increases expectations, with a consequent increase in placebo responses (Flaten et al. 1999; Luparello et al. 1970). Stronger placebo effects in the active arm compared to the placebo arm in randomized controlled trials would lead to an overestimation of the drug effect compared to placebo. This important problem needs more research.

Conclusions

Research on the effect of expectations on treatment outcomes shows that the benefit of treatment is increased when drug therapy (or other treatment) is paired with positive expectations about the therapy. One underlying mechanism could be the activation of homeostatic mechanisms that control health-related reactions, e.g., descending pain-inhibitory pathways.

References

Amanzio M, Benedetti F (1999) Neuropharmalogical dissection of placebo analgesia expectation-activated opioid systems versus conditioning-activated specific subsystems. J Neurosci 19:484–494

Apkarian AV, Bushnell MC, Treede R-D, Zubieta J-K (2005) Human brain mechanisms of pain perception and regulation in health and disease. Eur J Pain 9:463–484

Aslaksen PM, Flaten MA (2008) The roles of subjective and physiological stress in the effect of a placebo on experimentally induced pain. Psychosom Med 70:811–818

Aslaksen PM, Bystad M, Vambheim SM, Flaten MA (2011) Gender differences in placebo analgesia: event-related potentials and emotional modulation. Psychosom Med 73:193–199

Benedetti F et al (1998) The specific effects of prior opioid exposure on placebo analgesia and placebo respiratory depression. Pain 75:313–319

Benedetti F, Arduino C, Amanzio M (1999) Somatotopic activation of opioid systems by target-directed expectations of analgesia. J Neurosci 19:3639–3648

Benedetti F, Amanzio M, Vighetti S, Asteggiano G (2006) The biochemical and neuroendocrine bases of the hyperalgesic nocebo effect. J Neurosci 46:12014–12022

Bjørkedal E, Flaten MA (2011) Interaction between expectancies and drug effects: an experimental investigation of placebo analgesia with caffeine as an active placebo. Psychopharmacology 215:537–548

Bjørkedal E, Flaten MA (2012) Expectations of increased or decreased pain explain the effect of conditioned pain modulation in females. J Pain Res 5:289–300

Brodeur DW (1965) The effects of stimulant and tranquilizer placebos on healthy subjects in a real-life situation. Psychopharmacologia 7:444–452

Carlino E, Pollo A, Benedetti F (2011) Placebo analgesia and beyond: a melting pot of concepts and ideas for neuroscience. Curr Opin Anesthesiol 24(5):540–544

Eikelboom R, Stewart J (1982) Conditioning of drug-induced physiological responses. Psychol Rev 89:507–528

Eippert F, Finsterbusch J, Bingel U, Buchel C (2009a) Direct evidence for spinal cord involvement in placebo analgesia. Science 326(5951):404

Eippert F, Bingel U, Schoell ED, Yacubian J, Klinger R, Lorenz J, Büchel C (2009b) Activation of the opioidergic descending pain control system underlies placebo analgesia. Neuron 63:533–543

Fields HL, Basbaum AI, Heinricher MH (2006) Central nervous system mechanisms of pain modulation. In: McMahon SB, Koltzenburg M (eds) Textbook of pain, 5th edn. Elsevier, Edinburgh, pp 125–142

Flaten MA (1993) A comparison of electromyographic and photoelectric techniques in the study of classical eyeblink conditioning and startle reflex modification. J Psychophysiol 7:230–237

Flaten MA (1998) Information about drug effects modify arousal. An investigation of the placebo response. Nor J Psychiatry 52:147–151

Flaten MA (2009) Drug effects: agonistic and antagonistic processes. Scand J Psychol 50(6):652–659

Flaten MA (2010) Expectations of pharmacological treatment and their role for adjustment. In: Powell DA (ed) CNS mechanisms involved in learned autonomic adjustments. Research Signpost, Kerala, pp 115–131

Flaten MA, Blumenthal TD (1999) Caffeine-associated stimuli generate conditioned responses: an experimental investigation of the placebo effect. Psychopharmacology 145:105–112

Flaten MA, Simonsen T, Waterloo KK, Olsen H (1997) Pharmacological classical conditioning in humans. Hum Psychopharmacol Clin Exp 12:369–377

Flaten MA, Simonsen T, Olsen H (1999) Drug-related information generates placebo and nocebo responses that modify the drug response. Psychosom Med 61:250–255

Flaten MA, Åsli O, Blumenthal TD (2003) Expectancy and placebo responses to caffeine-associated stimuli. Psychopharmacology 169:198–204

Flaten MA, Simonsen T, Zahlsen K, Sager G, Aamo T, Olsen H (2004) Stimulant and relaxant drugs combined with stimulant and relaxant information: a study of active placebo. Psychopharmacology 176:426–434

Flaten MA, Aslaksen PM, Finset A, Simonsen T, Johansen O (2006) Cognitive and emotional factors in placebo analgesia. J Psychosom Res 61(1):81–89

Flor H, Grusser SM (1999) Conditioned stress-induced analgesia in humans. Eur J Pain 3(4):317–324

Goffaux P, Redmond WJ, Rainville P, Marchand S (2007) Descending analgesia - when the spine echoes what the brain expects. Pain 130:137–143

Granovsky Y, Granot M, Nir R-R, Yarnitsky D (2008) Objective correlate of subjective pain perception by contact heat-evoked potentials. J Pain 9:53–63

Kemeny ME, Schedlowski M (2007) Understanding the interaction between psychosocial stress and immune-related diseases: a stepwise progression. Brain Behav Immun 21:1009–1018

Kirsch I (1999) Response expectancy: an introduction. In: Kirsch I (ed) How expectancies shape experience. American Psychological Association, Washington DC, pp 3–13

Levine JD, Gordon NC (1984) Influence of the method of drug administration on analgesic response. Nature 312(5996):755–756

Levine JD, Gordon NC, Fields HL (1978) The mechanism of placebo analgesia. Lancet 23:654–657

Lipman JJ, Miller BE, Mays KS, Miller MN, North WC, Byrne WL (1990) Peak B endorphin concentration in cerebrospinal fluid: reduced in chronic pain patients and increased during the placebo response. Psychopharmacology 102:112–116

Lovallo WR (2005) Stress and health, 2nd edn. Sage, Thousand Oaks

Luparello TJ, Leist N, Lourie CH, Sweet P (1970) Interaction of psychologic stimuli and pharmacologic agents on airway reactivity in asthmatic subjects. Psychosom Med 32:509–513

Lyby PS, Aslaksen PM, Flaten MA (2010) Is fear of pain related to placebo analgesia? J Psychosom Res 68(4):369–377

Lyby PS, Aslaksen PM, Flaten MA (2011) Variability in placebo analgesia and the role of fear of pain - an ERP study. Pain 152(10):2405–2412

Lyby PS, Åsli O, Forsberg JT, Flaten MA (2012) Induced fear reduces the effectiveness of a placebo intervention on pain. Pain 153(5):1114–1121

Lyerly SB, Ross S, Krugman AD, Clyde DJ (1964) The effects of instructions upon performance and mood under amphetamine sulphate and chloral hydrate. J Abnorm Soc Psychol 68:321–327

Meissner K, Colloca L, Bingel U, Wager TD, Watson A, Flaten MA (2011) The placebo effect: advances from different methodological approaches. J Neurosci 31(45):16117–16124

Montgomery GH, Kirsch I (1997) Classical conditioning and the placebo effect. Pain 72:107–113

Petersen GL, Finnerup NB, Nørskov KN, Grosen K, Pilegaard HK, Benedetti F, Price DD, Jensen TS, Vase L (2012) Placebo manipulations reduce hyperalgesia in neuropathic pain. Pain 153 (6):1292–1300

Pollo A, Vighetti S, Rainero I, Benedetti F (2003) Placebo analgesia and the heart. Pain 102(1–2):125–133

Price DD, Milling LS, Kirsch I, Duff A, Montgomery GH, Nicholls SS (1999) An analysis of factors that contribute to the magnitude of placebo analgesia in an experimental paradigm. Pain 83:147–156

Rhudy JL, Williams AE, McCabe KM, Russell JL, Maynard LJ (2008) Emotional control of nociceptive reactions (ECON): do affective valence and arousal play a role? Pain 136(3):250–261

Scott DJ, Stohler CS, Egnatuk CM, Wang H, Koeppe RA, Zubieta J-K (2007) Individual differences in reward responding explain placebo-induced expectations and effects. Neuron 55:325–336

Siegel S (1975) Evidence from rats that morphine tolerance is a learned response. J Comp Physiol Psychol 89:498–506

Siegel S (1976) Morphine analgesic tolerance: its situation specificity supports a Pavlovian conditioning model. Science 193:323–325

ter Riet G, de Craen AJM, de Boer A, Kessels AGH (1998) Is placebo analgesia mediated by endogenous opioids? A systematic review. Pain 76:273–275

Vase L, Robinson ME, Verne GN, Price DD (2005) Increased placebo analgesia over time in irritable bowel syndrome (IBS) patients is associated with desire and expectation but not endogenous opioid mechanisms. Pain 115:338–347

Wager TD, Rilling JK, Smith EE, Sokolik A, Casey KL, Davidson RJ et al (2004) Placebo induced changes in fMRIin the anticipation and experience of pain. Science 303(5661):1162–1167

Wager TD, Matre D, Casey KL (2006) Placebo-effects in laser-evoked pain potentials. Brain Behav Immun 20:219–230

Wager TD, Scott DJ, Zubieta J-K (2007) Placebo effects on human mu-opioid activity during pain. PNAS 104:11056–11061

Watson A, El-Deredy W, Bentley DE, Vogt BA, Jones AKP (2006) Categories of placebo response in the absence of site-specific expectation of analgesia. Pain 126:115–122

Watson A, El-Deredy W, Vogt BA, Jones AKP (2007) Placebo analgesia is not due to compliance or habituation: EEG and behavioural evidence. Neuroreport 18:771–775

Zubieta J-K, Bueller JA, Jackson LR, Scott DJ, Xu YJ, Koeppe RA, Nichols TE, Stohler CS (2005) Placebo effects mediated by endogenous opioid activity on mu-opioid receptors. J Neurosci 25:7754–7762

How Positive and Negative Expectations Shape the Experience of Visceral Pain

Sigrid Elsenbruch

Contents

1 Introduction: Visceral Pain and Irritable Bowel Syndrome 98
2 Relevance of Placebo and Nocebo Effects in Visceral Pain 99
3 Mechanistic Studies 101
 3.1 Experimental Placebo Studies 102
 3.2 Experimental Nocebo Studies 106
4 Perspectives and Future Research Goals 109
 4.1 Learning/Conditioning 109
 4.2 Trait and State Emotions 110
 4.3 Sex/Gender Differences 112
References 114

Abstract

Knowledge from placebo and nocebo research aimed at elucidating the role of treatment expectations and learning experiences in shaping the response to visceral pain fills an important research gap. First, chronic abdominal pain, such as in irritable bowel syndrome (IBS), is highly prevalent, with detrimental individual and socioeconomic impact and limited effective treatment options. At the same time, IBS patients show high placebo response rates in clinical trials and benefit from placebo interventions. Second, psychological factors including emotions and cognitions in the context of visceral pain have been implicated in the pathophysiology of IBS and other conditions characterized by medically unexplained somatic symptoms. Hence, the study of nocebo and placebo effects in visceral pain constitutes a model to assess the contribution of psychological factors. Herein, the clinical relevance of visceral pain is introduced with a focus

S. Elsenbruch (✉)
Institute of Medical Psychology and Behavioral Immunobiology, University Hospital Essen, University of Duisburg-Essen, Essen, Germany
e-mail: Sigrid.Elsenbruch@uk-essen.de

F. Benedetti et al. (eds.), *Placebo*, Handbook of Experimental Pharmacology 225,
DOI 10.1007/978-3-662-44519-8_6

on IBS as a bio-psycho-social disorder, followed by a review of existing clinical and experimental work on placebo and nocebo effects in IBS and in clinically relevant visceral pain models. Finally, emerging research trends are highlighted along with an outlook regarding goals for ongoing and future research.

Keywords
Visceral pain • Functional gastrointestinal disorders • Irritable bowel syndrome • Visceral hyperalgesia

1 Introduction: Visceral Pain and Irritable Bowel Syndrome

Visceral pain is a common symptom of great clinical and socioeconomic significance in many areas of medicine. Patients experience acute, recurrent and/or chronic visceral pain in many medical disciplines, including internal medicine, gynaecology, visceral surgery, urology, and general medicine. Especially in patients presenting with chronic or recurrent abdominal pain, it is often difficult to identify an unequivocal organic cause, at least with established diagnostic tools. After exclusion of a number of common organic conditions such as inflammatory bowel disease (IBD) or esophagitis, chronic abdominal complaints are often classified as one of the *functional gastrointestinal disorders* (FGDs). The FGDs are considered an important public health problem because they are remarkably prevalent, can be disabling, and constitute a major individual, social, and economic burden (Agarwal and Spiegel 2011; Maxion-Bergemann et al. 2006). Irritable Bowel Syndrome (IBS), the most common FGD with prevalence rates of 8–23 % (Choung and Locke 2011; Talley 2008) is characterized by recurrent abdominal pain or discomfort in combination with disturbed bowel habits in the absence of identifiable organic cause. FGDs such as IBS are more prevalent in women (at least in Western countries) and often present with comorbid gastrointestinal, somatic, and psychological/psychiatric symptoms resulting in a significant overlap with other diagnoses. This overlap does not only exist with conditions associated with other, primarily gastrointestinal symptomatology, including chronic pelvic pain, faecal incontinence, or chronic constipation, but also with diagnoses involving extra-intestinal symptoms such as fibromyalgia and chronic fatigue syndrome (Choung and Locke 2011; Frissora and Koch 2005). All these conditions share a high incidence of psychiatric or psychological comorbidities, especially anxiety, depression, and somatization disorder with typical personality alterations including high neuroticism and catastrophizing and altered healthcare-seeking behaviour (Folks 2004; Whitehead et al. 2002). Finally, a history of abuse, early adverse life events and trauma has been linked to the onset of symptoms in a significant proportion of patients with medically unexplained bodily symptoms (Bradford et al. 2012).

Current etiological concepts for FGDs unequivocally assume bio-psycho-social models (Elsenbruch 2011; Tanaka et al. 2011). Consequently, it is assumed that the pathophysiology is multi-factorial encompassing biological, psychological, and social mechanisms mediated by the central, autonomic, neuroendocrine, and immune systems. Whereas earlier concepts focussed on the role of altered motility in FGDs, more recent evidence has led to a paradigm shift with a strong focus on altered afferent and central processing of painful stimuli. Specifically, visceral hyperalgesia (or hypersensitivity) and visceral hypervigilance constitute key concepts in current research on pathophysiological mechanisms. The role of central nervous system mechanisms along the "brain-gut axis" is increasingly appreciated, owing to accumulating evidence from brain imaging studies that the neural processing of painful visceral stimuli is altered in IBS together with long-standing knowledge regarding the contribution of stress and negative emotions to symptom frequency and severity. At the same time, there is growing evidence suggesting that peripheral and local immune mechanisms and disturbed neuro-immune communication could play a role in the pathophysiology of visceral hyperalgesia (Elsenbruch 2011).

Although little mortality is associated with FGDs including IBS, effective treatment is often difficult leading to (or exacerbating) multiple and costly medical procedures, decreased compliance and altered healthcare-seeking behaviour. Existing treatment options range from a number of symptom-oriented pharmacological options to psychological treatments, including psychotherapy and hypnotherapy (Enck et al. 2010). Given the bio-psycho-social disease model (Tanaka et al. 2011), it is recommended that these conditions are treated with interdisciplinary, personalized treatment approaches that require particular attention to the doctor–patient relationship (Palsson and Drossman 2005). Hence, apart from the high clinical relevance of visceral pain in FGDs, these conditions in general and IBS in particular can be viewed as "model conditions" in the development and testing of conceptual approaches aimed at understanding and improving the integration of the psychosocial context into treatment concepts. Indeed, within the field of clinically oriented placebo research, one primary goal is to integrate patient expectations and experiences into more "personalized" treatment approaches that integrate medical and psychological aspects. Interestingly, it is indeed in patients with IBS that the remarkable clinical effectiveness of a placebo-based intervention has been demonstrated in a clinical trial (Kaptchuk et al. 2008, 2010). Clearly, these seminal findings have catapulted visceral pain and IBS into the focus of placebo researchers both in basic and clinical sciences who strive to transfer knowledge from placebo research into clinical application.

2 Relevance of Placebo and Nocebo Effects in Visceral Pain

The seminal results of the above mentioned clinical trial with placebo acupuncture in IBS (Kaptchuk et al. 2008), together with another trial revealing the feasibility and clinical effectiveness of "open" placebo treatment without deception

(Kaptchuk et al. 2010), have impressively demonstrated the putative clinical potential of placebo treatment in IBS. Moreover, there exist at least two additional considerations that drive ongoing efforts to produce more experimental and clinical data on placebo and nocebo effects in visceral pain both in patients and healthy individuals. The first is the fact that IBS patients (and patients with other types of gastrointestinal conditions including IBD and GERD) demonstrate large placebo responses in clinical trials. For example, in a recent meta-analysis Ford et al. found that in 73 eligible RCTs including 8,364 patients with IBS allocated to placebo, pooled placebo response rate across all RCTs was 37.5 % (Ford and Moayyedi 2010). Similar results were reported in an earlier, smaller meta-analysis including 45 placebo-controlled RTCs (Patel et al. 2005). Herein, the population-weighted average placebo response rate was 40.2 % (Patel et al. 2005). Finally, in a meta-analysis of 19 randomized and placebo-controlled complementary and alternative medicine (CAM) trials, the pooled estimate of the placebo response rate was 42.6 %, and hence comparable when compared to "conventional" medical therapy trials (Dorn et al. 2007). However, it should also be noted that placebo response rates in functional bowel disorders (functional dyspepsia, irritable bowel syndrome) trials are similar to those in other pain conditions and are also comparable with other organic gastrointestinal diseases (duodenal ulcer, inflammatory bowel diseases) (Enck et al. 2012). Despite these facts, findings of high placebo response rates in FGDs, irrespective of differences or similarities with other conditions, have in fact contributed to a "negative image" of placebo effects as "nuisance" (Enck et al. 2008) which hinders rather than helps efforts to identify effective treatment options for FGDs. This negative view is only slowly being replaced by a more constructive appreciation of the chances associated with an improved understanding of psychological factors in general and placebo/nocebo knowledge in particular (Enck et al. 2013; Finniss et al. 2010; Price et al. 2008) with interesting contributions to our understanding of the pathophysiology and treatment of chronic abdominal pain (Elsenbruch 2011; Enck et al. 2012; Lu and Chang 2011). This is paralleled by more refined knowledge regarding the conceptualization, design, and analysis of clinical trials (Enck et al. 2013; Rief et al. 2011).

Secondly, there is a growing appreciation for the potential of placebo and nocebo research in interdisciplinary science aimed at elucidating the pathophysiology of chronic abdominal pain and IBS. Indeed, placebo analgesia and nocebo hyperalgesia constitute fruitful experimental models to assess the contribution of psychological factors in altered responses to visceral stimuli in general and visceral hyperalgesia in particular. In fact, it has been prominently noted already several years ago that "…these forms of hyperalgesia are also highly modifiable by placebo and nocebo factors […], synergistic interactions occur between placebo/nocebo factors and enhanced afferent processing so as to enhance, maintain, or reduce hyperalgesia in IBS" (Price et al. 2009), thereby catapulting placebo/nocebo issues "at the heart" of a multi-factorial, psychosocial disease model. Since then, a number of experimental studies assessing the mechanisms mediating placebo and nocebo effects in IBS and healthy volunteers have been accomplished. These findings, reviewed in the following section, have not only contributed to our understanding of placebo and nocebo effects in visceral pain, but have also highlighted the pivotal role of psychological factors in the response to visceral pain. Indeed, the study of nocebo and placebo effects in visceral pain constitutes a model to assess the contribution of psychological factors to the pathophysiology of IBS and other clinical conditions associated with chronic abdominal pain and medically unexplained bodily complaints (Elsenbruch 2011).

3 Mechanistic Studies

Knowledge about the neurobiology and neuropsychology underlying placebo and nocebo effects in visceral pain is steadily improving, but overall experimental evidence is much more limited in visceral pain when compared to somatic pain. Of note, separate studies in clinically relevant visceral pain models are important given significant differences between visceral and somatosensory signal processing both in the periphery and within the central nervous system. In fact, several fMRI studies support distinct processing of somatosensory and visceral pain in the human brain (Aziz et al. 2000; Dunckley et al. 2005a, 2007; Eickhoff et al. 2006). Similar differences also appear to exist within the brainstem (Dunckley et al. 2005b). Furthermore, attentional modulation of pain intensity perception for visceral and somatic pain, respectively, is reflected in different brain regions (Dunckley et al. 2007), which is interesting in the context of placebo-induced pain modulation. Finally, recent evidence showed that although statistically significant, the correlation between individual pain thresholds for visceral and somatic stimulation is relatively weak (Horing et al. 2013). Hence, although no studies exist that have directly compared the neural mechanisms mediating placebo analgesia in somatic vs. visceral pain models, it appears highly likely that the brain mechanisms differ. Therefore, studies on *visceral* placebo analgesia in no way duplicate but rather complement and extend findings from research using somatic pain models and/or address other chronic pain conditions. Using a barostat, pressure-controlled distensions of the rectum or oesophagus can be accomplished, and this procedure constitutes a clinically relevant, valid, and reliable visceral pain model. This paradigm represents the "gold standard" in the study of visceral sensitivity, very closely induces ("mimics") visceral discomfort or pain as well as urge-to-defecate (in the case of rectal distension), and is safely applicable in healthy subjects as well as patients. It also allows the determination of sensory thresholds for perception and pain such that individualized stimuli at pre-determined intensity levels for application in studies, including fMRI studies, can be chosen. Of note, with one exception of esophageal distensions (Lu et al. 2010), all experimental placebo and nocebo studies in the visceral pain field, reviewed below, have applied rectal distensions.

3.1 Experimental Placebo Studies

The group around D. Price was the first to conduct experimental placebo studies in the visceral pain field. Several studies, which all used the rectal distension model, were conducted within IBS patients (Price et al. 2007; Vase et al. 2003, 2005): The first study (Vase et al. 2003) documented that IBS patients reported significant reductions in rectal distension-induced pain intensity and pain unpleasantness in the placebo condition (i.e. verbal suggestions for pain relief regarding an inactive gel that was applied to the rectal balloon). The study also included conditions with rectal and oral lidocaine application, respectively. Interestingly, no differences were found between the placebo and either lidocaine condition and given previous findings by the same group showing that rectal lidocaine reversed visceral hyperalgesia (Verne et al. 2005), the authors concluded that "adding a verbal suggestion for pain relief can increase the magnitude of placebo analgesia to that of an active agent" (Vase et al. 2003). As this constituted the very first placebo study in the visceral pain field, this conclusion proved "prophetic" in the sense that today—a decade later—there is good evidence to support that placebo interventions may be used not only to enhance or complement conventional treatment approaches for IBS, but in fact to use them instead of pharmacological treatments (Kaptchuk et al. 2008, 2010). In a second study (Vase et al. 2005), the authors could again show a large placebo effect in a group of IBS patients in the same pain and placebo analgesia paradigm (i.e. rectal distensions delivered with instructions of pain relief). Interestingly, the placebo effect reportedly increased over time, while ratings of expected pain, desire for pain relief and anxiety decreased successively, resulting in more variable placebo responses during later parts of the experimental session. Based on these findings, the authors suggested that a reduction in negative emotions could play a role in placebo analgesia (Vase et al. 2005) (for a more detailed discussion of findings regarding the putative role of emotions, see Sect. 4). Further, the authors found no effect of naloxone treatment on the placebo response, indicating that herein the placebo effect was not associated with (or mediated by) endogenous opioids (Vase et al. 2005). This negative finding is interesting given broad evidence that somatic placebo analgesia involves the endogenous opioid system (Benedetti 1996; Benedetti et al. 2005; Eippert et al. 2009; Petrovic et al. 2002; Zubieta et al. 2005). This raises the question if indeed the mechanisms mediating placebo analgesia may be specific for pain modality and/or condition.

Owing to the growing appreciation of the crucial role of the brain in pain processing in general and placebo analgesia in particular, several groups have since then accomplished mechanistic placebo studies in visceral pain using brain imaging techniques. The first published brain imaging study on placebo effects in visceral pain was a positron emission tomography (PET) study (Lieberman et al. 2004). Herein, the brain response to rectal distensions in IBS patients was analyzed both before and after a 3-week placebo regimen. Increases in ventrolateral prefrontal cortex (VLPFC) activity from pre- to post-placebo treatment predicted self-reported symptom improvement, and this relationship was mediated by changes in dorsal anterior cingulate cortex (Lieberman et al. 2004). The second brain imaging study (Price et al. 2007) used fMRI to assess rectal distension-induced brain activation in patients with IBS in the same rectal placebo paradigm described above (Vase et al. 2003, 2005). The results revealed large reductions in pain ratings and in distension-induced brain activation within pain-related regions (i.e. thalamus, somatosensory cortices, insula, and anterior cingulate cortex) in the placebo condition. The authors noted that decreases in activity were related to suggestion (i.e. expectation) and a second factor ("habituation/attention/conditioning") (Price et al. 2007). Two re-analyses (Craggs et al. 2007, 2008) of this first fMRI study in IBS patients (Price et al. 2007) were subsequently carried out: One connectivity analysis described the interactions of neural networks during placebo analgesia using structural equation models (Craggs et al. 2007), the other focussed on the temporal characteristics of neural networks activated during placebo analgesia (Craggs et al. 2008).

The above studies were carried out exclusively in patients with IBS. Placebo analgesia and its underlying neural mechanisms were first described for healthy humans in a study utilizing an esophageal distension pain model (Lu et al. 2010). The authors reported large reductions of pain extent and pain ratings, along with reduced brain activity in the visceral pain matrix (i.e. thalamus, somatosensory cortices, insula, prefrontal cortex, and anterior cingulate cortex) in the placebo condition in healthy subjects (Lu et al. 2010). Interestingly, this was also the first study to pay attention to pain anticipation, which appears to play a significant role in subsequent responses to pain. Herein, the VLPFC was associated with increased activity during anticipation of visceral pain, which was interpreted as evidence in support of "top-down control" in the modulation of the pain experience (Lu et al. 2010). Utilizing the rectal distension model, our group has implemented several expectation-induction and learning procedures to study visceral placebo (along with nocebo) responses utilizing behavioural, peripheral, and central measures including fMRI (Benson et al. 2012; Elsenbruch et al. 2012a, b; Kotsis et al. 2012; Schmid et al. 2013, 2014; Theysohn et al. 2014). In this series of studies, our first main goal was to clarify the role of expectation in visceral placebo analgesia in healthy volunteers (Elsenbruch et al. 2012a). To do so, we delivered visceral pain stimuli in three expectation conditions designed to vary the level of expectancy regarding the intravenous administration of a supposed analgesic drug which was in reality saline. In a within-subject design with a counterbalanced order of conditions, participants were told that they had a 100, 50, or 0 % chance of receiving the active drug. The results revealed that the expectation of pain relief effectively reduced perceived painfulness of visceral stimuli in a "dose-dependent" manner, i.e. the greater the expectation of analgesia, the more pronounced the placebo analgesic effect. Analysis of blood-oxygen-level-dependent (BOLD) responses during cued pain anticipation and painful stimulation revealed that placebo analgesia was associated with activity changes in the thalamus, prefrontal, and somatosensory cortices in placebo responders when comparing the 100 % and 0 % expectation conditions (Elsenbruch et al. 2012a). Expectation-induced changes in cortical activation were particularly pronounced for the pain anticipation phase, underscoring the pivotal role of pain anticipation in central pain modulation during placebo-induced positive expectation, consistent with findings in the oesophageal placebo analgesia study reviewed above (Lu et al. 2010). In a follow-up analysis of the 50 % expectation condition of this dataset, we could show that *perceived* treatment allocation affected behavioural and neural responses to placebo treatment (Kotsis et al. 2012). Given a 50 % probability of receiving active treatment, the magnitude of placebo-induced subjective pain relief and pain-induced neural activation was significantly greater in subjects who believed to be in the active treatment group. These findings have interesting implications for clinical trials where patients are typically told that they have a 50 % chance of receiving the active drug. Our most recent work has subsequently focussed on implementing and testing experimental paradigms to study nocebo hyperalgesia in parallel to placebo analgesia in visceral pain, in order to be able to better understand these opposite effects and their underlying mechanisms at the behavioural and neural levels. These studies (Elsenbruch et al. 2012b; Schmid et al. 2013) are summarized in detail below (see Sect. 3.2).

All studies reviewed thus far were conducted either exclusively within IBS patients or exclusively within healthy controls, which precludes an assessment of possible alterations in the neural response during placebo analgesia in patients with chronic abdominal pain. To this date, there exist only two studies (Lee et al. 2012; Schmid et al. 2014), one of them from our group, that *directly* compared placebo analgesia responses in IBS patients and a healthy control group. In the first study (Lee et al. 2012), placebo analgesia was induced by a combination of verbal suggestions and a prior learning experience (i.e. "pre-conditioning") involving a technical manipulation to simulate a potent analgesic effect. The results of this study revealed comparable placebo analgesia responses in IBS patients compared to healthy controls in subjective parameters, including pain ratings. Interestingly, greater anxiety responses were negatively correlated with the magnitude of placebo-induced subjective pain reduction, which led the authors to suggest that higher affective disturbances in IBS patients may predict a weak placebo effect. Furthermore, despite comparable placebo responses at the behavioural level, there was greater activity in affective and cognitive brain regions, including the insula, cingulate cortex, and VLPFC in IBS patients during placebo analgesia, suggesting altered neural processing of placebo-induced changes in pain perception in IBS (Lee et al. 2012). These data are supported by our own recently published data (Schmid et al. 2014) showing similar behavioural placebo analgesia but altered neural modulation in IBS patients not only when compared to healthy controls but also to patients with ulcerative colitis in remission, suggesting a specific deficit in endogenous pain inhibition due to affective disturbances in IBS (Fig. 1).

Together, these studies impressively demonstrate that placebo-induced cognitive and/or learning processes are highly relevant for central and behavioural pain responses not only in patients but also in healthy controls. At the same time, there exist several areas where more knowledge is urgently needed (for details, see Sect. 4). In this context of mechanistic studies, two aspects appear most important: First, more studies comparing the mechanisms mediating placebo responses in patient groups compared to healthy control groups are clearly needed to

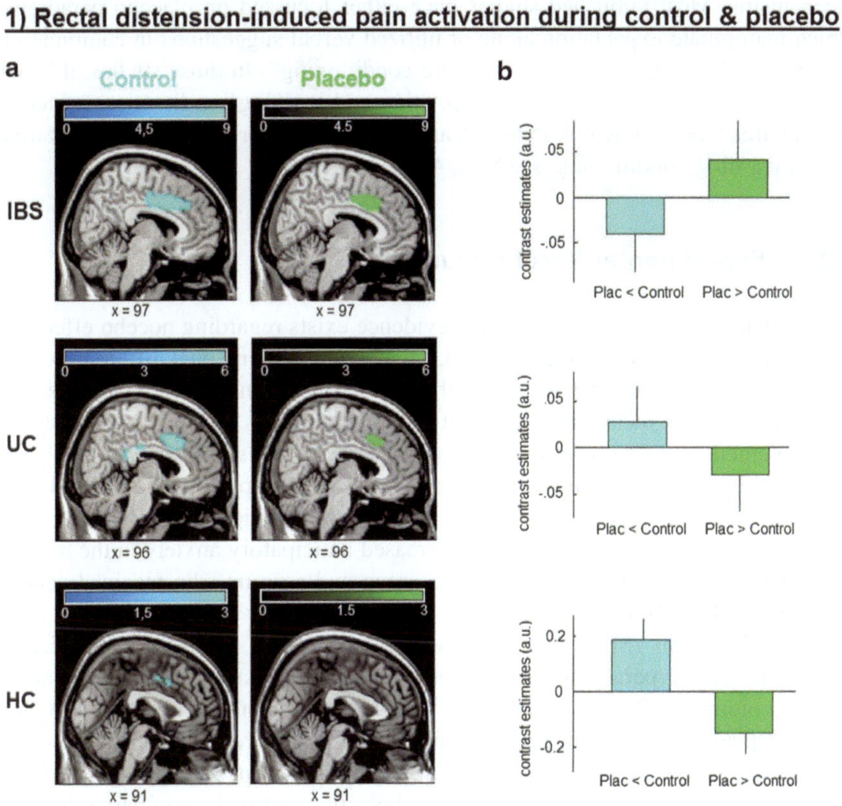

Fig. 1 (a) Rectal distension-induced neural activation in the midcingulate cortex in the control condition (i.e. neutral expectations induced by instructions of receiving saline, *left column*, activation shown in *blue colour*) and placebo condition (i.e. expectation of pain-relief induced by deceptive instructions of receiving a spasmolytic drug, *right column*, activation shown in *green colour*) in patients with irritable bowel syndrome (IBS, *upper row*), patients with ulcerative colitis in remission (UC, *middle row*), and healthy controls (HC, *lowest row*). Results of within-group analyses on the contrast (placebo > off) using one-sample *t*-tests revealing significantly reduced pain-induced neural activation in the placebo condition in UC and HC but not in IBS, resulting in significant group differences upon two-sample *t*-tests (not shown). Images overlaid on a structural T_1-weighted MRI used for spatial normalization and thresholded at $p < 0.05$ uncorrected using an anatomical mask for visualization purposes; colour bars indicate *t*-score. (**b**) Plots of contrast estimates of changes in pain-related neural activation in the respective differential contrast within each group for the cingulate cortex, a.u., arbitrary units [Adapted from Schmid et al. (2014)]

complement and extend the only two existing study (Lee et al. 2012; Schmid et al. 2014). Second, although expectation and conditioning have been identified as the two major neuropsychological mechanisms mediating placebo and nocebo effects, in the above reviewed experimental research on visceral placebo analgesia there exists virtually no data addressing the putative role of conditioning/learning mechanisms. Most published studies have either focussed on placebo paradigms which manipulate expectation alone or utilized verbal suggestions in combination with a prior learning experience (i.e. "pre-conditioning") In those studies, it is not possible to disentangle effects of expectation and learning/conditioning, which is another area where more research is clearly needed (for more details on the putative role of learning/conditioning, see Sect. 4).

3.2 Experimental Nocebo Studies

To this date, virtually no experimental evidence exists regarding nocebo effects in visceral pain. To close this research gap, we recently implemented different experimental approaches to investigate nocebo effects in a clinically relevant visceral pain paradigm (i.e. rectal distensions) in healthy volunteers. In a behavioural study, we implemented a combination of negative verbal suggestions about (supposed) pain sensitization and a prior learning experience of surreptitiously enhanced pain intensity (i.e. "pre-conditioning"). The results revealed significantly greater pain ratings (i.e. nocebo hyperalgesia) and increased anticipatory anxiety in the nocebo group when compared to both a placebo group and a group who received neutral instructions (Elsenbruch et al. 2012b).

In a subsequent fMRI study, we assessed the neural mechanisms mediating visceral nocebo hyperalgesia along with placebo analgesia in a separate group of healthy volunteers (Schmid et al. 2013). To do so, effects of negative (nocebo) and positive (placebo) treatment expectations following intravenous application of an inert substance on the response to painful rectal distensions were analysed in two groups: Whereas the placebo group received positive instructions of pain relief due to the supposed application of a spasmolytic drug with analgesic properties, the nocebo group was instructed about an increase in pain due to the application of the opioid antagonist naloxone. In reality, only saline was administered in all groups. Within each group, there a control condition was implemented (in counterbalanced order) during which participants received truthful neutral instructions of saline application, allowing us to directly contrast positive and negative expectations, respectively, with neutral expectations in analyses of BOLD responses. As expected, results in the placebo group revealed significantly reduced rectal-distension induced perceived pain (Fig. 2a) along with a reduction of pain-induced neural activation within the insula (Fig. 2b). Interestingly, the nocebo group showed increased perceived pain (Fig. 2a), which was paralleled by increased insula activation during painful stimulation when comparing negative and neutral expectations (Fig. 2c) (Schmid et al. 2013).

Given that the insula is crucial for interoception, multi-modal sensory integration as well as pain-related decision making and emotional awareness (Craig 2003; Linnman et al. 2011; Wiech et al. 2010), these findings are an important step in identifying the brain mechanism(s) mediating visceral pain modulation by expectations. Since our insula finding during nocebo hyperalgesia is in line with existing brain imaging data on nocebo hyperalgesia for somatic pain (Bingel

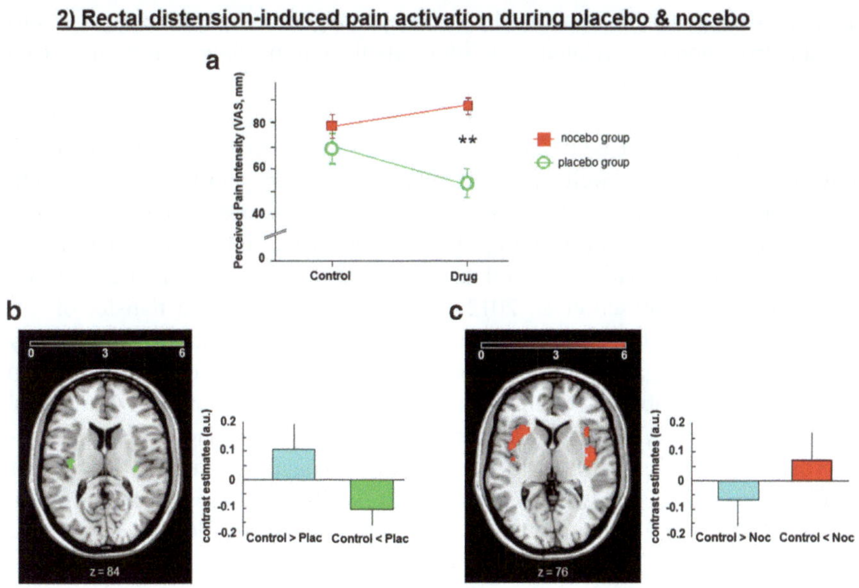

Fig. 2 (**a**) Visual analogue scale (VAS, 0–100 mm) ratings of perceived pain intensity in response to rectal distensions in the placebo and nocebo groups during neutral expectations (control: truthful instructions of saline administration) and deceptive drug-specific expectations (i.e. placebo group: instructions of a spasmolytic drug; nocebo group: instructions of an opioid antagonist). The placebo and nocebo groups differed significantly in perceived pain ratings during drug-specific expectations (**results of post-hoc independent samples t-test: $p < 0.001$). Data are shown as mean ± standard error of the mean (SEM). (**b**) Rectal distension-induced modulation of neural activation by deceptive verbal suggestions within the placebo group revealing significantly reduced activation of the insula during positive (placebo) compared to neutral expectations (control). (**c**) Within the nocebo group, insula activation was significantly increased during negative (Noc) when compared to neutral expectations (control). *Left columns* in **b, c**: Images overlaid on a structural T_1-weighted MRI used for spatial normalization and thresholded at $p < 0.01$ using an anatomical mask and uncorrected for visualization purposes. *Colour bars* indicate t score. *Right columns* in **b, c**: Plots of parameter estimates of changes in pain-related neural activation in differential contrasts within the insula, a.u., arbitrary units [Adapted from Schmid et al. (2013)]

et al. 2011; Kong et al. 2008), one may conclude that there may exist at least some shared brain regions for central pain modulation by cognitions and/or emotions, irrespective of pain modality. Interestingly, we previously documented more pronounced insula modulation in a negative emotional context, induced by psychosocial stress, in IBS patients (Elsenbruch et al. 2010), supporting a role of the insula in pain modulation also in patients with chronic abdominal pain.

Our nocebo results are especially interesting in light of recent evidence that negative treatment expectancy abolished opioid analgesia in a somatic pain model (Bingel et al. 2011). Together, these findings strongly underscore that negative expectations induced by verbal suggestions shape the response to pain, irrespective of the presence of an actual analgesic drug. Furthermore, they extend our own previous work in which we induced nocebo hyperalgesia by negative expectations resulting from non-drug-related suggestions about an impending worsening of pain together with surreptitiously increased distension pressures ("pre-conditioning", see above) (Elsenbruch et al. 2012b). Together, these data support that nocebo effects in experimental pain can occur as a result of verbal suggestions in the context of active as well as inert pharmacological substances (Benedetti et al. 2006; Bingel et al. 2011; Scott et al. 2008), and also in situations that induce negative expectations not resulting from drug-related information but rather disease-related information as well as from learning or conditioning (Colloca et al. 2008, 2010; Jensen et al. 2012; Kong et al. 2008). For a transfer of this knowledge into clinical application, it is important to appreciate that in daily clinical routine, negative expectations regarding worsening of symptoms can occur through a number of possible factors which may or may not involve an actual drug or medication (Colloca and Miller 2011). In fact, there is increasing appreciation that it is the *entire* context surrounding medical encounters that shapes patients' expectations and hence placebo and nocebo responses in daily clinical practice (Colloca and Miller 2011). By inference, attempts to systematically reduce or minimize nocebo effects in clinical settings will have to address treatment-specific as well as non-treatment directed negative expectations, which could pose a challenge that researchers will have to overcome once more data becomes available describing nocebo effects in clinical settings—which is thus far not available in the field of visceral pain.

In conclusion, taking together results from experimental pain research in visceral and other pain models unequivocally underscores the "power" of positive and negative expectancies in shaping the response to pain not only at the behavioural level but also within the brain. Indeed, brain imaging studies have made a fundamental contribution to leaving behind earlier criticism that placebo or nocebo responses are merely the result of a response bias (Price et al. 2008) and moving to discerning the neural mechanisms mediating placebo/nocebo-induced alterations in endogenous pain inhibition. At the same time, the role of genetic (Hall et al. 2012) and peripheral mechanisms, including neuroendocrine and immune mediators (Elsenbruch et al. 2012b; Kokkotou et al. 2010), is beginning to emerge. Continuing this work in the visceral pain field will be vital for us to gain a more complete picture encompassing the complex interactions between the central nervous system and the periphery during visceral placebo and nocebo responses as a basis for much needed clinically oriented research not only in IBS patients but also in other gastrointestinal conditions such as inflammatory bowel diseases (Bonaz and Bernstein 2013).

4 Perspectives and Future Research Goals

4.1 Learning/Conditioning

More research is needed to discern the role of learning/conditioning processes in placebo and especially in nocebo effects in visceral pain models and in patients with chronic visceral pain. As reviewed above, there currently exists only very limited experimental and clinical evidence on nocebo effects in visceral pain, and the few existing studies have primarily focused on the role of negative expectations. However, nocebo findings in somatic pain models support a role of learning/conditioning processes in nocebo effects, although available research does remain scarce and heterogeneous also in the somatic pain field. For example, Colloca et al. showed that one session of conditioning (i.e. pairing coloured lights to stimuli that were surreptitiously increased or reduced in intensity) was sufficient to induce nocebo responses to non-painful and painful stimuli (Colloca et al. 2010), but these responses extinguished rapidly. Four sessions of conditioning led to more robust nocebo responses that did not extinguish as rapidly, supporting that the "strength" of learning is related to the magnitude of the nocebo response. On the other hand, in an earlier study, the same group reported that a preconditioning procedure did not increase allodynia and hyperalgesia induced by verbal suggestions alone (Colloca et al. 2008), leading the authors to conclude that learning may be less in important in nocebo hyperalgesia when compared to its role in placebo analgesia. More recently, Jensen et al. implemented visual cues indicating high and low pain to induce nocebo and placebo responses (Jensen et al. 2012). While visual cues were designed to be clearly visible in one experiment, there occurred non-conscious (masked) exposure to the same cues in a second experiment. The results revealed significant nocebo effects for both clearly visible as well as masked visual cues, supporting that nocebo effects can be induced without conscious awareness of the predictive cues (Jensen et al. 2012).

These findings provide an interesting link to another learning-based experimental paradigm that is well-established in the field of learning and memory, namely fear conditioning. Fear conditioning and its extinction is based on the principles of classical conditioning and is an established translational model both in the context of anxiety disorders, drug addiction and relapse, and (chronic) pain (Milad and Quirk 2012). During fear conditioning, neutral stimuli are repeatedly paired with aversive unconditioned stimuli (US). In many studies, the US is a painful stimulus, most commonly electric shock, which is consistently paired with neutral visual stimuli during a learning/conditioning phase (Sehlmeyer et al. 2009). As a result of contingent pairing of neutral stimuli and US, the previously neutral stimuli turn into predictive cues that are now fear-provoking conditioned stimuli (CS) even when presented alone. When applied to the putative conditioning/learning mechanisms mediating nocebo hyperalgesia, this model may prove useful and relevant since conditioned pain-related anticipatory fear likely contributes to hyperalgesia and its underlying central mechanisms. In other words, one may reconceptualise the CSs as pain-signalling predictive cues capable of eliciting (or enhancing) nocebo responses. Indeed, the concept that classical conditioning is relevant for placebo effects is well-established in classically conditioned immunosuppression, but thus far it has not been systematically studied in the context of nocebo hyperalgesia in pain models. There does, however, exist evidence from human and animal research supporting a link between conditioned fear and hyperalgesia. For example, effects of conditioned fear on somatic pain thresholds have been documented in healthy volunteers (Williams and Rhudy 2007). In animal models of visceral hypersensitivity, learned associations between predictive contextual cues and painful stimuli were reportedly relevant for the development of visceral hypersensitivity (Tyler et al. 2007) and for the retrieval of visceral pain-conditioned passive avoidance (Wang et al. 2011). In IBS patients, conditioning led to reduced pain thresholds (Nozu et al. 2006) and a role of interoceptive fear conditioning in visceral pain has been proposed (De Peuter et al. 2011). Hence, associative learning and extinction processes appear to be involved in hyperalgesia and may thereby contribute to nocebo responses in pain. Although the vast majority of existing human fear conditioning studies utilized non-visceral USs (e.g. electric shock), it is possible to implement fear conditioning with oesophageal or rectal distensions as effective US (Kattoor et al. 2013, 2014; Schmid et al. 2013; Yágüez et al. 2005). Based on these initial studies, more research is needed to provide data addressing the putative role of learning/conditioning in nocebo effects. In doing so, it will be important to disentangle classically conditioned processes that may operate without conscious awareness from primarily "conscious" learning experiences that trigger cognitions because of expectations.

Finally, in the context of learning mechanisms, recent advances have been made pointing to the role of social/observational learning in nocebo hyperalgesia, which may also prove relevant for patients with visceral pain (Swider and Babel 2013; Vögtle et al. 2013). Herein, it will be necessary to develop and test appropriate paradigms to assess observational/social learning in nocebo (as well as placebo) effect using clinically relevant visceral pain models.

4.2 Trait and State Emotions

The putative relevance of psychological trait and state variables is only beginning to be understood, and there is growing evidence from the somatic pain field to suggest a role of emotions in nocebo as well as placebo responses (Flaten et al. 2011). In nocebo effects, the role of negative emotions, especially anxiety and stress, has previously been documented in several pain models, including experimental ischemic arm pain (Benedetti et al. 2006; Johansen et al. 2003), painful mechanical and/or electrical stimulation (Colloca et al. 2008, 2010; van Laarhoven et al. 2011), and heat pain (Kong et al. 2008) in healthy subjects, as well as in patients with postoperative pain (Benedetti et al. 1997). Given the small number of available nocebo studies in visceral pain models, it is difficult to ascertain if negative emotions play a similar role in visceral nocebo hyperalgesia. In our own first nocebo study implementing the rectal distension pain model in healthy individuals, we observed significantly greater anticipatory state anxiety in the nocebo group (Elsenbruch et al. 2012b). On the other hand, our subsequent fMRI nocebo study revealed significant nocebo hyperalgesia in the nocebo group in the absence of obvious changes in state anxiety or tension (Schmid et al. 2013). Based on these negative findings, we concluded that it is possible for nocebo hyperalgesia to occur in the absence of increased negative emotions.

The notion that placebo effects could be mediated at least in part by reduced negative emotions has been put forward (Flaten et al. 2011) based on the previously established connection between placebo analgesia and reward processing (Petrovic et al. 2005). However, with one recent exception (Lyby et al. 2012), no studies exist thus far which have directly manipulated emotions in order to directly test for changes in placebo and/or nocebo responses. In this study, the authors tested effects of experimentally-induced fear (i.e. anticipation of electric shock) on subsequent placebo analgesia in a somatic pain model. The results supported that induced fear abolished placebo analgesia, especially in participants with high fear of pain (Lyby et al. 2012). These findings fit together nicely with correlative evidence from the study by Lee et al. (see above) showing that within IBS patients, high scores on the Hospital Anxiety and Depression Scale correlated significantly and negatively with indicators of the placebo analgesia response (Lee et al. 2012). These findings led the authors to suggest that affective disturbances (. . .) "might partially predict a weak placebo effect in IBS patients" (Lee et al. 2012), which is clearly supported by our own recent IBS data showing a correlation of weaker placebo analgesia with higher depression scores on the Hospital Anxiety and depression scale (Schmid et al. 2014). Given proper replication also in other pain models and conditions, these findings have important implications for attempts to bring experimental findings from the placebo field into the clinic. Herein, it will then be important to incorporate and systematically take into account emotional state and trait variables of the patient, including fear of pain (Lyby et al. 2011). This would be especially important in clinical settings that are per se anxiety provoking, such as in the context of receiving treatment following a frightening diagnosis or awaiting a potentially painful treatment such as a surgical intervention. Clearly, more knowledge about effects of trait and state emotions on placebo analgesia and nocebo hyperalgesia is needed to improve our understanding of inter-individual differences in placebo and nocebo responses. Given effects of positive and negative expectations on drug efficacy (Bingel et al. 2011) and effects of patients–provider interaction on the magnitude of placebo effects (Kaptchuk et al. 2008), a more refined understanding about the role of emotional context factors will be crucial for optimizing doctor-patient communication irrespective of treatment with a "real" drug or a placebo. This can ultimately lead to more effective and "personalized" (placebo) treatments while minimizing unwanted nocebo effects.

4.3 Sex/Gender Differences

Knowledge regarding possible sex and/or gender differences in placebo analgesia is scarce, and virtually non-existent in nocebo hyperalgesia (Bjorkedal and Flaten 2012; Swider and Babel 2013). Indeed, a recent review concluded that "studies are urgently required in order to better understand the role of sex-gender on placebo mechanism and its impact on randomized clinical trials outcomes" (Franconi et al. 2012). Attempts to draw conclusions about possible sex differences in placebo/nocebo responses in visceral pain are further complicated by the fact that the role of sex or gender in the response to and central processing of visceral stimuli themselves (without additional placebo/nocebo modulation) remains incompletely understood. Although sex differences in the prevalence of the functional gastrointestinal disorders including IBS are well-documented (Chang et al. 2006b; Fillingim et al. 2009; Mayer et al. 2004; Mogil and Bailey 2010; Unruh 1996), experimental evidence is scares and heterogeneous. Studies on sex differences in visceral sensitivity revealed conflicting and even contradictory results. For example, in healthy subjects, results demonstrated no sex differences (Kern et al. 2001; Sloots et al. 2000; Soffer et al. 2000), reduced perception but normal pain thresholds in females (Kim et al. 2006), and increased discomfort thresholds in females (Chang et al. 2006a). For patients with IBS, current evidence is similarly conflicting (Berman et al. 2000; Chang et al. 2006a; Mertz et al. 1995). At the level of neural processing of visceral stimuli, the few available brain imaging studies supported sex differences in IBS patients (Berman et al. 2000; Labus et al. 2008; Naliboff et al. 2003). In healthy subjects, on the other hand, the few existing fMRI studies revealed contradictory results. Whereas Berman et al. found a trend for greater activation in males in the insula, anterior, and midcingulate cortex compared to females (Berman et al. 2006), Kern et al. reported the opposite result, i.e. an activation of the insula and anterior cingulate cortex (ACC) only in females, but not in males (Kern et al. 2001). In age- and BMI-matched healthy subjects, we assessed sex differences in rectal sensory and pain thresholds along with the neural response to painful rectal stimuli (Benson et al. 2012). Our analysis of rectal thresholds revealed no differences between males and females. At the level of the brain, males and females demonstrated a largely comparable pattern of neural activation in the majority of pain-processing brain regions, although there was a tendency for females to show a slightly different activation of prefrontal regions during cued anticipation and pain (Benson et al. 2012). Building on our fear conditioning work with rectal pain as unconditioned stimulus (Gramsch et al. 2014; Kattoor et al. 2013, 2014), we recently documented sex differences in the neural mechanisms mediating fear conditioning, extinction, and reinstatement in healthy males and females (Benson et al. 2014), with interesting implications for the putative role of learned pain-related fear in nocebo hyperalgesia (Elsenbruch 2011).

Given these complex findings, it is not surprising that current evidence from placebo research is similarly conflicting. There exists some evidence from experimental placebo studies and clinical trials focusing on somatic pain (Aslaksen et al. 2011; Averbuch and Katzper 2001; Butcher and Carmody 2012; Compton et al. 2003; Flaten et al. 2006), nausea (Klosterhalfen et al. 2009; Weimer et al. 2012), and IBS symptoms (Kelley et al. 2009). Herein, there are results supporting either greater placebo responses in men (Aslaksen et al. 2011; Butcher and Carmody 2012; Compton et al. 2003; Flaten et al. 2006) or in women (Kelley et al. 2009), or suggest no sex differences (Averbuch and Katzper 2001). Hence, it remains elusive if one sex shows larger placebo responses and may hence be considered to be more "placebo-prone". Of note, in this context it is crucial not only to consider the sex of the participant but also that of the investigator which reportedly plays a role at least in somatic pain responses (albeit without placebo modulation) (Aslaksen et al. 2007; Gijsbers and Nicholson 2005). For example, in a heat pain experiment, Aslaksen et al. reported that investigator x subject sex interaction influenced pain ratings with lower pain reports in male subjects given female investigators (Aslaksen et al. 2007). In addition, a recent study on nocebo hyperalgesia induced by social observational learning revealed that the magnitude of nocebo hyperalgesia was greater after a male model was observed, regardless of the sex of the subject (Swider and Babel 2013). Taken together, these initial results clearly indicate that there exist complex interactions between sex, sex hormones and gender (roles) that are likely to contribute to placebo and nocebo effects, which need to be addressed in future studies.

Conclusions

Results from clinical and experimental research in the field of visceral pain complement and extend findings from other pain modalities and in chronic somatic pain conditions. Together, this growing body of evidence unequivocally underscores the "power" of positive and negative expectancies and learning experiencing in shaping the response to pain not only at the behavioural level but also within the brain, with profound clinical implications. Indeed, brain imaging studies have made a fundamental contribution to leaving behind earlier criticism that placebo or nocebo responses may merely reflect response bias and moving to discerning the neural mechanisms mediating placebo/nocebo-induced alterations in endogenous pain inhibition. At the same time, the role of peripheral mechanisms, including mediators of the HPA axis and the autonomic nervous system, is beginning to be understood such that a more complete picture encompassing the complex interactions between the central nervous system and the periphery during placebo and nocebo responses is beginning to emerge. Within a bio-psycho-social conceptualization of placebo and nocebo mechanisms, emotions constitute primary targets for future research aimed at elucidating the *modulators* of placebo and nocebo responses both in experimental and clinical studies in the context of visceral pain and beyond.

References

Agarwal N, Spiegel BM (2011) The effect of irritable bowel syndrome on health-related quality of life and health care expenditures. Gastroenterol Clin North Am 40:11–19

Aslaksen PM, Myrbakk IN, Hoifodt RS, Flaten MA (2007) The effect of experimenter gender on autonomic and subjective responses to pain stimuli. Pain 129:260–268

Aslaksen PM, Bystad M, Vambheim SM, Flaten MA (2011) Gender differences in placebo analgesia: event-related potentials and emotional modulation. Psychosom Med 73:193–199

Averbuch M, Katzper M (2001) Gender and the placebo analgesic effect in acute pain. Clin Pharmacol Ther 70:287–291

Aziz Q, Thompson DG, Ng VW, Hamdy S, Sarkar S, Brammer MJ, Bullmore ET, Hobson A, Tracey I, Gregory L, Simmons A, Williams SC (2000) Cortical processing of human somatic and visceral sensation. J Neurosci 20:2657–2663

Benedetti F (1996) The opposite effects of the opiate antagonist naloxone and the cholecystokinin antagonist proglumide on placebo analgesia. Pain 64:535–543

Benedetti F, Amanzio M, Casadio C, Oliaro A, Maggi G (1997) Blockade of nocebo hyperalgesia by the cholecystokinin antagonist proglumide. Pain 71:135–140

Benedetti F, Mayberg HS, Wager TD, Stohler CS, Zubieta JK (2005) Neurobiological mechanisms of the placebo effect. J Neurosci 25:10390–10402

Benedetti F, Amanzio M, Vighetti S, Asteggiano G (2006) The biochemical and neuroendocrine bases of the hyperalgesic nocebo effect. J Neurosci 26:12014–12022

Benson S, Kotsis V, Rosenberger C, Bingel U, Forsting M, Schedlowski M, Gizewski ER, Elsenbruch S (2012) Behavioural and neural correlates of visceral pain sensitivity in healthy men and women: does sex matter? Eur J Pain 16:349–358

Benson S, Kattoor J, Kullmann JS, Hofmann S, Engler H, Forsting M, Gizewski ER, Elsenbruch S (2014) Towards understanding sex differences in visceral pain: enhanced reactivation of classically-conditioned fear in healthy women. Neurobiol Learn Mem 109:113–121

Berman S, Munakata J, Naliboff BD, Chang L, Mandelkern M, Silverman D, Kovalik E, Mayer EA (2000) Gender differences in regional brain response to visceral pressure in IBS patients. Eur J Pain 4:157–172

Berman SM, Naliboff BD, Suyenobu B, Labus JS, Stains J, Bueller JA, Ruby K, Mayer EA (2006) Sex differences in regional brain response to aversive pelvic visceral stimuli. Am J Physiol Regul Integr Comp Physiol 291:R268–R276

Bingel U, Wanigasekera V, Wiech K, Ni Mhuircheartaigh R, Lee MC, Ploner M, Tracey I (2011) The effect of treatment expectation on drug efficacy: imaging the analgesic benefit of the opioid remifentanil. Sci Transl Med 3:70ra14

Bjorkedal E, Flaten MA (2012) Expectations of increased and decreased pain explain the effect of conditioned pain modulation in females. J Pain Res 5:289–300

Bonaz BL, Bernstein CN (2013) Brain-gut interactions in inflammatory bowel disease. Gastroenterology 144:36–49

Bradford K, Shih W, Videlock EJ, Presson AP, Naliboff BD, Mayer EA, Chang L (2012) Association between early adverse life events and irritable bowel syndrome. Clin Gastroenterol Hepatol 10:385–390

Butcher BE, Carmody JJ (2012) Sex differences in analgesic response to ibuprofen are influenced by expectancy: a randomized, crossover, balanced placebo-designed study. Eur J Pain 16:1005–1013

Chang L, Mayer EA, Labus JS, Schmulson M, Lee OY, Olivas TI, Stains J, Naliboff BD (2006a) Effect of sex on perception of rectosigmoid stimuli in irritable bowel syndrome. Am J Physiol Regul Integr Comp Physiol 291:R277–R284

Chang L, Toner BB, Fukudo S, Guthrie E, Locke GR, Norton NJ, Sperber AD (2006b) Gender, age, society, culture, and the patient's perspective in the functional gastrointestinal disorders. Gastroenterology 130:1435–1446

Choung RS, Locke GR III (2011) Epidemiology of IBS. Gastroenterol Clin North Am 40:1–10

Colloca L, Miller FG (2011) The nocebo effect and its relevance for clinical practice. Psychosom Med 73:598–603

Colloca L, Sigaudo M, Benedetti F (2008) The role of learning in nocebo and placebo effects. Pain 136:211–218

Colloca L, Petrovic P, Wager TD, Ingvar M, Benedetti F (2010) How the number of learning trials affects placebo and nocebo responses. Pain 151:430–439

Compton P, Charuvastra VC, Ling W (2003) Effect of oral ketorolac and gender on human cold pressor pain tolerance. Clin Exp Pharmacol Physiol 30:759–763

Craggs JG, Price DD, Verne GN, Perlstein WM, Robinson MM (2007) Functional brain interactions that serve cognitive-affective processing during pain and placebo analgesia. NeuroImage 38:720–729

Craggs JG, Price DD, Perlstein WM, Verne GN, Robinson ME (2008) The dynamic mechanisms of placebo induced analgesia: evidence of sustained and transient regional involvement. Pain 139:660–669

Craig AD (2003) Interoception: the sense of the physiological condition of the body. Curr Opin Neurobiol 13:500–505

De Peuter S, Van Diest I, Vansteenwegen D, Van denBergh O, Vlaeyen JW (2011) Understanding fear of pain in chronic pain: interoceptive fear conditioning as a novel approach. Eur J Pain 15:889–894

Dorn SD, Kaptchuk TJ, Park JB, Nguyen LT, Canenguez K, Nam BH, Woods KB, Conboy LA, Stason WB, Lembo AJ (2007) A meta-analysis of the placebo response in complementary and alternative medicine trials of irritable bowel syndrome. Neurogastroenterol Motil 19:630–637

Dunckley P, Wise RG, Aziz Q, Painter D, Brooks J, Tracey I, Chang L (2005a) Cortical processing of visceral and somatic stimulation: differentiating pain intensity from unpleasantness. Neuroscience 133:533–542

Dunckley P, Wise RG, Fairhurst M, Hobden P, Aziz Q, Chang L, Tracey I (2005b) A comparison of visceral and somatic pain processing in the human brainstem using functional magnetic resonance imaging. J Neurosci 25:7333–7341

Dunckley P, Aziz Q, Wise RG, Brooks J, Tracey I, Chang L (2007) Attentional modulation of visceral and somatic pain. Neurogastroenterol Motil 19:569–577

Eickhoff SB, Lotze M, Wietek B, Amunts K, Enck P, Zilles K (2006) Segregation of visceral and somatosensory afferents: an fMRI and cytoarchitectonic mapping study. NeuroImage 31:1004–1014

Eippert F, Bingel U, Schoell ED, Yacubian J, Klinger R, Lorenz J, Büchel C (2009) Activation of the opioidergic descending pain control system underlies placebo analgesia. Neuron 63:533–543

Elsenbruch S (2011) Abdominal pain in irritable bowel syndrome: a review of putative psychological, neural and neuro-immune mechanisms. Brain Behav Immun 25:386–394

Elsenbruch S, Rosenberger C, Bingel U, Forsting M, Schedlowski M, Gizewski ER (2010) Patients with irritable bowel syndrome have altered emotional modulation of neural responses to visceral stimuli. Gastroenterology 139:1310–1319

Elsenbruch S, Kotsis V, Benson S, Rosenberger C, Reidick D, Schedlowski M, Bingel U, Theysohn N, Forsting M, Gizewski ER (2012a) Neural mechanisms mediating the effects of expectation in visceral placebo analgesia: an fMRI study in healthy placebo responders and nonresponders. Pain 153:382–390

Elsenbruch S, Schmid J, Basler M, Cesko E, Schedlowski M, Benson S (2012b) How positive and negative expectations shape the experience of visceral pain: an experimental pilot study in healthy women. Neurogastroenterol Motil 24:914–e460

Enck P, Benedetti F, Schedlowski M (2008) New insights into the placebo and nocebo responses. Neuron 59:195–206

Enck P, Junne F, Klosterhalfen S, Zipfel S, Martens U (2010) Therapy options in irritable bowel syndrome. Eur J Gastroenterol Hepatol 22:1402–1411

Enck P, Horing B, Weimer K, Klosterhalfen S (2012) Placebo responses and placebo effects in functional bowel disorders. Eur J Gastroenterol Hepatol 24:1–8

Enck P, Bingel U, Schedlowski M, Rief W (2013) The placebo response in medicine: minimize, maximize or personalize? Nat Rev Drug Discov 12:191–204

Fillingim RB, King CD, Ribeiro-Dasilva MC, Rahim-Williams B, Riley JL (2009) Sex, gender, and pain: a review of recent clinical and experimental findings. J Pain 10:447–485

Finniss DG, Kaptchuk TJ, Miller F, Benedetti F (2010) Biological, clinical, and ethical advances of placebo effects. Lancet 375:686–695

Flaten MA, Aslaksen PM, Finset A, Simonsen T, Johansen O (2006) Cognitive and emotional factors in placebo analgesia. J Psychosom Res 61:81–89

Flaten MA, Aslaksen PM, Lyby PS, Bjorkedal E (2011) The relation of emotions to placebo responses. Philos Trans R Soc Lond B Biol Sci 366:1818–1827

Folks DG (2004) The interface of psychiatry and irritable bowel syndrome. Curr Psychiatry Rep 6:210–215

Ford AC, Moayyedi P (2010) Meta-analysis: factors affecting placebo response rate in the irritable bowel syndrome. Aliment Pharmacol Ther 32:144–158

Franconi F, Campesi I, Occhioni S, Antonini P, Murphy MF (2012) Sex and gender in adverse drug events, addiction, and placebo. Handb Exp Pharmacol 107–126

Frissora CL, Koch KL (2005) Symptom overlap and comorbidity of irritable bowel syndrome with other conditions. Curr Gastroenterol Rep 7:264–271

Gijsbers K, Nicholson F (2005) Experimental pain thresholds influenced by sex of experimenter. Percept Mot Skills 101:803–807

Gramsch C, Kattoor J, Icenhour A, Forsting M, Schedlowski M, Gizewski ER, Elsenbruch S (2014) Learning and remembering pain-related fear: neural mechanisms mediating rapid differential conditioning, extinction and memory retrieval processes in human visceral pain. Neurobiol Learn Mem (in revision)

Hall KT, Lembo AJ, Kirsch I, Ziogas DC, Douaiher J, Jensen KB, Conboy LA, Kelley JM, Kokkotou E, Kaptchuk TJ (2012) Catechol-O-methyltransferase val158met polymorphism predicts placebo effect in irritable bowel syndrome. PLoS ONE 7:e48135

Horing B, Kugel H, Brenner V, Zipfel S, Enck P (2013) Perception and pain thresholds for cutaneous heat and cold, and rectal distension: associations and disassociations. Neurogastroenterol Motil 25:e791–e802

Jensen KB, Kaptchuk TJ, Kirsch I, Raicek J, Lindstrom KM, Berna C, Gollub RL, Ingvar M, Kong J (2012) Nonconscious activation of placebo and nocebo pain responses. Proc Natl Acad Sci USA 109:15959–15964

Johansen O, Brox J, Flaten MA (2003) Placebo and nocebo responses, cortisol, and circulating beta-endorphin. Psychosom Med 65:786–790

Kaptchuk TJ, Kelley JM, Conboy LA, Davis RB, Kerr CE, Jacobson EE, Kirsch I, Schyner RN, Nam BH, Nguyen LT, Park M, Rivers AL, McManus C, Kokkotou E, Drossman DA, Goldman P, Lembo AJ (2008) Components of placebo effect: randomised controlled trial in patients with irritable bowel syndrome. BMJ 336:999–1003

Kaptchuk TJ, Friedlander E, Kelley JM, Sanchez MN, Kokkotou E, Singer JP, Kowalczykowski M, Miller FG, Kirsch I, Lembo AJ (2010) Placebos without deception: a randomized controlled trial in irritable bowel syndrome. PLoS ONE 5:e15591

Kattoor J, Gizewski ER, Kotsis V, Benson S, Gramsch C, Theysohn N, Maderwald S, Forsting M, Schedlowski M, Elsenbruch S (2013) Fear conditioning in an abdominal pain model: neural responses during associative learning and extinction in healthy subjects. PLoS ONE 8:e51149

Kattoor J, Thurling M, Gizewski ER, Forsting M, Timmann D, Elsenbruch S (2014) Cerebellar contributions to different phases of visceral aversive extinction learning. Cerebellum 13:1–8

Kelley JM, Lembo AJ, Ablon JS, Villanueva JJ, Conboy LA, Levy R, Marci CD, Kerr CE, Kirsch I, Jacobson EE, Riess H, Kaptchuk TJ (2009) Patient and practitioner influences on the placebo effect in irritable bowel syndrome. Psychosom Med 71:789–797

Kern MK, Jaradeh S, Arndorfer RC, Jesmanowicz A, Hyde J, Shaker R (2001) Gender differences in cortical representation of rectal distension in healthy humans. Am J Physiol Gastrointest Liver Physiol 281:G1512–G1523

Kim HS, Rhee PL, Park J, Lee JH, Kim YH, Kim JJ, Rhee JC (2006) Gender-related differences in visceral perception in health and irritable bowel syndrome. J Gastroenterol Hepatol 21:468–473

Klosterhalfen S, Kellermann S, Braun S, Kowalski A, Schrauth M, Zipfel S, Enck P (2009) Gender and the nocebo response following conditioning and expectancy. J Psychosom Res 66:323–328

Kokkotou E, Conboy LA, Ziogas DC, Quilty MT, Kelley JM, Davis RB, Lembo AJ, Kaptchuk TJ (2010) Serum correlates of the placebo effect in irritable bowel syndrome. Neurogastroenterol Motil 22:285–e81

Kong J, Gollub RL, Polich G, Kirsch I, Laviolette P, Vangel M, Rosen B, Kaptchuk TJ (2008) A functional magnetic resonance imaging study on the neural mechanisms of hyperalgesic nocebo effect. J Neurosci 28:13354–13362

Kotsis V, Benson S, Bingel U, Forsting M, Schedlowski M, Gizewski ER, Elsenbruch S (2012) Perceived treatment group affects behavioral and neural responses to visceral pain in a deceptive placebo study. Neurogastroenterol Motil 24:935–e462

Labus JS, Naliboff BN, Fallon J, Berman SM, Suyenobu B, Bueller JA, Mandelkern M, Mayer EA (2008) Sex differences in brain activity during aversive visceral stimulation and its expectation in patients with chronic abdominal pain: a network analysis. NeuroImage 41:1032–1043

Lee HF, Hsieh JC, Lu CL, Yeh TC, Tu CH, Cheng CM, Niddam DM, Lin HC, Lee FY, Chang FY (2012) Enhanced affect/cognition-related brain responses during visceral placebo analgesia in irritable bowel syndrome patients. Pain 153:1301–1310

Lieberman MD, Jarcho JM, Berman S, Naliboff BD, Suyenobu BY, Mandelkern M, Mayer EA (2004) The neural correlates of placebo effects: a disruption account. NeuroImage 22:447–455

Linnman C, Rougemont-Bucking A, Beucke JC, Zeffiro TA, Milad MR (2011) Unconditioned responses and functional fear networks in human classical conditioning. Behav Brain Res 221:237–245

Lu CL, Chang FY (2011) Placebo effect in patients with irritable bowel syndrome. J Gastroenterol Hepatol 26(Suppl 3):116–118

Lu HC, Hsieh JC, Lu CL, Niddam DM, Wu YT, Yeh TC, Cheng CM, Chang FY, Lee SD (2010) Neuronal correlates in the modulation of placebo analgesia in experimentally-induced esophageal pain: a 3 T-fMRI study. Pain 148:75–83

Lyby PS, Aslaksen PM, Flaten MA (2011) Variability in placebo analgesia and the role of fear of pain–an ERP study. Pain 152:2405–2412

Lyby PS, Forsberg JT, Asli O, Flaten MA (2012) Induced fear reduces the effectiveness of a placebo intervention on pain. Pain 153:1114–1121

Maxion-Bergemann S, Thielecke F, Abel F, Bergemann R (2006) Costs of irritable bowel syndrome in the UK and US. Pharmacoeconomics 24:21–37

Mayer EA, Berman S, Chang L, Naliboff BD (2004) Sex-based differences in gastrointestinal pain. Eur J Pain 8:451–463

Mertz H, Naliboff B, Munakata J, Niazi N, Mayer EA (1995) Altered rectal perception is a biological marker of patients with irritable bowel syndrome. Gastroenterology 109:40–52

Milad MR, Quirk GJ (2012) Fear extinction as a model for translational neuroscience: ten years of progress. Annu Rev Psychol 63:129–151

Mogil JS, Bailey AL (2010) Sex and gender differences in pain and analgesia. Prog Brain Res 186:141–157

Naliboff BD, Berman S, Chang L, Derbyshire SW, Suyenobu B, Vogt BA, Mandelkern M, Mayer EA (2003) Sex-related differences in IBS patients: central processing of visceral stimuli. Gastroenterology 124:1738–1747

Nozu T, Kudaira M, Kitamori S, Uehara A (2006) Repetitive rectal painful distention induces rectal hypersensitivity in patients with irritable bowel syndrome. J Gastroenterol 41:217–222

Palsson OS, Drossman DA (2005) Psychiatric and psychological dysfunction in irritable bowel syndrome and the role of psychological treatments. Gastroenterol Clin North Am 34:281–303
Patel SM, Stason WB, Legedza A, Ock SM, Kaptchuk TJ, Conboy L, Canenguez K, Park JK, Kelly E, Jacobson E, Kerr CE, Lembo AJ (2005) The placebo effect in irritable bowel syndrome trials: a meta-analysis. Neurogastroenterol Motil 17:332–340
Petrovic P, Kalso E, Petersson KM, Ingvar M (2002) Placebo and opioid analgesia– imaging a shared neuronal network. Science 295:1737–1740
Petrovic P, Dietrich T, Fransson P, Andersson J, Carlsson K, Ingvar M (2005) Placebo in emotional processing–induced expectations of anxiety relief activate a generalized modulatory network. Neuron 46:957–969
Price DD, Craggs J, Verne GN, Perlstein WM, Robinson ME (2007) Placebo analgesia is accompanied by large reductions in pain-related brain activity in irritable bowel syndrome patients. Pain 127:63–72
Price DD, Finniss DG, Benedetti F (2008) A comprehensive review of the placebo effect: recent advances and current thought. Annu Rev Psychol 59:565–590
Price DD, Craggs JG, Zhou Q, Verne GN, Perlstein WM, Robinson ME (2009) Widespread hyperalgesia in irritable bowel syndrome is dynamically maintained by tonic visceral impulse input and placebo/nocebo factors: evidence from human psychophysics, animal models, and neuroimaging. NeuroImage 47:995–1001
Rief W, Bingel U, Schedlowski M, Enck P (2011) Mechanisms involved in placebo and nocebo responses and implications for drug trials. Clin Pharmacol Ther 90:722–726
Schmid J, Theysohn N, Gass F, Benson S, Gramsch C, Forsting M, Gizewski ER, Elsenbruch S (2013) Neural mechanisms mediating positive and negative treatment expectations in visceral pain: a functional magnetic resonance imaging study on placebo and nocebo effects in healthy volunteers. Pain 154:2372–2380
Schmid J, Langhorst J, Gass F, Theysohn N, Benson S, Engler H, Gizewski ER, Forsting M, Elsenbruch S (2014) Placebo analgesia in patients with functional and organic abdominal pain: a fMRI study in irritable bowel syndrome, ulcerative colitis and healthy volunteers. Gut (in press)
Scott DJ, Stohler CS, Egnatuk CM, Wang H, Koeppe RA, Zubieta JK (2008) Placebo and nocebo effects are defined by opposite opioid and dopaminergic responses. Arch Gen Psychiatry 65:220–231
Sehlmeyer C, Schöning S, Zwitserlood P, Pfleiderer B, Kircher T, Arolt V, Konrad C (2009) Human fear conditioning and extinction in neuroimaging: a systematic review. PLoS ONE 4: e5865
Sloots CE, Felt-Bersma RJ, Cuesta MA, Meuwissen SG (2000) Rectal visceral sensitivity in healthy volunteers: influences of gender, age and methods. Neurogastroenterol Motil 12:361–368
Soffer EE, Kongara K, Achkar JP, Gannon J (2000) Colonic motor function in humans is not affected by gender. Dig Dis Sci 45:1281–1284
Swider K, Babel P (2013) The effect of the sex of a model on nocebo hyperalgesia induced by social observational learning. Pain 154:1312–1317
Talley NJ (2008) Functional gastrointestinal disorders as a public health problem. Neurogastroenterol Motil 20(Suppl 1):121–129
Tanaka Y, Kanazawa M, Fukudo S, Drossman DA (2011) Biopsychosocial model of irritable bowel syndrome. J Neurogastroenterol Motil 17:131–139
Theysohn N, Schmid J, Icenhour A, Mewes C, Forsting M, Gizewski ER, Schedlowski M, Elsenbruch S, Benson S (2014) Are there sex differences in placebo analgesia during visceral pain processing? - a fMRI study in healthy subjects. Neurogastroenterol Motil (in revision)
Tyler K, Moriceau S, Sullivan RM, Greenwood-van Meerveld B (2007) Long-term colonic hypersensitivity in adult rats induced by neonatal unpredictable vs predictable shock. Neurogastroenterol Motil 19:761–768
Unruh AM (1996) Gender variations in clinical pain experience. Pain 65:123–167

van Laarhoven AI, Vogelaar ML, Wilder-Smith OH, Van Riel PL, van de Kerkhof PC, Kraaimaat FW, Evers AW (2011) Induction of nocebo and placebo effects on itch and pain by verbal suggestions. Pain 152:1486–1494

Vase L, Robinson ME, Verne GN, Price DD (2003) The contributions of suggestion, desire, and expectation to placebo effects in irritable bowel syndrome patients. An empirical investigation. Pain 105:17–25

Vase L, Robinson ME, Verne GN, Price DD (2005) Increased placebo analgesia over time in irritable bowel syndrome (IBS) patients is associated with desire and expectation but not endogenous opioid mechanisms. Pain 115:338–347

Verne GN, Sen A, Price DD (2005) Intrarectal lidocaine is an effective treatment for abdominal pain associated with diarrhea-predominant irritable bowel syndrome. J Pain 6:493–496

Vögtle E, Barke A, Kroner-Herwig B (2013) Nocebo hyperalgesia induced by social observational learning. Pain 154:1427–1433

Wang Z, Bradesi S, Charles JR, Pang RD, Maarek JM, Mayer EA, Holschneider DP (2011) Functional brain activation during retrieval of visceral pain-conditioned passive avoidance in the rat. Pain 152:2746–2756

Weimer K, Schulte J, Maichle A, Muth ER, Scisco JL, Horing B, Enck P, Klosterhalfen S (2012) Effects of ginger and expectations on symptoms of nausea in a balanced placebo design. PLoS ONE 7:e49031

Whitehead WE, Palsson O, Jones KR (2002) Systematic review of the comorbidity of irritable bowel syndrome with other disorders: what are the causes and implications? Gastroenterology 122:1140–1156

Wiech K, Lin CS, Brodersen KH, Bingel U, Ploner M, Tracey I (2010) Anterior insula integrates information about salience into perceptual decisions about pain. J Neurosci 30:16324–16331

Williams AE, Rhudy JL (2007) The influence of conditioned fear on human pain thresholds: does preparedness play a role? J Pain 8:598–606

Yágüez L, Coen S, Gregory LJ, Amaro E, Altman C, Brammer MJ, Bullmore ET, Williams SC, Aziz Q (2005) Brain response to visceral aversive conditioning: a functional magnetic resonance imaging study. Gastroenterology 128:1819–1829

Zubieta JK, Bueller JA, Jackson LR, Scott DJ, Xu Y, Koeppe RA, Nichols TE, Stohler CS (2005) Placebo effects mediated by endogenous opioid activity on mu-opioid receptors. J Neurosci 25:7754–7762 van Laarhoven AI, Vogelaar ML, Wilder-Smith OH, van Riel PL, van de Kerkhof PC, Kraaimaat FW, Evers AW (2011) Induction of nocebo and placebo effects on itch and pain by verbal suggestions. Pain 152:1486–1494

[illegible]

[illegible]

[illegible]

[illegible]

[illegible]

[illegible]

[illegible]

[illegible]

Placebo Effects in Idiopathic and Neuropathic Pain Conditions

Lene Vase, Gitte Laue Petersen, and Karen Lund

Contents

1 Introduction 122
2 Placebo Analgesia Effects in Idiopathic Pain 123
  2.1 Placebo Effects in Idiopathic Pain 123
  2.2 The Contribution of Expectations and Emotional Feelings 125
  2.3 Neurobiological Underpinnings 126
3 Placebo Analgesia Effects in Neuropathic Pain 128
  3.1 The Magnitude of Placebo and Nocebo Effects in Neuropathic Pain 128
  3.2 The Contribution of Psychological Factors 129
4 Can Knowledge from Placebo Mechanisms Studies Improve RCTs 131
  4.1 Are Placebo Effects and Drug Effects Additive? 131
  4.2 Can Knowledge from Placebo Mechanisms Studies Improve Randomized Clinical Trials? 133
References 134

Abstract

The magnitude of placebo analgesia effect appears to be large in chronic pain patients experiencing hyperalgesic states. So far, placebo effects have primarily been investigated in idiopathic pain conditions, such as irritable bowel pain syndrome, but more recently they have also been investigated in neuropathic pain patients, in which the underlying nerve injury is known. Expected pain levels and emotional feelings are central to placebo effects in both types of pain.

L. Vase (✉) • G.L. Petersen
Department of Psychology and Behavioral Sciences, School of Business and Social Sciences, Aarhus University, Aarhus, Denmark

Danish Pain Research Center, Aarhus University Hospital, Aarhus, Denmark
e-mail: lenevase@psy.au.dk

K. Lund
Danish Pain Research Center, Aarhus University Hospital, Aarhus, Denmark

F. Benedetti et al. (eds.), *Placebo*, Handbook of Experimental Pharmacology 225,
DOI 10.1007/978-3-662-44519-8_7

They appear to help patients to engage in a mindset for pain relief and activate the pain-modulating system. Furthermore, expectations, emotional feelings, and the experience of pain seem to interact over time, thereby maintaining or enhancing the pain-relieving effect. Expectations and emotional feelings also contribute to the effect of active drugs, and recent studies indicate that drug effects and placebo effects interact in ways that may complicate the interpretations of the findings from clinical trials. It is suggested that expectations and emotional feelings may act as additional or alternative measures in the testing of new pharmacological agents, thereby improving the understanding of the interaction between pharmacological effects and placebo effects, which may have far-reaching implications for research and clinical practice.

Keywords
Placebo effect • Expectations • Emotional feelings and additivity

1 Introduction

Traditionally, placebo agents have been used as control conditions for active treatments in randomized clinical trials (Andersen 1997; Harrington 1997). During the last decades, however, experimental studies have shown that placebo effects may be large and clinically relevant (Benedetti 2009; Price et al. 2008; Vase et al. 2002, 2009), and the psychological and neurophysiological factors underlying these effects have been specified to a higher extent (Benedetti 2009). Placebo effects appear to be related to patients' *perception or direct experience* of a treatment, i.e., seeing, smelling, and hearing verbal information about the treatment as well as actively integrating this sensory information with memories of previous experiences and current expectations, and recent studies have illustrated how the patient's perception of a treatment contributes to both placebo effects and active drug effects (Benedetti 2009; Lund et al. 2014; Price et al. 2008; Vase et al. 2003, 2005, 2011). Currently, clinical trials' ability to differentiate between drug effects and placebo effects is debated (Dworkin et al. 2012). The advanced knowledge of placebo mechanisms could help improve the information obtained from clinical trials (Vase and Petersen 2013), which may ultimately improve the understanding of the factors that contribute to the optimization of pharmacological and nonpharmacological treatments, and thereby enhance treatment outcomes in clinical practice.

Meta-analyses have shown that the magnitude of placebo analgesia effects is highly variable (Hróbjartsson and Gøtzsche 2001, 2004, 2010; Price et al. 2003; Vase et al. 2002, 2009), and the largest placebo effects were found in patients with hyperalgesia (Vase et al. 2009). A hyperalgesic state is characterized by an increased pain response to stimuli that normally provoke pain, and it is believed to be related to sensitization of the nociceptive pain processing system (IAPS guidelines; Price 1999). Thus, based on these findings it would be interesting to investigate placebo effects in relation to hyperalgesic states. The majority of research on placebo effects in chronic pain conditions and hyperalgesic states has been conducted in patients with irritable bowel syndrome (IBS) (Conboy et al. 2006; Craggs et al. 2007, 2008; Kaptchuk et al. 2010; Price et al. 2007; Vase et al. 2003, 2005; Verne et al. 2003). Although, IBS is a good model for studying placebo effects, IBS can be considered an idiopathic pain in so far as the pain has no apparent underlying cause (Piche et al. 2011). In order to fully understand the mechanisms underlying placebo analgesia effects in hyperalgesic states and chronic pain conditions, it may be helpful to investigate placebo effects in a chronic pain condition such as neuropathic pain, in which the pain is caused by a (known) lesion or disease of the somatosensory nervous system (Jensen et al. 2011).

So far, the majority of placebo research has investigated either the psychological or the neurophysiological mechanisms underlying placebo effects. However, as placebo effects obviously involve both psychological and neurophysiological factors, it would be interesting to increase our understanding of how these factors influence each other in specific placebo effects (Price and Barrell 2012). Moreover, as some studies investigate placebo effects and drug effects in the same study, it is relevant to examine the relationship between placebo effects and drug effects in order to improve the test of pharmacological agents and to optimize treatment effects in clinical practice (Amanzio et al. 2001; Benedetti 2009; Vase et al. 2003, 2005).

In this chapter we will review central studies of placebo effects in chronic pain conditions involving hyperalgesic states. First, we will look at placebo effects observed in IBS, which can be considered an idiopathic pain, and subsequently we will look at placebo effects in neuropathic pain following thoracotomy. Special focus will be given to how both psychological and neurophysiological mechanisms contribute to these placebo effects and how they possibly relate to each other. Finally, it will be debated whether the knowledge obtained from studies of placebo mechanisms can be utilized to improve our understanding of the effects of active drugs and placebos in randomized clinical trials and in clinical practice.

2 Placebo Analgesia Effects in Idiopathic Pain

2.1 Placebo Effects in Idiopathic Pain

Several studies have investigated placebo analgesia effects in IBS patients within traditional designs typically including an active treatment, a placebo treatment, and a no treatment condition or group (Craggs et al. 2007, 2008; Kaptchuk et al. 2010, ; Lieberman et al. 2004; Price et al. 2008; Vase et al. 2003, 2005; Verne et al. 2003). These studies have shown a significant and often large placebo analgesia effect as indicated by the difference in pain levels between the placebo treatment and the no

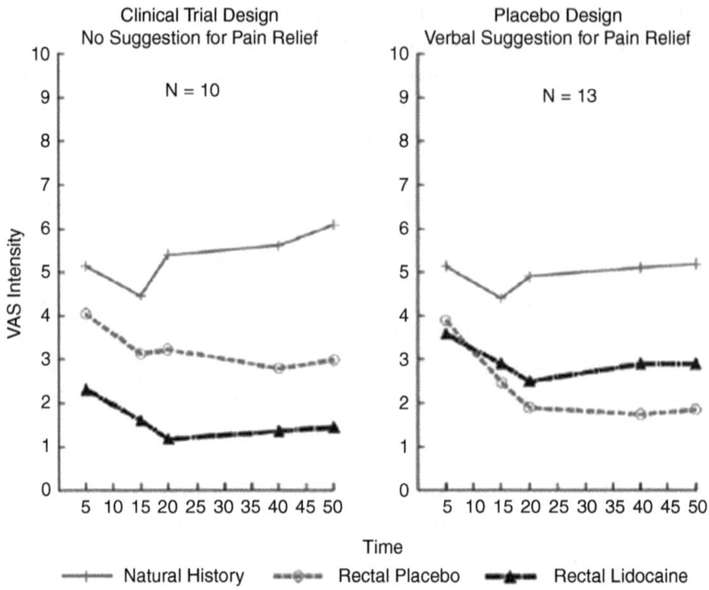

Fig. 1 Comparison of natural history, rectal placebo, and rectal lidocaine scores on visceral pain intensity ratings during a 50-min session within a clinical trial design, where no suggestions for pain relief are given (*left*) and within a placebo design with verbal suggestions for pain relief (*right*)

treatment group or condition (Benedetti 2009; Price et al. 2008; Vase et al. 2002, 2004).

In two early studies of placebo effects in IBS patients, 10 and 13 chronic IBS patients, respectively, were exposed to rectal stimulation and tested under rectal lidocaine, rectal placebo (lubricant), and no treatment conditions in a crossover fashion (Vase et al. 2003; Verne et al. 2003). These studies showed that suggestions for pain relief may influence the magnitude of the placebo effect (Vase et al. 2004). In the first study, patients received an informed consent form stating that they "may receive an active pain reducing mediation or an inert placebo agent." This study found a significant pain-relieving effect of rectal lidocaine compared with rectal placebo and a significant pain-relieving effect of rectal placebo compared with the untreated natural history condition. The second study was conducted in a similar manner, the main difference being that in this study patients were told that "the agent you have just been given is known to significantly reduce pain in some patients" (Vase et al. 2003). A much larger placebo effect was found in this second study, and the magnitude of the placebo effect was so high that there was no longer a significant difference between the effect of rectal lidocaine and rectal placebo (Fig. 1). Hence, these two studies suggest that it is possible to increase the effect of placebo analgesia to the level of an active agent by adding an overt suggestion for pain relief.

2.2 The Contribution of Expectations and Emotional Feelings

Verbal suggestions for pain relief are likely to influence the magnitude of placebo analgesia effects through patients' expectations of pain relief. Several studies have shown that expected pain levels contribute to placebo analgesia effects (Benedetti 2009; Montgomery and Kirsch 1997; Price et al. 1999). In the study of IBS patients described above, the patients were asked to rate their expected pain levels and desire for pain relief on well-validated visual analog scales immediately after each of the three conditions and just before any analgesic effects could take place (Vase et al. 2003). The combination of ratings of the expected pain level and the desire for pain relief accounted for 77 and 81 % of the variance in the pain ratings during the placebo and lidocaine conditions, respectively. These strong correlations show that expected pain levels and the desire for relief are central to the experience of pain relief during placebo as well as active treatment.

In a subsequent similar study of IBS patients, the temporal development of the placebo analgesia effect as well as the temporal changes in expected pain levels, the desire for pain relief, and anxiety levels were investigated (Vase et al. 2005). In this study, IBS patients were also asked to rate their expected pain levels, desire for pain relief, and anxiety levels at the beginning of the study and then again halfway through the study. The study showed an increasing placebo analgesia effect during the 40 min of investigation, with a markedly increasing placebo effect during the first 20 min and a plateauing placebo effect during the last 20 min. Interestingly, the expected pain levels, the desire for pain relief, and anxiety *decreased* from the early part (first 20 min) to the late part (last 20 min) of the session and the three psychological variables came to account for considerably more of the variance in the placebo response and in the response to lidocaine treatment over time. These findings may be interpreted as follows: In the beginning of the experiment, IBS patients had a mild to moderate desire for pain relief and expected a reduction in pain as a result of the suggestion for pain relief and the administration of an agent. This psychological mindset is likely to have contributed to the actual experience of some pain relief during the first part of the session. The actual experience of pain relief could then have led to the further reductions in anxiety and expected pain levels in the second part of the session, and these changes may have contributed to a further self-reinforcing pain reduction in the late part of the study. Thus, combinations of the expectations and an overall reduction in negative emotions are likely to have contributed to an increase in the placebo effect over time (Fig. 2). These findings have been supported by more recent studies showing that expectations of pain relief may lead to a reduction of anxiety, and this may be a central component of the placebo effect (Flaten et al. 2011). Taken together these studies illustrate that expectations and emotional feelings are embedded in active and placebo treatments, and that the dynamic interactions between these parameters contribute to a self-reinforcing analgesic effect over time (Craggs et al. 2008; Price et al. 2007; Vase et al. 2004, 2011).

In the study by Vase et al. (2005), the patients were interviewed about their direct experience of receiving a treatment and of possible pain relief following the

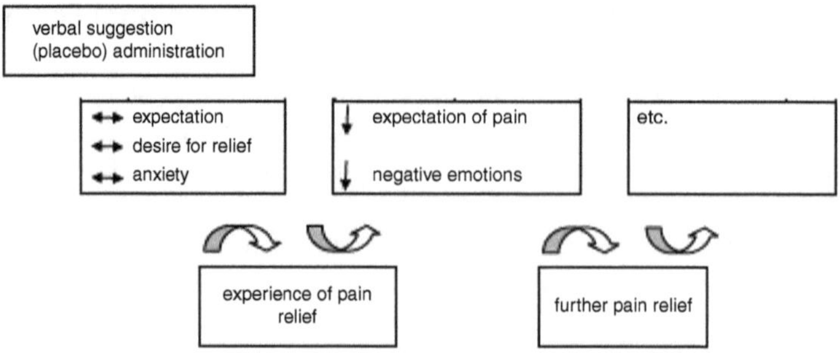

Fig. 2 An illustration of how expectations and emotional feelings may contribute to a self-reinforcing placebo analgesia effect. Following the verbal suggestions given for pain relief, patients have moderate expectations of pain relief and moderate levels of negative emotional feelings. These psychological factors are likely to contribute to the actual experience of pain relief during the first 20 min of the study. The actual experience of pain relief appears to contribute to lower expectations of pain and lower levels of anxiety and desire in the late part of the study, which are likely to contribute to the even further experience of pain relief in the last 20 min of the experiment. In this manner the placebo analgesia effect may be self-reinforcing

last session in which they received either a placebo or an active lidocaine treatment (Vase et al. 2011). The patients were asked about their perception of the description of the agent, their focus of attention, and their ongoing thoughts and feelings during the early part (first 20 min) and the late part (the last 20 min) of the session. Patients reported similar experiences during placebo treatment and active lidocaine treatment, which is probably related to similar magnitudes of analgesia in the two conditions as well as absence of side effects. The data illustrate that the relationship with the healthcare provider and the verbal suggestions given for pain relief were important for the perception of the treatment. These factors seemed to help patients to be actively engaged in generating a placebo effect in the beginning of the placebo condition, in which several of the patients reported listening to the doctor's verbal suggestions and focusing on how this matched their bodily sensations. This is illustrated by a patient saying: "I am paying attention to what he [the doctor] says and . . .um. . .trying to get in touch with how I feel physically." These factors also appear to contribute to the patients' feeling of calmness. Once this analgesic effect had been established, however, the patients appeared to go into a mode of either maintaining the effect or focusing on other things, possibly because their pain was no longer salient for them. This can be illustrated by a patient saying: "I am staring at the wall . . .um. . .thinking about stuff I need to buy this afternoon."

2.3 Neurobiological Underpinnings

Changes at the psychological level have been shown to be associated with neurophysiological changes. Functional magnetic resonance imaging (fMRI) studies have demonstrated that during the period in which patients *anticipate* pain (relief) in a placebo treatment, there is an increased activity in brain areas such as the orbitofrontal cortex (OBC) and the dorsolateral prefrontal cortex (DLPFC), regions known to be involved in expectations and emotional factors (Wager et al. 2004). During the actual *experience* of pain and pain reduction, however, studies of, for example, IBS patients have shown that there is decreased activity in pain-processing areas of the brain such as the thalamus, somatosensory cortices, the anterior insular cortex, and the anterior cingulate cortex (Price et al. 2007; Wager et al. 2004). Thus, patients' expectations of pain relief and their emotional feelings in relation to a treatment seem to be related to both the reduction in afferent processing of pain and in the generation of analgesia.

In an fMRI study of IBS patients using a similar design to that described above, both decreased (Price et al. 2007) and increased (Craggs et al. 2008) neural activity during placebo analgesia was investigated. Interestingly, during the early phase of the study (first three inflations), there was an increased activity in areas of the temporal lobe (involved in memory), the precuneus (involved in associative thinking), and the amygdala (involved in descending modulation of pain). Similar to the magnitude of the placebo effect itself, the neural activity in these three areas was much greater during the early part of the session, in which the placebo effect increased over time and reached peak levels. These findings may be interpreted as follows: During the early part of the placebo condition (the first 3–4 min), patients were likely to make associations between the verbal suggestions given for pain relief, internal cues that suggested whether or not the agent was working, and their expectations about future experiences. These associations require memory, somatic focus, and comparison of present experience to expectations of pain following placebo suggestion. Once the placebo effect was established, however, there was much less activity in these brain areas during the remainder of the experimental session. Thus, the placebo effect may persist beyond the time of activation of the brain structures that induced it. These findings and conceptualizations are also consistent with a self-enhancing feedback mechanism (Price et al. 2007; Verne et al. 2012).

Taken together, the combination of the psychophysiological data, the interviews, and the brain imaging data gives a picture of patients actively engaging in generating a mindset for pain reduction and a corresponding active engagement of a descending pain control system following placebo and active drug administration. Both psychological and neural generation of analgesia occurs early in the placebo process and once analgesia is established, it may persist beyond the duration of factors that generate it.

3 Placebo Analgesia Effects in Neuropathic Pain

3.1 The Magnitude of Placebo and Nocebo Effects in Neuropathic Pain

Recently, placebo analgesia effects have been investigated in chronic neuropathic pain conditions, in which the underlying nerve damage/injury is specified (Petersen et al. 2012). In these studies, patients who had developed chronic neuropathic pain following thoracotomy were exposed to a placebo manipulation via an open versus hidden administration of lidocaine, controlled for the natural history of the pain and tested with quantitative sensory testing in an area close to the surgery site.

In the first study, 19 patients went through 3 randomized sessions: (1) open administration of lidocaine, (2) hidden administration of lidocaine, and (3) no treatment (Fig. 3). In the open condition, lidocaine was applied to a disinfection napkin in full view of the patients, and the patients were told: "The agent you have just been given is known to powerfully reduce pain in some patients." In the hidden condition, lidocaine was applied to the disinfection napkin without the patients' knowledge and the patients were told: "This is a control condition for the active medication." In the control condition, patients were not given any medication on the disinfection napkin and they were told: "We will test your response to different types of stimuli in order to get a better understanding of how (your) pain is processed." Prior to each test condition there was a baseline condition in order to control for the daily variability in the patients' pain. Hence, the placebo effect was calculated as the difference in pain between the baseline-open versus baseline-hidden administrations of lidocaine controlled for the baseline-control natural history of pain. A disinfection napkin is typically used to disinfect the test area in quantitative sensory testing studies, so the open and hidden administrations of lidocaine could easily be embedded in this procedure. In addition, pilot testing had indicated that the administration was double blind. In each session/condition, the patients' spontaneous pain levels were measured and evoked pain was tested via

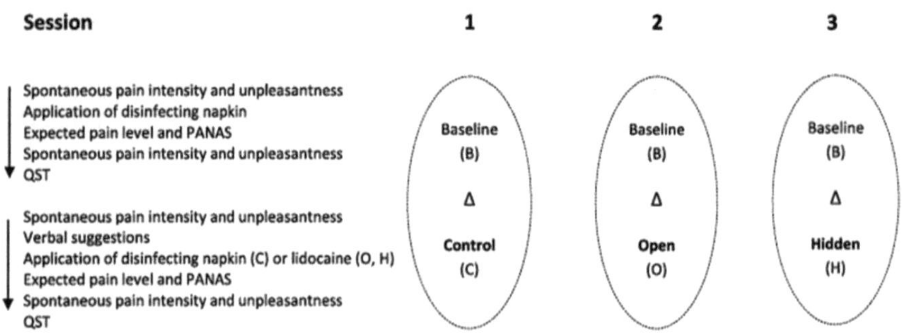

Fig. 3 Study design. Each patient goes through three randomized sessions on separate days. Each session includes a baseline condition and a control condition (1), open condition (2), or hidden condition (3). The exact test in each session is listed chronologically to the *left*. QST is an abbreviation of Quantitative Sensory Testing

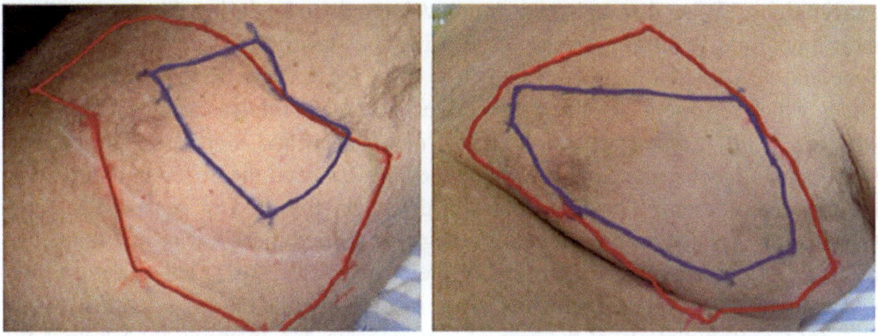

Fig. 4 The area of hyperalgesia in one of the patients. The placebo effect is defined as the difference between ΔB-O versus ΔB-H. The red line indicates the area of hyperalgesia determined at baseline, and the blue line indicates the area of hyperalgesia determined in O and H, respectively. In this patient, the area of hyperalgesia is markedly reduced in the open condition compared with baseline, whereas the area of hyperalgesia is only slighted reduced in the hidden compared with the baseline condition. Thus, this patient experiences a large placebo effect

brush and cold allodynia, heat pain tolerance, area of pinprick hyperalgesia, and wind-up-like pain after pinprick stimulation.

There was a large and significant placebo effect on the area of hyperalgesia (Fig. 4), but there was not a significant placebo effect on spontaneous pain levels. In this study, the patients had to experience a pain intensity of at least 3 on a 0–10 numerical rating scale to be included, but the patients experienced pain levels around 2 on average on the actual test days. Hence, the lack of a placebo effect on spontaneous pain could be due to a floor effect.

To investigate this further, a new study similar to the one described above was conducted in patients with chronic neuropathic pain following thoracotomy (unpublished observations). In this study, care was taken to ensure that all patients experienced pain intensity levels above 3 on a numerical rating scale on each test day. This study showed large and significant placebo effects on spontaneous pain and evoked types of pain. Interestingly, the placebo effect on spontaneous pain was investigated at the beginning of the session and approximately 30 min later, and the latter magnitude was larger than the former (unpublished observations), thereby suggesting that the magnitude of placebo effects in neuropathic pain may also increase over time.

3.2 The Contribution of Psychological Factors

In the studies of placebo effects in neuropathic pain, the patients' expected pain levels and emotional feelings were assessed immediately after the disinfection napkin was applied and before the potential administration of lidocaine had taken effect. In the first study, patients were asked one general question in relation to their expectations: "What do you expect your level of pain to be during this session?" and their emotional feelings were assessed using the Positive Affect Negative Affect Schedule (PANAS) (Watson et al. 1988). The PANAS is divided into *negative affect*, which represents levels of subjective distress and unpleasurable engagement, whereas *positive affect* reflects levels of enthusiasm and alertness. The patients were asked to assess the extent to which they experienced the different emotions in the present moment. The large placebo effect on the area of hyperalgesia was significantly related to low levels of negative affect, but it was not related to positive affect or expected pain levels (Petersen et al. 2012). The reason why it was not related to expected pain levels was most likely that the expectation measure was not targeted directly at the area of hyperalgesia, for example, by asking if the patients expected the treatment to reduce the area of hyperalgesia.

In the second study by Petersen et al. (unpublished observations), measures of expected pain levels were specifically targeted at each pain measure by asking: "What do you expect your pain level to be" in relation to spontaneous pain and each of the evoked pain measures In addition, the measures of emotional feelings were changed, so patients rated the intensity of emotional feelings on a visual analogue scale and qualitatively described these emotions as these ratings may more directly relate to their actual immediate experience (Price and Barrell 2012).

In this study, the placebo effects were related to the expected pain. Also, patients reported a much higher intensity of positive than negative emotions following the open administration of lidocaine.

The finding that patients had high levels of positive emotions and low levels of negative emotions during the open administration of lidocaine is partly in contrast to previous studies using the PANAS, in which only low levels of negative affect were significantly related to pain in the placebo condition whereas high levels of positive affect was non- significantly related to placebo (Scott et al. 2007; Petersen et al. 2012). One explanation of these seemingly contradictory results may be that during open administrations of treatments, patients do not experience the positive emotions predefined in the PANAS but instead positive emotions that are captured by the open reports of the immediate experience. Another implication of the finding is that patients' expectations of a treatment effect are not neutral, but co-exist with emotional feelings. This may be especially important to keep in mind when dealing with chronic pain patients. Furthermore, in this study, patients also reported that they were focused on sensations related to the perception of the treatment and on monitoring how their body responded to the treatment. These observations are in accordance with the above-mentioned findings from the interviews and brain imaging studies of IBS patients (Craggs et al. 2007; Price et al. 2008; Vase et al. 2011), and they suggest that patients are actively engaged in initiating the pain relief experienced in open administrations of treatments.

4 Can Knowledge from Placebo Mechanisms Studies Improve RCTs

4.1 Are Placebo Effects and Drug Effects Additive?

The basic assumption in randomized clinical trials and meta-analyses hereof is that the effect of the placebo agent and the effect of the active agent are additive. In other words, if the pain level following placebo administration is subtracted from the pain level following active administration, the effect of the active medication can be deduced. However, as illustrated in the studies above, psychological factors such as expectations and emotional feelings that contribute to the placebo effect also contribute to the effect of the active drug (Benedetti 2009; Vase et al. 2003, 2005). Moreover, meta-analyses investigating the efficacy of pain medication in clinical trials have questioned whether the additivity assumption is correct (Finnerup et al. 2010; Katz et al. 2008; Khan et al. 2003; Moerman 2000). Furthermore, increasing analgesic effects have been observed following administration of inactive placebo treatments in clinical trials that do not include a natural history control group/condition, and these large effects appear to be an obstacle to obtain approval of supposedly new active medications, and even approval of medications previously approved (Dworkin et al. 2012; Silberman 2009; Usdin 2011). This has led to a renewed focus on the best ways to conduct clinical trials and to test assay sensitivity, i.e., the ability of a clinical trial to distinguish an effective treatment from a less effective or ineffective treatment (Dworkin et al. 2012).

Recently, the additivity assumption has been directly investigated in an experimental study (Lund et al. 2014). Forty-eight healthy volunteers were exposed to a randomized, double-blind, within-subjects balanced, placebo design in which they received active drugs and inactive placebo agents along with either correct or incorrect verbal suggestions leaving four treatment conditions (Fig. 5). Pain was induced via hypertonic saline injection into the masseter muscle, and the active drug and placebo agents were lidocaine and saline injections, respectively. In condition A (control condition), the participants received an injection with hypertonic saline and were told that they received hypertonic saline. In condition B (active drug condition), they received an injection with hypertonic saline plus lidocaine but were told that they only received hypertonic saline. In condition C

		Information	
		No drug	Drug
Administration	No drug	A - Control	C - Placebo
	Drug	B - Active	D - Total

Fig. 5 Study design. The order of injections was randomized

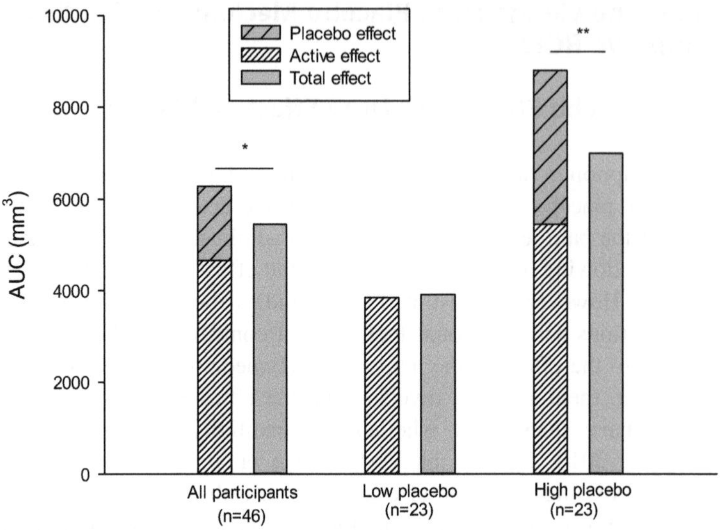

Fig. 6 Mean area under the curve (AUC) for the sum of the active effect and the placebo effect and the total effect for all participants and for the groups with low and high placebo effects. $*p < 0.05$, $**p < 0.01$

(placebo condition), participants received an injection with hypertonic saline plus placebo but were told that they received hypertonic saline along with a pain killer. In condition D (total treatment effect, i.e., administration of active drug along with verbal suggestions for pain relief), participants received an injection with hypertonic saline and lidocaine and were told that they receive hypertonic saline and a painkiller.

The total treatment effect was smaller than the sum of the drug effect and the placebo effect (D < B + C). Interestingly, the difference between the total treatment effect versus the sum of the drug effect and the placebo effect increased with the increasing magnitude of the placebo effect (Fig. 6). Hence, for participants with a low placebo effect the total treatment effect was not different from the sum of the drug effect and the placebo effect, but for participants with a high placebo effect, there was a significant difference. The implication of this for clinical trials is that the drug effect size may be underestimated in studies in which the placebo response is large, thereby contributing to problems with low assay sensitivity. Another noteworthy finding of the study was that the effect of the active drug also tended to be higher for participants with a high placebo effect than for participants with a low placebo effect, and a positive correlation between placebo and drug effects indicated that attempts to decrease placebo effects and responses may also decrease drug effects. Thus, taken together these and other findings (Hammami et al. 2010) illustrate the dilemmas and shortcomings of the randomized clinical trial. If the placebo response is large, the assay sensitivity appears to be low and the active drug effect may thus be underestimated. On the other hand, if the placebo response is reduced, the magnitude of the active drug effect may also be reduced.

4.2 Can Knowledge from Placebo Mechanisms Studies Improve Randomized Clinical Trials?

Given the challenges and apparent shortcomings of the randomized clinical trial, it is discussed whether alterations in factors related to patient characteristics, study design, study sites, and outcome measures may represent ways of improving the information obtained from clinical trials. However, the studies presented in this chapter illustrate that placebo effects and drug effects influence each other (Lund et al. 2014), and that psychological factors such as expectations of pain relief and emotional feelings contribute to both placebo and drug effects. Hence, an additional or alternative way of improving the understanding of the extent to which placebo factors contribute to the pain relief following active drug and placebo administration in clinical trials may be to directly ask patients about their expectations and emotional feelings (Price and Vase 2013; Vase and Petersen 2013). These measures are simple to administer and they could form a valuable adjunct measure in standard clinical trials. Also, in meta-analyses of randomized clinical trials it might be possible to make approximations of expectations and emotional feelings by looking at randomization rate, strength of active medication, dosing regimen as well as frequency and type of interaction with healthcare professionals. These parameters could be used as predictors in meta-analyses of clinical trials, whereby it may be possible to find new ways of explaining the variability in analgesic effects in clinical trials (Vase and Petersen 2013). Such an approach has successfully been applied to the understanding of how verbal suggestions for pain relief may influence adverse events in the placebo arm of clinical trials (Amanzio et al. 2009), and it may have far-reaching implications for the way of testing pain medication and for the optimization of placebo factors in clinical practice (Price and Vase 2013). Hence, in future studies it may be recommendable to use current knowledge from placebo mechanism studies to improve the design and interpretation of clinical pain trials.

Conclusion

The magnitude of placebo effects in hyperalgesic states is large. Psychological factors such as expectations and emotional feelings are central to these placebo effects, and they seem to help patients engage actively in a mindset for generating pain-reducing effects through activation of the descending pain control system. The factors appear to interact over time, thereby maintaining or enhancing the pain-relieving effects. Psychological factors that contribute to placebo effects also contribute to active drug effects, which shows that drug effects and placebo effects are not independent of each other. Experimental studies have indicated that the magnitude of the placebo effect influences the magnitude of the active drug effect, thereby complicating the conclusions that can be drawn from clinical trials testing active drugs against placebos. Based on the knowledge obtained from placebo mechanism studies, expectations and emotional feelings can be proposed as additional or alternative means of assessing the placebo component of pharmacological trials, which may improve our understanding of the relationship between pharmacological effects and placebo effects. This may be helpful in the clinical testing of new pharmacological agents by ensuring that (only) agents that have true effects over and above a placebo are approved for pain treatment. It may also be helpful in clinical settings, as a better understanding of how placebo factors interact with pharmacological effects may help improve the placebo component of active drugs in clinical practice, thereby enhancing the overall treatment outcome.

Acknowledgment "The research leading to these results is part of the Europain Collaboration, which has received support from the Innovative Medicines Initiative Joint Undertaking, under Grant Agreement No. 115007, resources of which are composed of financial contribution from the European Union's Seventh Framework Programme (FP7/2007-2013) and EFPIA companies' in kind contribution" www.imi.europa.eu. The funding source had the possibility to comment on the design and reporting of the study, but had no role in the collection, analysis and interpretation of data and decision to submit for publication. Ethical approval: This study was approved by the ethical committee of the Central Denmark Region

References

Amanzio M, Pollo A, Maggi G, Benedetti F (2001) Response variability to analgesics: a role for non-specific activation of endogenous opioids. Pain 90:205–215

Amanzio M, Corrazini LL, Vase L, Benedetti F (2009) A systematic review of adverse events in placebo groups of anti-migraine clinical trials. Pain 146:261–269

Andersen LO (1997) Placebo – historisk og kulturelt. In: Andersen OK, Claësson MH, Hjóbjartsson A, Sørensen AN (eds) Placebo. Historie, biologi, og effekt. Akademisk Forlag A/S, Copenhagen, pp 69–120

Benedetti F (2009) Placebo effects. Understanding the mechanisms in health and disease. Oxford University Press, Oxford

Bingel U, Wanigasekera V, Wiech K, Ni Mhuircheartaigh R, Lee MC, Ploner M et al (2011) The effect of treatment expectation on drug efficacy: imaging the analgesic benefit of the opioid remifentanil. Sci Transl Med 3:70ra14

Conboy LA, Wasserman RH, Jacobson EE, Davis RB, Legedza AT, Park M et al (2006) Investigating placebo effects in irritable bowel syndrome: a novel research design. Contemp Clin Trials 27:123–134

Craggs JG, Price DD, Verne GN, Perlstein WM, Robinson MM (2007) Functional brain interactions that serve cognitive-affective processing during pain and placebo analgesia. NeuroImage 38:720–729

Craggs JG, Price DD, Perlstein WM, Verne GN, Robinson ME (2008) The dynamic mechanisms of placebo induced analgesia: evidence of sustained and transient regional involvement. Pain 139:660–669

Dworkin RH, Turk DC, Peirce-Sandner S, Burke LB, Farrar JT, Gilron I et al (2012) Considerations for improving assay sensitivity in chronic pain clinical trials: IMMPACT recommendations. Pain 153:1148–1158

Finnerup NB, Sindrup SH, Jensen TS (2010) The evidence for pharmacological treatment of neuropathic pain. Pain 150:573–581

Flaten MA, Aslaksen PM, Lyby PS, Bjørkedal E (2011) The relation of emotions to placebo analgesia. Philos Trans R Soc Lond B Biol Sci 366:1818–1827

Hammami MM, Al-Gaai EA, Alvi S, Hammami MB (2010) Interaction between drug and placebo effects: a cross-over balanced placebo design trial. Trials 11:1–10

Harrington A (1997) The placebo effect. An interdisciplinary exploration. Harvard University Press, Cambridge

Hróbjartsson A, Gøtzsche PC (2001) Is the placebo powerless? An analysis of clinical trials comparing placebo with no treatment. N Engl J Med 344:1594–1602

Hróbjartsson A, Gøtzsche PC (2004) Is the placebo powerless? Update of a systematic review with 52 new randomised trials comparing placebo with no treatment. J Intern Med 256:91–100

Hróbjartsson A, Gøtzsche PC (2010) Placebo interventions for all clinical conditions. Cochrane Database Syst Rev 1, CD003974

Jensen TS, Baron R, Haanpaa M, Kalso E, Loeser JD, Rice AS et al (2011) A new definition of neuropathic pain. Pain 152:2204–2205

Kaptchuk TJ, Friedlander E, Kelley JM, Sanchez MN, Kokkotou E, Singer JP et al (2010) Placebos without deception: a randomized controlled trial in irritable bowel syndrome. PLoS ONE 5: e15591

Katz J, Finnerup NB, Dworkin RH (2008) Clinical trial outcome in neuropathic pain: relationship to study characteristics. Neurology 70:263–272

Khan A, Detke M, Khan SR, Mallinckrodt C (2003) Placebo response and antidepressant clinical trial outcome. J Nerv Ment Dis 191:211–218

Lieberman MD, Jarcho JM, Berman S, Naliboff BD, Suyenobu BY, Mandelkern M et al (2004) The neural correlates of placebo effects: a disruption account. NeuroImage 22:447–455

Lund K, Vase L, Petersen GL, Jensen TS, Finnerup NB (2014) Randomised controlled trials may underestimate drug effects: balanced placebo trial design. PLoS ONE 9(1):e84104. doi:10.1371/journal.pone.0084104. eCollection 2014

Moerman DE (2000) Cultural variations in the placebo effect: ulcers, anxiety, and blood pressure. Med Anthropol Q 14:51–72

Montgomery G, Kirsch I (1997) Classical conditioning and the placebo effect. Pain 72:107–113

Petersen GL, Finnerup NB, Nørskov KN, Grosen K, Pilegaard HK, Benedetti F et al (2012) Placebo manipulations reduce hyperalgesia in neuropathic pain. Pain 153:1292–1300

Piche M, Bouin M, Arsenault M, Poitras P, Rainville P (2011) Decreased pain inhibition in irritable bowel syndrome depends on altered descending modulation and higher-order brain processes. Neuroscience 195C:166–175

Price DD (1999) Psychological mechanisms of pain and analgesia. IASP Press, Seattle

Price DD, Barrell JJ (2012) Inner experience and neuroscience: merging both perspectives. MIT Press, Cambridge

Price DD, Vase L (2013) The contributions of desire, expectation and reduced negative emotions to placebo anti-hyperalgesia in irritable bowel syndrome. In: Colloca L, Flaten M, Meissner K (eds) Placebo and pain. From bench to bedside. Elsevier, Amsterdam, pp 215–226

Price DD, Milling LS, Kirsch I, Duff A, Montgomery GH, Nicholls SS (1999) An analysis of factors that contribute to the magnitude of placebo analgesia in an experimental paradigm. Pain 83:147–156

Price DD, Riley JL III, Vase L (2003) Reliable differences in placebo effects between clinical analgesic trials and studies of placebo analgesia mechanisms. Pain 104:715–716

Price DD, Craggs J, Verne GN, Perlstein WM, Robinson ME (2007) Placebo analgesia is accompanied by large reductions in pain-related brain activity in irritable bowel syndrome patients. Pain 127:63–72

Price DD, Finniss DG, Benedetti F (2008) A comprehensive review of the placebo effect: recent advances and current thought. Annu Rev Psychol 59:565–590

Scott DJ, Stohler CS, Egnatuk CM, Wang H, Koeppe RA, Zubieta JK (2007) Individual differences in reward responding explain placebo-induced expectations and effects. Neuron 55:325–336

Silberman S (2009) Placebos are getting more effective. Drugmakers are desperate to know why. Wired Magazine. http://www.wired.com. Accessed 24 Aug 2009

Usdin S (2011) Product discovery and development: shaking the cup for pain. BioCentury 19:13–14

Vase L, Petersen GL (2013) Conceptualizations and magnitudes of placebo analgesia effects across meta-analyses and experimental studies. In: Colloca L, Flaten M, Meissner K (eds) Placebo and pain. From bench to bedside. Elsevier, Amsterdam, pp 203–213

Vase L, Riley JL III, Price DD (2002) A comparison of placebo effects in clinical analgesic trials versus studies of placebo analgesia. Pain 99:443–452

Vase L, Robinson ME, Verne GN, Price DD (2003) The contribution of suggestion, desire, and expectation to placebo effects in irritable bowel syndrome patients. An empirical investigation. Pain 105:17–25

Vase L, Price DD, Verne GN, Robinson ME (2004) The contribution on changes in expected pain levels and desire for pain relief to placebo analgesia. In: Price DD, Bushnell MC (eds) Psychological methods of pain control: basic science and clinical perspectives. IASP Press, Seattle, pp 207–234

Vase L, Robinson ME, Verne GN, Price DD (2005) Increased placebo analgesia over time in irritable bowel syndrome (IBS) patients is associated with desire and expectation but not endogenous opioid mechanisms. Pain 115:338–347

Vase L, Petersen GL, Riley JL 3rd, Price DD (2009) Factors contributing to large analgesic effects in placebo mechanism studies conducted between 2002 and 2007. Pain 145:36–44

Vase L, Nørskov KN, Petersen GL, Price DD (2011) Patients' direct experiences as central elements of placebo analgesia. Philos Trans R Soc Lond B Biol Sci 366:1913–1921

Verne GN, Robinson ME, Vase L, Price DD (2003) Reversal of visceral and cutaneous hyperalgesia by local rectal anesthesia in irritable bowel syndrome (IBS) patients. Pain 105:223–230

Verne GN, Price DD, Callam CS, Zhang B, Peck J, Zhou Q (2012) Viscerosomatic facilitation in a subset of IBS patients, an effect mediated by N-Methyl-D-aspartate receptors. J Pain 13:901–909

Wager TD, Rilling JK, Smith EE, Sokolik A, Casey KL, Davidson RJ et al (2004) Placebo-induced changes in fMRI in the anticipation and experience of pain. Science 303:1162–1167

Watson D, Clark LA, Tellegen A (1988) Development and validation of brief measures of positive and negative affect: the PANAS scales. J Pers Soc Psychol 54:1063–1070

Part II

Emerging Models

Great Expectations: The Placebo Effect in Parkinson's Disease

Sarah Christine Lidstone

Contents

1 Parkinson's Disease as a Model for Studying the Placebo Effect 140
2 Evidence for the Placebo Effect in Parkinson's Disease 141
3 Placebos as Rewards 142
4 The Importance of Expectation 143
5 Implications and Future Directions 144
References 145

Abstract

Our understanding of the neural mechanisms underlying the placebo effect has increased exponentially in parallel with the advances in brain imaging. This is of particular importance in the field of Parkinson's disease, where clinicians have described placebo effects in their patients for decades. Significant placebo effects have been observed in clinical trials for medications as well as more invasive surgical trials including deep-brain stimulation and stem-cell implantation. In addition to placebo effects occurring as a byproduct of randomized controlled trials, investigation of the placebo effect itself in the laboratory setting has further shown the capacity for strong placebo effects within this patient population. Neuroimaging studies have demonstrated that placebos stimulate the release of dopamine in the striatum of patients with Parkinson's disease and can alter the activity of dopamine neurons using single-cell recording. When taken together with the findings from other medical conditions discussed elsewhere in this publication, a unified mechanism for the placebo effect in Parkinson's disease is emerging that blends expectation-induced neurochemical changes and disease-specific nigrostriatal dopamine release.

S.C. Lidstone (✉)
Neurology Residency Program, University of Toronto, Toronto, ON, Canada
e-mail: lidstone@interchange.ubc.ca

F. Benedetti et al. (eds.), *Placebo*, Handbook of Experimental Pharmacology 225,
DOI 10.1007/978-3-662-44519-8_8

Keywords
Placebo effect • Parkinson's disease • Expectation • Dopamine and reward

1 Parkinson's Disease as a Model for Studying the Placebo Effect

The primary neuropathology of Parkinson's disease is the selective loss of dopaminergic neurons in the midbrain that project to the motor areas of the striatum (nigrostriatal pathway). It is diagnosed based on the presence of the classic motor symptoms of tremor, cogwheel rigidity, slowness of movement (bradykinesia), and postural instability. The goal of pharmacological therapy—either dopamine replacement with levodopa or dopamine receptor agonists—is to alleviate the disabling motor symptoms. Less well-recognized but equally disabling are the autonomic, mood, sleep, and cognitive symptoms of Parkinson's disease which generally do not respond to dopamine replacement and are treated with adjunctive therapies (Calne et al. 2008).

Parkinson's disease is an excellent model to study the placebo effect. Firstly, and most generally, it is a true patient population and thus clinical improvements (whether they be attributable to active medication or placebo effects) have direct relevance to the clinical realm and need not be extrapolated. This is in contrast to studies using healthy control subjects, who cannot fully represent the myriad of complex psychosocial factors underlying the experience of living with a chronic disease, which strongly influences expectation. Unique to Parkinson's disease is that the deficits occur primarily in the motor system, thus the placebo effect is represented by improvement in motor function (although any symptom patients experience is subject to a placebo response, including mood, autonomic, or any other aspect of their illness causing reduced quality of life). In an experimental design, the patients' neurological status can therefore be assessed objectively following active treatment or placebo administration by a blinded examiner trained to perform a neurological exam. This is in contrast to experimental placebo analgesia or depression in which patients are often required to use visual analog scales to quantify reductions in pain or changes in mood. This being said, it is equally important to emphasize that the clinical scales used for measuring motor function are subjective themselves. Also, patients may be less prone to report clinical changes than the clinicians are to observe them (Freed et al. 2001). Finally, in addition to the clinical placebo effect (i.e., improvement in motor symptoms), the neurochemical/neurophysiological response to placebo can be measured directly. Endogenous dopamine release can be quantified using [^{11}C] raclopride positron emission tomography, and the activity of dopaminergic neurons in the subthalamic nucleus can be recorded intraoperatively during STN deep brain stimulation surgery (Benedetti et al. 2004). Together, these techniques have provided valuable insights into the mechanisms of the placebo effect in Parkinson's disease and have extended to other conditions as well.

2 Evidence for the Placebo Effect in Parkinson's Disease

Clinical trials for oral anti-Parkinson's medications demonstrate significant clinical improvement in 14–21 % of patients receiving placebo, which can be sustained to 6 months (Goetz et al. 2002a, b). In a double-blind trial of the dopamine agonist pergolide, significant improvement was seen in both the pergolide-treated group (30 % after 24 weeks) and the placebo group (23 % after 24 weeks) (Diamond et al. 1985). Finally, a meta-review demonstrated that 12 of 36 articles reported a 9–59 % improvement in patient motor symptoms following placebo (Shetty et al. 1999). Surgical trials also demonstrate substantial placebo effects, consistent with the observation that stronger interventions result in stronger placebo effects (Benedetti, et al. 2004; Benedetti 2012). Patients who underwent intrastriatal implantation of fetal porcine ventral mesencephalic tissue had the same the degree of improvement at 18 months as those in the sham group (Watts et al. 2001). In a human fetal transplantation trial for Parkinson's, there was no significant clinical benefit of the transplant compared to sham surgery (Olanow et al. 2003). In another study, at 18-month post-transplant, quality of life outcomes were better predicted by which treatment the patient thought she/he was assigned to rather than the actual treatment assignment (Freed et al. 2001; McRae et al. 2004).

Experiments aimed at studying the placebo effect itself have further demonstrated clinical improvement following placebo administration. Patients with subthalamic nucleus deep-brain stimulators as treatment for Parkinson's demonstrate improved motor performance when they believe their stimulators are turned on and perform worse than baseline when they believe their stimulators are turned off, compared to the conditions in which they were blind to stimulator function (Mercado et al. 2006). In an elegant series of studies using an overt-covert experimental design, Benedetti and colleagues demonstrated that sham STN-DBS improves bradykinesia as measured by hand velocity (Benedetti et al. 2003; Pollo et al. 2002).

Placebos have also been shown to stimulate the release of dopamine in the dorsal and ventral striatum (de la Fuente-Fernandez et al. 2001, 2002; Lidstone et al. 2010; Strafella et al. 2006). This is thought to represent the "disease-specific" component of the placebo effect in Parkinson's disease and is remarkable considering that patients must lose upwards of 80 % of their dopamine-producing cells before their symptoms become clinically apparent. Using [^{11}C] raclopride positron emission tomography, de la Fuente-Fernandez and colleagues demonstrated that a placebo injection stimulates the robust release of endogenous dopamine, in quantities comparable to the response to amphetamine in subjects with an intact dopamine system (de la Fuente-Fernandez et al. 2001). Furthermore, the dopamine release was greater in those patients who reported clinical improvement (i.e., placebo responders). Dopamine release has also been shown in response to sham repetitive transcranial magnetic stimulation in Parkinson's patients (Strafella et al. 2006). These results suggest that the biochemical basis for the placebo effect in Parkinson's is to replace the depleted striatal dopamine. These results are corroborated by an electrophysiology study performed in PD patients undergoing STN-DBS surgery, in which it was shown that a placebo (saline injection) evoked changes in neuronal firing in the subthalamic nucleus in placebo responders (Benedetti et al. 2004; Lanotte et al. 2005). The neurons displayed a decrease in mean discharge frequency and a shift from bursting to non-bursting activity in response to placebo, which was correlated with a reduction in upper limb rigidity.

3 Placebos as Rewards

Dopamine is hypothesized to play a prominent role in all placebo effects through its key involvement in reward processing (Lidstone and Stoessl 2007). Dopamine is a neuromodulator of all thalamocortical-basal ganglia loops underlying cognitive, motor, and emotional processing (Haber and Fudge 1997). It is synthesized by a population of neurons localized in the ventral midbrain that project to the basal ganglia and forebrain in a topographic distribution, thereby modulating excitatory and inhibitory neural transmission. In the motor system, dopamine depletion such as occurs in Parkinson's disease results in overall hypoactivity of the circuit, resulting in the clinical syndrome of bradykinesia and rigidity. The mesolimbic projections to the ventral striatum (nucleus accumbens), ventral prefrontal cortex, anterior cingulate cortex, and other limbic areas represent a major component of motivation and reward processing.

"Rewards" are defined as stimuli which, when administered to an organism following a correct or desired response, produce repeated approach behaviors or the repetition of responses (Bishop et al. 1963; Olds and Milner 1954). Thus, a reward is an operational concept used to describe the positive value that an organism attributes to an object, behavior, or internal physical state (Breiter and Rosen 1999). The ability of an organism to detect, approach, and interact with (i.e., consume, in the case of food rewards) the rewarding stimuli in its environment is a fundamental component of goal-directed behavior and requires the integration of cognitive, motivational, and motor circuits, in which dopamine plays a crucial modulatory role. The majority of dopamine neurons show phasic activation in response to primary liquid and food rewards, visual, auditory, and somatosensory reward-predicting stimuli, and intense, novel stimuli (Horvitz 2000; Schultz 2000; Ljungberg et al. 1992). Rather than signaling the absolute presence of a reward, dopamine neuron activity codes the discrepancy between the predicted reward and the actual reward, which is termed the "prediction error." (Mirenowicz and Schultz 1994; Schultz 1998) Thus, dopamine neurons are activated when rewards occur without being predicted or are better than expected and are depressed when predicted rewards are omitted or are worse than predicted. These responses of dopamine neurons are stronger to either rewards or reward-predicting stimuli that are associated with higher reward magnitude, probability, and expected reward value (Fiorillo et al. 2003; Schultz 1998, 2001; Tobler et al. 2005). In humans, increases in striatal dopamine release have been demonstrated in response to primary food reward (Small et al. 2003) and monetary rewards (Koepp et al. 1998; Zald et al. 2004). Dopamine neurons also demonstrate sustained activations during the interval between a reward-predicting cue and the delivery of the reward, which is thought to encode the uncertainty associated with reward expectation (Fiorillo et al. 2003). This represents the organism's natural environment, in which rewards occur with some degree of uncertainty. If the reward value is held constant, and if an animal is trained to associate certain conditioned stimuli with discrete probabilities of reward delivery, more than one third of dopamine neurons show a relatively slow, sustained, and moderate activation between the onset of the reward-predicting stimulus and the delivery of the reward. These tonic dopamine responses are maximally active at a probability of 0.5 ($p = 0.5$), decline both at $p = 0.25$ and $p = 0.75$, and are virtually zero at both extremes of the probability distribution ($p = 0$ and $p = 1$) (Fiorillo et al. 2003). This response reflects the uncertainty associated with reward expectation, as uncertainty can be expressed as the variance of the probability distribution, which is an inverted-U-shaped function with a peak at $p = 0.5$ (intuitively, it can be understood that an outcome is most uncertain when the likelihood of its occurrence is 50 %, and most certain to occur or not occur, at 100 and 0 %, respectively). These findings have been extended to humans using fMRI (Dreher et al. 2006).

The dopaminergic reward circuits are the same, fundamental neural pathways that have been shown to be involved in the mechanism of the placebo effect. The anticipation of therapeutic benefit in response to placebo can easily be conceptualized as a form of reward expectation, particularly in patients suffering from a chronic illness (de la Fuente-Fernandez et al. 2002, 2004; Lidstone et al. 2010). The relief of discomfort from unpleasant symptoms (i.e., removal of pain or suffering) is also a form of reward expectation, for potentially increasing or prolonging survival. Unsurprisingly, placebos have been shown to activate reward circuitry in both pain and Parkinson's disease, including stimulation of dopamine release in the ventral striatum (de la Fuente-Fernandez et al. 2002; Scott et al. 2008; Strafella et al. 2006).

4 The Importance of Expectation

As previously mentioned, patients' expectations play a central role in the mechanism of the placebo effect. Expectation is now recognized as a major driving force for the downstream physiological changes underlying placebo responses across most medical conditions and experimental paradigms (Benedetti 2013). An expectation can be loosely defined as a person's subjective sense of the probability of some future event. As it applies to the placebo effect, this can be conceptualized as two distinct entities depending on the situation. In a clinical encounter, an expected efficacy is produced when the patient believes that the treatment they are receiving will alleviate their symptoms. In a clinical trial, an expectation of perceived treatment is generated depending on whether the patient believes they have been assigned to active treatment or placebo. In both cases, the expectation of therapeutic benefit and symptom alleviation is produced. Interestingly, a placebo effect is absent in patient populations with frontal lobe pathology such as Alzheimer's disease (Benedetti et al. 2006), which is attributed to the inability to generate and/or maintain cognitive expectations (Benedetti 2010).

Manipulation of expectation has been shown to affect the clinical motor performance of patients with Parkinson's disease (Benedetti et al. 2003, 2004; Colloca et al. 2004; Mercado et al. 2006; Pollo et al. 2002). The relationship between the strength of expectation of improvement generated by a placebo and the resulting placebo effect was studied in Parkinson's disease (Lidstone et al. 2010). The outcome measures were dopamine release ("biochemical" placebo effect), the objective clinical symptoms, and the patients' subjective feeling of improvement/worsening. Patients were given a specific numeric probability that they were receiving active medication, in order to capture the distribution of the probability curve: 25, 50, 75, or 100 %, but in all cases they received placebo. Dopamine release was measured using [^{11}C] raclopride positron emission tomography and results compared to the response to active medication. Striatal dopamine release was significantly increased when the stated probability of receiving active medication was 75 %, i.e., some degree of uncertainty but reasonable sure they would receive medication and hence symptom relief. Those patients also demonstrated the greatest clinical benefit as measured by a modified version of the Unified Parkinson's Disease Rating Scale, motor component (tremor, rigidity, and bradykinesia in the supine position). Importantly, patients who had a more robust dopaminergic response to active treatment also had stronger placebo-induced dopamine release, indicating that prior treatment experience was the major determinant of dopamine release in the dorsal striatum. However, expectation of clinical improvement (i.e., the probability) was additionally required to drive dopamine release in the ventral striatum, indicating the involvement of reward expectation pathways in the placebo response (Lidstone et al. 2010). We concluded that these results illustrated a dissociation between the different dopamine circuits involved in the placebo effect in Parkinson's disease: a permissive, or reward-expectation component, driven by expectation and mesolimbic dopamine release, and a disease-specific component, represented by nigrostriatal dopamine release in the motor striatum, aimed at replenishing the depleted dopamine that occurs in the disease state.

5 Implications and Future Directions

This two-component model of the mechanism of the placebo effect could conceptually extend to other disease states and be used as a framework for further investigation and hypothesis generation. In this view, we have proposed that all placebo effects are created by (at least) two separate but related components: a generalized, fundamental reward-expectation component, driven by mesolimbic dopaminergic systems, and a disease-specific component responsible for the specific physiologic improvement (Lidstone and Stoessl 2007). This disease-specific component is unique to the medical condition experienced by the patient and is responsible for the clinical improvement, and can be conceptualized as an "effector" physiological response, such as the release of endogenous opioids in placebo analgesia, or serotonin in depression and so forth. In support of this view, dopamine release in the ventral striatum has been demonstrated in experimental placebo analgesia, in addition to endogenous opioid release (Scott et al. 2007, 2008). That both components are mediated by dopamine in Parkinson's disease (i.e. the reward expectation and physiological dopamine depletion in the motor striatum) and can be measured by PET further illustrates how powerful this patient population is as a model for studying the mechanism of the placebo effect. Future studies should be directed towards applying these results to the clinical context, particularly in a disease population such as Parkinson's disease where patients take multiple doses of medication per day that are associated with long-term side effects, such as disabling dyskinesias. Elucidating the factors responsible for maximizing endogenous dopamine release, such as the expectation of benefit, could serve as another avenue of potential adjunctive treatment in the management of this chronic disease.

Conclusion

A growing body of literature supports the existence and beneficial effects of placebo effects in Parkinson's disease. What was previously noted anecdotally in clinics, or obscuring the results of clinical trials, has evolved as a legitimate area of study and possibly future treatment in its own right. Studying the placebo effect enables researchers and clinicians to work together to understand the neural mechanisms at the core of the physician–patient relationship, bridging the laboratory and the clinic in order to explore new avenues for patient-centered care. Equally as important are the contributions that research in this area provide to the knowledge of basic neuroscience. The concept of adding scientific rigor to understanding the intricacies of human relationships and their impact on health outcomes is an exciting and compelling area of future study.

References

Benedetti F (2010) No prefrontal control, no placebo response. Pain 148(3):357–358

Benedetti F (2012) Placebo-induced improvements: how therapeutic rituals affect the patient's brain. J Acupunct Meridian Stud 5(3):97–103

Benedetti F (2013) Placebo and the new physiology of the doctor-patient relationship. Physiol Rev 93(3):1207–1246

Benedetti F, Pollo A, Lopiano L, Lanotte M, Vighetti S, Rainero I (2003) Conscious expectation and unconscious conditioning in analgesic, motor, and hormonal placebo/nocebo responses. J Neurosci 23(10):4315–4323

Benedetti F, Colloca L, Torre E, Lanotte M, Melcarne A, Pesare M, Bergamasco B, Lopiano L (2004) Placebo-responsive Parkinson patients show decreased activity in single neurons of subthalamic nucleus. Nat Neurosci 7(6):587–588

Benedetti F, Arduino C, Costa S, Vighetti S, Tarenzi L, Rainero I, Asteggiano G (2006) Loss of expectation-related mechanisms in Alzheimer's disease makes analgesic therapies less effective. Pain 121(1–2):133–144

Bishop MP, Elder ST, Heath RG (1963) Intracranial self-stimulation in man. Science 140:394–396

Breiter HC, Rosen BR (1999) Functional magnetic resonance imaging of brain reward circuitry in the human. Ann N Y Acad Sci 877:523–547

Calne SM, Lidstone SC, Kumar A (2008) Psychosocial issues in young-onset Parkinson's disease: current research and challenges. Parkinsonism Relat Disord 14(2):143–150

Colloca L, Lopiano L, Lanotte M, Benedetti F (2004) Overt versus covert treatment for pain, anxiety, and Parkinson's disease. Lancet Neurol 3(11):679–684

de la Fuente-Fernandez R, Ruth TJ, Sossi V, Schulzer M, Calne DB, Stoessl AJ (2001) Expectation and dopamine release: mechanism of the placebo effect in Parkinson's disease. Science 293 (5532):1164–1166

de la Fuente-Fernandez R, Phillips AG, Zamburlini M, Sossi V, Calne DB, Ruth TJ, Stoessl AJ (2002) Dopamine release in human ventral striatum and expectation of reward. Behav Brain Res 136(2):359–363

de la Fuente-Fernandez R, Schulzer M, Stoessl AJ (2004) Placebo mechanisms and reward circuitry: clues from Parkinson's disease. Biol Psychiatry 56(2):67–71

Diamond SG, Markham CH, Treciokas LJ (1985) Double-blind trial of pergolide for Parkinson's disease. Neurology 35(3):291–295

Dreher JC, Kohn P, Berman KF (2006) Neural coding of distinct statistical properties of reward information in humans. Cereb Cortex 16:561–573

Fiorillo CD, Tobler PN, Schultz W (2003) Discrete coding of reward probability and uncertainty by dopamine neurons. Science 299(5614):1898–1902

Freed CR, Greene PE, Breeze RE, Tsai WY, DuMouchel W, Kao R, Dillon S, Winfield H, Culver S, Trojanowski JQ, Eidelberg D, Fahn S (2001) Transplantation of embryonic dopamine neurons for severe Parkinson's disease. N Engl J Med 344(10):710–719

Goetz CG, Leurgans S, Raman R (2002a) Objective changes in motor function during placebo treatment in PD. Neurology 54(3):710–714

Goetz CG, Leurgans S, Raman R (2002b) Placebo-associated improvements in motor function: comparison of subjective and objective sections of the UPDRS in early Parkinson's disease. Mov Disord 12(2):283–288

Haber SN, Fudge JL (1997) The primate substantia nigra and VTA: integrative circuitry and function. Crit Rev Neurobiol 11(4):323–342

Horvitz JC (2000) Mesolimbocortical and nigrostriatal dopamine responses to salient non-reward events. Neuroscience 96:651–656

Koepp MJ, Gunn RN, Lawrence AD, Cunningham VJ, Dagher A, Jones T, Brooks DJ, Bench CJ, Grasby PM (1998) Evidence for striatal dopamine release during a video game. Nature 393:266–268

Lanotte M, Lopiano L, Torre E, Bergamasco B, Colloca L, Benedetti F (2005) Expectation enhances autonomic responses to stimulation of the human subthalamic limbic region. Brain Behav Immun 19(6):500–509

Lidstone SC, Stoessl AJ (2007) Understanding the placebo effect: contributions from neuroimaging. Mol Imaging Biol 9(4):176–185

Lidstone SC, Schulzer M, Dinelle K, Mak E, Sossi V, Ruth TJ, Fuente-Fernandez R, Phillips AG, Stoessl AJ (2010) Effects of expectation on placebo-induced dopamine release in Parkinson disease. Arch Gen Psychiatry 67(8):857–865

Ljungberg T, Apicella P, Schultz W (1992) Responses of monkey dopamine neurons during learning of behavioral reactions. J Neurophysiol 67:145–163

McRae C, Cherin E, Yamazaki TG, Diem G, Vo AH, Russell D, Ellgring JH, Fahn S, Greene P, Dillon S, Winfield H, Bjugstad KB, Freed CR (2004) Effects of perceived treatment on quality of life and medical outcomes in a double-blind placebo surgery trial. Arch Gen Psychiatry 61 (4):412–420

Mercado R, Constantoyannis C, Mandat T, Kumar A, Schulzer M, Stoessl AJ, Honey CR (2006) Expectation and the placebo effect in Parkinson's disease patients with subthalamic nucleus deep brain stimulation. Mov Disord 21(9):1457–1461

Mirenowicz J, Schultz W (1994) Importance of unpredictability for reward responses in primate dopamine neurons. J Neurophysiol 72:1024–1027

Olanow CW, Goetz CG, Kordower JH, Stoessl AJ, Sossi V, Brin MF, Shannon KM, Nauert GM, Perl DP, Godbold J, Freeman TB (2003) A double-blind controlled trial of bilateral fetal nigral transplantation in Parkinson's disease. Ann Neurol 54(3):403–414

Olds J, Milner P (1954) Positive reinforcement produced by electrical stimulation of septal area and other regions of rat brain. J Comp Physiol Psychol 47:419–427

Pollo A, Torre E, Lopiano L, Rizzone M, Lanotte M, Cavanna A, Bergamasco B, Benedetti F (2002) Expectation modulates the response to subthalamic nucleus stimulation in Parkinsonian patients. NeuroReport 13(11):1383–1386

Schultz W (1998) Predictive reward signal of dopamine neurons. J Neurophysiol 80(1):1–27

Schultz W (2000) Multiple reward signals in the brain. Nat Rev Neurosci 1:199–207

Schultz W (2001) Reward signaling by dopamine neurons. Neuroscientist 7(4):293–302

Scott DJ, Stohler CS, Egnatuk CM, Wang H, Koeppe RA, Zubieta JK (2007) Individual differences in reward responding explain placebo-induced expectations and effects. Neuron 55(2):325–336

Scott DJ, Stohler CS, Egnatuk CM, Wang H, Koeppe RA, Zubieta JK (2008) Placebo and nocebo effects are defined by opposite opioid and dopaminergic responses. Arch Gen Psychiatry 65 (2):220–231

Shetty N, Friedman JH, Kieburtz K, Marshall FJ, Oakes D (1999) The placebo response in Parkinson's disease. Parkinson Study Group. Clin Neuropharmacol 22(4):207–212

Small DM, Jones-Gotman M, Dagher A (2003) Feeding-induced dopamine release in dorsal striatum correlates with meal pleasantness ratings in healthy human volunteers. NeuroImage 19:1709–1715

Strafella AP, Ko JH, Monchi O (2006) Therapeutic application of transcranial magnetic stimulation in Parkinson's disease: the contribution of expectation. NeuroImage 31(4):1666–1672

Tobler PN, Fiorillo CD, Schultz W (2005) Adaptive coding of reward value by dopamine neurons. Science 307(5715):1642–1645

Watts RL, Freeman TB, Hauser RA, Bakay RAE, Ellias SA, Stoessl AJ, Eidelberg D, Fink JS (2001) A double-blind, randomised, controlled, multicentor clinical trial of the safety and efficacy of stereotaxic intrastriatal implantation of fetal porcine ventral mesencephalic tissue (NeurocellTM-PD) vs. imitation surgery in patients with Parkinson's disease (PD). Parkinson ism Relat Disord 7(S87)

Zald DH, Boileau I, El Dearedy W, Gunn R, McGlone F, Dichter GS, Dagher A (2004) Dopamine transmission in the human striatum during monetary reward tasks. J Neurosci 24:4105–4112

Mercado R, Constantoyannis C, Mandat T, Kumar A, Schulzer M, Stoessl AJ, Honey CR (2006) Expectation and the placebo effect in Parkinson's disease patients with subthalamic nucleus deep brain stimulation. Mov Disord 21(9):1457–1461

[illegible] (1984) [illegible] dependent [illegible] J Neurology [illegible]

[illegible] Stoessl [illegible] (2005) [illegible] that [illegible] in Parkinson's disease. Ann [illegible]

[illegible]

[illegible] patients [illegible]

[illegible]

[illegible]

[illegible]

[illegible]

The Effects of Placebos and Nocebos on Physical Performance

Elisa Carlino, Alessandro Piedimonte, and Elisa Frisaldi

Contents

1 Introduction 150
2 The Role of Positive and Negative Expectations 150
3 The Role of Learning 152
4 The Central Governor of Fatigue 153
References 156

Abstract

In this chapter we present and discuss recent studies on the mechanisms underlying placebo and nocebo effects in physical performance, showing how expectations and both pharmacological and nonpharmacological preconditioning procedures can be very effective in inducing placebo responses, with important implications for sport competitions. Furthermore, we place these findings within the biological model of central governor of fatigue, whose main goal is to protect our body from damage. A crucial aspect of this emerging field of placebo studies is related to the limit beyond which these procedures can be called doping in all respects.

Keywords

Placebo • Nocebo • Physical performance • Motor system • Sport • Fatigue • Central governor • Expectation • Conditioning • Doping • Human

E. Carlino (✉) • A. Piedimonte • E. Frisaldi
Neuroscience Department and National Institute of Neuroscience Labs, University of Turin Medical School, Corso Raffaello 30, 10125 Torino, Italy
e-mail: elisa.carlino@unito.it

F. Benedetti et al. (eds.), *Placebo*, Handbook of Experimental Pharmacology 225,
DOI 10.1007/978-3-662-44519-8_9

1 Introduction

Placebos are traditionally used in the clinical context: on one hand, they are widely used in clinical trials to test the efficacy of medical treatments (Enck et al. 2013); on the other hand, they are used in clinical practice to ameliorate patients' symptoms (Carlino et al. 2012). Besides their clinical use, placebo and its opposite nocebo effects are good examples of how mental activity may affect several physiological functions (Benedetti et al. 2011; Benedetti 2014; Pollo et al. 2011). For this reason, they have been studied with the aim to investigate their role in symptoms relief across a variety of conditions, such as pain, Parkinson's disease, and depression (Benedetti 2008, 2013; Frisaldi et al. 2013; Kirsch 2009, 2010).

In this chapter we will move away from the clinical perspective and we will analyze placebo and nocebo effects in motor performance. Several studies have been performed to investigate different parameters, such as perceived fatigue, time to exhaustion, and power to lift a weight.

An important basic question is related to the biological and psychological mechanisms underlying placebo effects in motor performance. From a biological perspective, central mechanisms would play a role in muscle performance and fatigue, as postulated early in the 1910s by Krogh and Lindhard (1913) and also by Mosso (1915) through the concept of "central governor." From a psychological perspective, the expectation of improvement or worsening, as well as the exposure to previous effective treatments (conditioning), have been found to play a crucial role in affecting motor performance (Beedie and Foad 2009). Within this context, the very special interaction between the athlete and his/her coach probably acts in a way that is similar to the doctor-patient relationship (Benedetti 2013). In the following sections, we will first review the recent studies on the mechanisms underlying placebo and nocebo effects in physical performance, and then we will place these findings within the biological model of central governor of fatigue.

2 The Role of Positive and Negative Expectations

Expectation plays a major role in enhancing motor performance. It can be viewed as a product of a cognitive engagement involving the anticipation of an event and is often induced by verbal suggestions (Tracey 2010). The study of expectations in sports has a long history: one of the first studies investigated the role of placebos related to anabolic steroids in a group of athletes (Ariel and Saville 1972). After the administration of an oral placebo resembling the usual steroid pill, the athletes showed greater performance in different lifting exercises compared to a baseline period in which the normal performance was assessed. However, even though this study clearly showed the psychophysiological benefits of expectancy in motor performance, it still lacked the specific assessment of placebo effects because of the absence of a natural history control group (i.e., a group that received neither a placebo nor a real steroid) as part of the experimental design.

In another and more recent study on professional power lifters, subjects received a placebo instead of the real anabolic steroid along with positive information about the increase in motor performance (Maganaris et al. 2000). The role of expectations was also investigated by disclosing the real nature of the placebo (saccharine) during the experiment. The results indicated a significant improvement in motor performance only when power lifters believed that a steroid had been administered, while this improvement drastically diminished when they were informed about the absence of a real active principle. By using the same experimental paradigm, expectancies have been investigated not only in professional athletes but also in college students. For example, Kalasountas et al. (2007) found that students in the placebo group significantly improved their motor performance (consisting of different lifting exercises) compared to the control group: however, after the placebo was disclosed (thus the expectancy component was removed), control and placebo groups showed approximately the same performance.

A balanced placebo design was used in an experiment aimed at evaluating carbohydrate supplementation in an endurance cycling performance (Clark et al. 2000). In this study, more than 40 competitive endurance cyclists were first asked to record a baseline performance, after ingesting water and performing two 40 km time trials; then they were divided in two groups: the placebo group, in which a noncaloric placebo drink was administered, and the group which received a carbohydrate drink. These two groups were further divided in three subgroups according to the given information: the first subgroup was told that the drink was a carbohydrate supplement; the second was told that the drink was a placebo; the third group knew that there was a 50:50 chance of receiving either a placebo or the carbohydrate. All subgroups were asked to complete a second performance trial. The placebo group (get placebo, told carbohydrate) showed a net measurable placebo effect, corresponding to a mean power increase of approximately 4 % and a speed increase of 1.5 % with respect to the baseline performance. No significant effect of the sole carbohydrate administration was found (get carbohydrate, told placebo) and only a small increase in mean power and speed was found in the group that received the carbohydrate and knew about its administration. Interestingly, no improvement was observed in the group in which participants were uncertain about receiving a placebo or the carbohydrate, which suggests that uncertain information can weaken positive expectations.

Professional cycling has also been the focus of other studies. In a dose–response study, trained cyclists completed a baseline performance trial and were evenly divided in three experimental groups (Beedie et al. 2006). All groups received a placebo drink without caffeine but participants of each group were informed that they would receive either a placebo (caffeine-free), a low dose of caffeine (4.5 mg/kg), or a high dose of caffeine (9 mg/kg). Accordingly, the cycling performance proportionally increased, with a power increase ranging from 1.3 % when cyclists thought they had received less caffeine to 3.1 % when they thought they have received more caffeine. Moreover, when cyclists were informed that they had received a placebo, their performance decreased by 1 %. In another study Beedie (2007) administered a hypothetical ergogenic aid to cyclists along with opposite verbal instructions of either increase or decrease in performance. In a sprint paradigm, the positive expectation group showed a significant trend of higher speed, whereas the negative expectation group ran 1.57 % slower than baseline. Interestingly, the same authors (Beedie et al. 2008) also showed that different personality traits, such as extroversion and openness, may improve placebo responsiveness in motor performance.

3 The Role of Learning

Today there is compelling experimental evidence that the placebo effect is a learning phenomenon, at least in most circumstances (Colloca and Miller 2011), for example in pain (Amanzio and Benedetti 1999). Moreover, placebo effects were found to be more robust and long lasting after a conditioning procedure compared to expectation alone (Colloca and Benedetti 2006; Colloca et al. 2008). These learning effects are particularly relevant to the world of sport, particularly within the context of doping. Indeed, an important ethical question is whether the use of illegal drugs during training (pharmacological preconditioning) can be concealed by replacing them with a placebo during competition.

In a study, Benedetti et al. (2007) used a paradigm of pharmacological preconditioning in a simulation of a sport competition in which different teams were involved in a pain endurance task. After repeated administrations of morphine in the pre-competition training phase, its replacement with a placebo on the competition day induced an opioid-mediated increase in pain endurance, which was crucial to win the competition. According to the drugs list of the World Anti-Doping Agency (WADA 2014), performance-enhancing supplements can be categorized into those which are illegal all the times, and those which are legal during the training but illegal during the competition. Therefore, although in the experiment by Benedetti et al. (2007) the use of morphine in the training phase should be considered legal, it is not clear whether it should also be considered ethical.

Similar findings were obtained with a nonpharmacological conditioning procedure (Pollo et al. 2008) in which the effects of an ergogenic placebo on the quadriceps muscle, which is responsible of the extension of the leg relative to the thigh, were studied. A placebo, which the subjects believed to be caffeine at high doses, was administered twice in two different sessions. Each time the weight to be lifted with the quadriceps was reduced surreptitiously so as to make the subjects believe that the "ergogenic agent" was effective. After this conditioning procedure, the load was restored to the original weight, and both muscle work and fatigue were assessed after placebo administration. A robust placebo effect occurred, with a significant increase in muscle work and a decrease in muscle fatigue.

Moreover, Pollo et al. (2012) showed that it is possible to negatively modulate the performance of subjects carrying out a muscle exercise to volitional maximum effort by employing discouraging suggestions and negative conditioning. In this study, the authors observed a significant decrease in the work performed under volitional maximal effort in the nocebo group compared to a significant increase of about 15 % observed in the control group. In an attempt to evaluate whether a negative conditioning can strengthen the effect of expectation elicited by verbal suggestion, the authors coupled the application of a sham electrical stimulation of the quadriceps muscles with the surreptitious increase of the weight to lift (procedural conditioning). In this case also, a sharp difference between groups was found, with controls improving about 29 % and nocebo subjects showing no changes in work performed. These findings may have profound implications for training strategies, because negative expectations may counteract the positive effects of training programs.

In a more recent study on the performance of the quadriceps muscles (Carlino et al. 2014), the contribution of four different verbal suggestions was investigated, i.e., 100, 50, 25, and 0 %, where the percentages refer to the amount of expectations. In the 100 % group, the participants were given decaffeinated coffee, but they were told it was high dose caffeine. In the 50 % group, the subjects had to choose between two cups of decaffeinated coffee, but they were told that only one cup contained a high dose of caffeine, whereas the other was decaffeinated. In the 25 % group, the subjects were presented with four different cups of decaffeinated coffee, but they were told that only one cup contained the high dose of caffeine. The 0 % group received only one cup, along with the information that it contained a decaffeinated drink. Only the 50 % group showed an increase in motor performance whereas the 25 and 0 % groups showed no effect. It is worth noting that these two latter groups showed an increase in performance if, and only if, a preconditioning procedure was carried out, thus showing that conditioning can enhance the effects of expectations.

4 The Central Governor of Fatigue

Fatigue can be considered an emotion in all respects, thus it can undergo a complex modulation by psychological factors, such as motivation, degree of self-belief, sense of rivalry, fear, memory of prior activities (St Clair Gibson et al. 2003; Noakes 2012). Its main goal is to protect our body from damage. The factors controlling muscle fatigue are not completely understood. The Hill's original peripheral model (Hill 1924) claimed that exercise is regulated by the failure of the cardiac output to provide the muscles with an adequate oxygen supply, and in this context metabolites accumulation (e.g., lactate) or depletion (e.g., glycogen, ATP) play a crucial role.

For more than a century Hill's model supplanted the preexisting idea that both brain and muscles alter their function during exercise (Krogh and Lindhard 1913; Mosso 1915). Indeed, according to the central model, motor performance is not limited by a failure of homeostasis in key organs but rather it is regulated at early stages in order to ensure that exercise is completed before harm develops. The idea is that of a central governor of fatigue which integrates metabolic and sensory peripheral factors (such as heart and respiratory rate, lactate, carbohydrate availability, and mechanical strain) with psychological and motivational factors (St Clair Gibson et al. 2001, 2003, 2006; Lambert et al. 2005) and regulates exercise specifically to ensure its completion whilst homeostasis is retained in all bodily systems.

Because of the complex nature of the sensation of fatigue, the regions of the brain from which it originates have been difficult to assess. Neuroimaging investigation found increased thalamo-insular activation in motor fatigue (Hilty et al. 2011), an effect that is similar to other sensations whose function is to alert the organism to urgent homeostatic imbalance, like air hunger (Banzett et al. 2000; Brannan et al. 2001; Evans et al. 2002; Liotti et al. 2001) and hunger for food (Tataranni et al. 1999).

The current evidence (Tanaka and Watanabe 2012) proposes that physical fatigue is regulated through the balance between two systems, one inhibitory and the other facilitatory, which provide continuous inputs to the motor areas. During physical fatigue, sensory input from the peripheral system activates the inhibition system to limit motor output from the motor cortex; this system includes a neural pathway interconnecting spinal cord, thalamus, secondary somatosensory cortex, medial and posterior insular cortex, anterior cingulated cortex (ACC), and motor cortex (Noakes 2012). In contrast, a motivational input activates the facilitation system to increase motor output from motor cortex; this system includes basal ganglia, thalamus, the limbic system, ACC, prefrontal cortex, orbitofrontal cortex, and motor cortex (Fig. 1).

Within the context of the central governor model and the central regulation of fatigue, placebos and nocebos might affect the output of the central governor by altering the individual evaluation of the ongoing muscles performance. Thus, placebos could act as a cue signaling the central governor to release its brake, while nocebos would induce a stronger limitation of muscle activity. Albeit speculative, this opposing action by placebo and nocebo would be in line with what was demonstrated for other systems regulating pain. Moreover, networks involved in homeostatic regulation of several functions and in pain processing seem to partly overlap, especially regarding ACC, insular, and thalamic regions (Craig 2003; Hilty et al. 2011). Interestingly, in a recent study it was found that the improvement of motor performance after a placebo treatment occurred along with enhanced excitability of the motor evoked potentials and decreased duration of the cortical silent period (Fiorio et al. 2014). This finding hints at a top-down, cognitive enhancement of corticospinal excitability as a neural signature of placebo modulation of motor performance, thus supporting an effect of placebos on motor output.

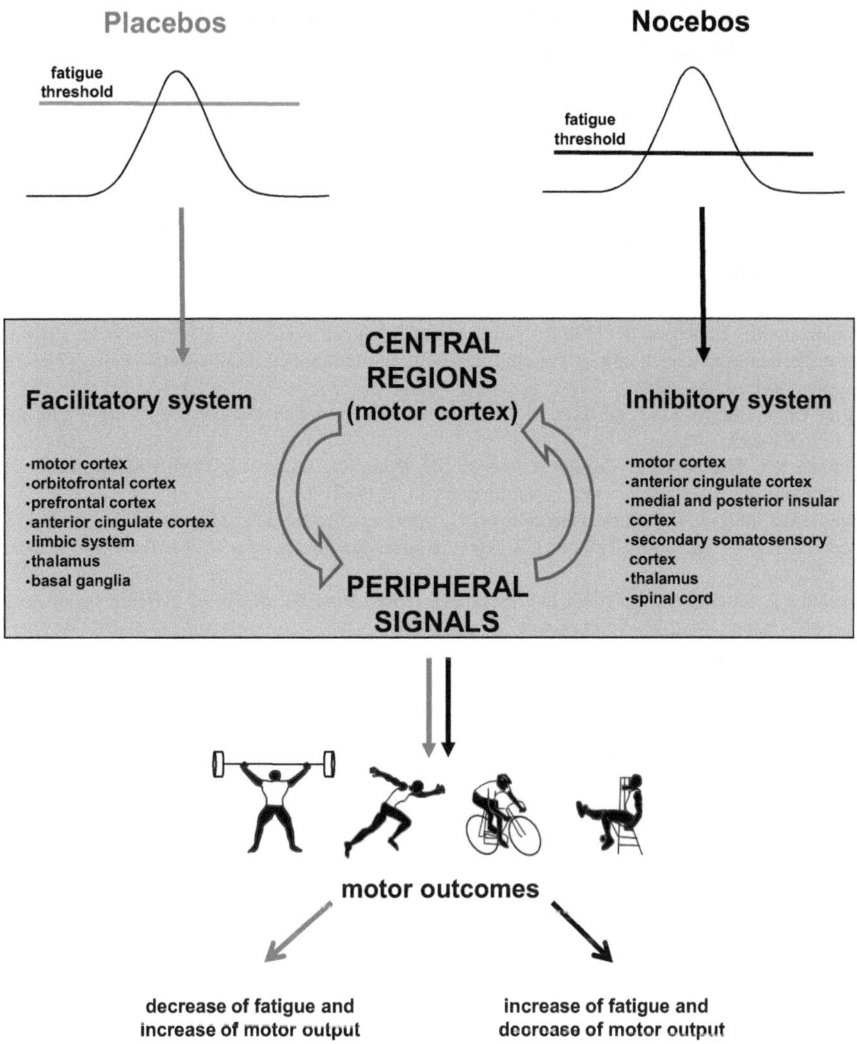

Fig. 1 The central governor of fatigue. According to the central model, physical fatigue is regulated through the balance between an inhibitory system and a facilitatory system. Within this model, placebos and nocebos might act on the balance between these two systems by altering the individual evaluation of the ongoing muscles performance: on one hand, placebos could act to increase fatigue threshold with the consequent increase of motor output and decrease of perceived fatigue; on the other hand, nocebos could act to decrease fatigue threshold with the consequent decrease of motor output and increase of perceived fatigue

Conclusions

Placebo and nocebo effects have important implications for physical performance of athletes and for training strategies (Beedie and Foad 2009; Pollo et al. 2011). However, an important ethical debate is related to the limit beyond which these effects can be called "doping" in all respects. Both pharmacological and nonpharmacological preconditioning procedures can be very effective in inducing robust placebo responses. These should be included in the discussion revolving around the world of sport, in which ergogenic substances and procedures must be used in full respect of WADA World Anti-Doping Code.

References

Amanzio M, Benedetti F (1999) Neuropharmacological dissection of placebo analgesia: expectation-activated opioid systems versus conditioning-activated specific subsystems. J Neurosci 19:484–494

Ariel G, Saville W (1972) Effect of anabolic steroids on reflex components. J Appl Physiol 32:795–797

Banzett RB, Mulnier HE, Murphy K, Rosen SD, Wise RJ, Adams L (2000) Breathlessness in humans activates insular cortex. Neuroreport 11:2117–2120

Beedie CJ (2007) Placebo effects in competitive sport: qualitative data. J Sports Sci Med 6:21–28

Beedie CJ, Foad AJ (2009) The placebo effect in sports performance: a brief review. Sports Med 39:313–329

Beedie CJ, Stuart EM, Coleman DA, Foad AJ (2006) Placebo effects of caffeine on cycling performance. Med Sci Sports Exerc 38:2159–2164

Beedie CJ, Foad AJ, Coleman DA (2008) Identification of placebo responsive participants in 40 km laboratory cycling performance. J Sports Sci Med 7:166–175

Benedetti F (2008) Mechanisms of placebo and placebo-related effects across diseases and treatments. Annu Rev Pharmacol Toxicol 48:33–60

Benedetti F (2013) Placebo and the new physiology of the doctor-patient relationship. Physiol Rev 93:1207–1246

Benedetti F (2014) Placebo effects: understanding the mechanisms in health and disease, 2nd edn. Oxford University Press, Oxford, UK

Benedetti F, Pollo A, Colloca L (2007) Opioid-mediated placebo responses boost pain endurance and physical performance: is it doping in sport competitions? J Neurosci 27:11934–11939

Benedetti F, Carlino E, Pollo A (2011) How placebos change the patient's brain. Neuropsychopharmacology 36:339–354

Brannan S, Liotti M, Egan G, Shade R, Madden L, Robillard R, Abplanalp B, Stofer K, Denton D, Fox PT (2001) Neuroimaging of cerebral activations and deactivations associated with hypercapnia and hunger for air. Proc Natl Acad Sci USA 98:2029–2034

Carlino E, Pollo A, Benedetti F (2012) The placebo in practice: how to use it in clinical routine. Curr Opin Supp Palliat Care 6:220–225

Carlino E, Benedetti F, Pollo A (2014) The effects of manipulating verbal suggestion on physical performance. Z Pers 222:154–164

Clark VR, Hopkins WG, Hawley JA, Burke LM (2000) Placebo effect of carbohydrate feedings during a 40-km cycling time trial. Med Sci Sports Exerc 32:1642–1647

Colloca L, Benedetti F (2006) How prior experience shapes placebo analgesia. Pain 124:126–133

Colloca L, Miller FG (2011) How placebo responses are formed: a learning perspective. Philos Trans R Soc Lond B Biol Sci 366:1859–1869

Colloca L, Tinazzi M, Recchia S, Le Pera D, Fiaschi A, Benedetti F, Valeriani M (2008) Learning potentiates neurophysiological and behavioral placebo analgesic responses. Pain 139:306–314

Craig AD (2003) A new view of pain as a homeostatic emotion. Trends Neurosci 26:303–307

Enck P, Bingel U, Schedlowski M, Rief W (2013) The placebo response in medicine: minimize, maximize or personalize? Nat Rev Drug Discov 12:191–204

Evans KC, Banzett RB, Adams L, McKay L, Frackowiak RSJ, Corfield DR (2002) BOLD fMRI identifies limbic, paralimbic, and cerebellar activation during air hunger. J Neurophysiol 88:1500–1511

Fiorio M, Emadi Andani M, Marotta A, Classen J, Tinazzi M (2014) Placebo-induced changes in excitatory and inhibitory corticospinal circuits during motor performance. J Neurosci 34:3993–4005

Frisaldi E, Carlino E, Lanotte M, Lopiano L, Benedetti F (2013) Characterization of the thalamic-subthalamic circuit involved in the placebo response through single-neuron recording in Parkinson patients. Cortex. doi:10.1016/j.cortex.2013.12.003, Epub ahead of print

Hill AV (1924) Muscular activity and carbohydrate metabolism. Science 60:505–514

Hilty L, Jäncke L, Luechinger R, Boutellier U, Lutz K (2011) Limitation of physical performance in a muscle fatiguing handgrip exercise is mediated by thalamo-insular activity. Hum Brain Mapp 32:2151–2160

Kalasountas V, Reed J, Fitzpatrick J (2007) The effect of placebo-induced changes in expectancies on maximal force production in college students. J Appl Sport Psychol 19:116–124

Kirsch I (2009) Antidepressants and the placebo response. Epidemiol Psichiatr Soc 18:318–322

Kirsch I (2010) Review: benefits of antidepressants over placebo limited except in very severe depression. Evid Based Mental Health 13:49

Krogh A, Lindhard J (1913) The regulation of respiration and circulation during the initial stages of muscular work. J Physiol Lond 47:112–136

Lambert EV, St Clair Gibson A, Noakes TD (2005) Complex systems model of fatigue: integrative homoeostatic control of peripheral physiological systems during exercise in humans. Br J Sports Med 39:52–62

Liotti M, Brannan S, Egan G, Shade R, Madden L, Abplanalp B, Robillard R, Lancaster J, Zamarripa FE, Fox PT, Denton D (2001) Brain responses associated with consciousness of breathlessness (air hunger). Proc Natl Acad Sci USA 98:2035–2040

Maganaris CN, Collins DJ, Sharp M (2000) Expectancy effects and strength training: do steroids make a difference? http://www.e-space.mmu.ac.uk/e-space/handle/2173/13266

Mosso A (1915) Fatigue. Allen & Unwin Ltd., London

Noakes TD (2012) Fatigue is a brain-derived emotion that regulates the exercise behavior to ensure the protection of whole body homeostasis. Front Physiol 3:82

Pollo A, Carlino E, Benedetti F (2008) The top-down influence of ergogenic placebos on muscle work and fatigue. Eur J Neurosci 28:379–388

Pollo A, Carlino E, Benedetti F (2011) Placebo mechanisms across different conditions: from the clinical setting to physical performance. Philos Trans R Soc Lond B Biol Sci 366:1790–1798

Pollo A, Carlino E, Vase L, Benedetti F (2012) Preventing motor training through nocebo suggestions. Eur J Appl Physiol 112:3893–3903

St Clair Gibson A, Schabort EJ, Noakes TD (2001) Reduced neuromuscular activity and force generation during prolonged cycling. Am J Physiol Regul Integr Comp Physiol 281:R187–R196

St Clair Gibson A, Baden DA, Lambert MI, Lambert EV, Harley YXR, Hampson D, Russell VA, Noakes TD (2003) The conscious perception of the sensation of fatigue. Sports Med 33:167–176

St Clair Gibson A, Lambert EV, Rauch LHG, Tucker R, Baden DA, Foster C, Noakes TD (2006) The role of information processing between the brain and peripheral physiological systems in pacing and perception of effort. Sports Med 36:705–722

Tanaka M, Watanabe Y (2012) Supraspinal regulation of physical fatigue. Neurosci Biobehav Rev 36:727–734

Tataranni PA, Gautier JF, Chen K, Uecker A, Bandy D, Salbe AD, Pratley RE, Lawson M, Reiman EM, Ravussin E (1999) Neuroanatomical correlates of hunger and satiation in humans using positron emission tomography. Proc Natl Acad Sci USA 96:4569–4574

Tracey I (2010) Getting the pain you expect: mechanisms of placebo, nocebo and reappraisal effects in humans. Nat Med 16:1277–1283

Learned Placebo Responses in Neuroendocrine and Immune Functions

Laura Wendt, Antje Albring, and Manfred Schedlowski

Contents

1 Introduction 161
2 Prerequisites of Learned Immune Responses 161
  2.1 Multidirectional Communication Between the CNS, the Neuroendocrine, and the Peripheral Immune System 161
  2.2 Associative Learning Rules in Behaviorally Conditioned Immune Responses 163
3 Mechanisms Mediating the Behaviorally Conditioned Immune Response 164
4 Clinical Relevance of Learned Immune Responses in Experimental Animals 165
5 Learned Placebo Responses in Immune Functions Within Humans 166
6 Learned Placebo Responses in Neuroendocrine Functions Within Humans 170
7 Clinical Implications of Learned Placebo Responses in Neuroendocrine and Immune Functions: Reproducibility, Predictability and Extinction 172
8 Future Direction 173
References 174

Abstract

The phenomenon of learned placebo responses in neuroendocrine and immune functions is a fascinating example of communication between the brain and both the endocrine and peripheral immune systems. In this chapter, we will give a short overview of afferent and efferent communication pathways, as well as the central mechanisms, which steer the behavioral conditioned immune response. Subsequently, we will focus on data that provides evidence for learned immune responses in experimental animals and learned neuroendocrine and immune placebo responses in humans. Finally, we will take a critical look at these learning protocols, to determine whether or not they can be considered a viable additional treatment option to pharmacological regimens in clinical routine. This

L. Wendt • A. Albring • M. Schedlowski (✉)
Institute of Medical Psychology and Immunobiology, University Hospital Essen, University of Duisburg-Essen, Hufelandstrasse 55, D-45122 Essen, Germany
e-mail: Manfred.schedlowski@uk-essen.de

F. Benedetti et al. (eds.), *Placebo*, Handbook of Experimental Pharmacology 225,
DOI 10.1007/978-3-662-44519-8_10

is fundamental, since there are still a number of issues, which need to be solved, such as the potential reproducibility, predictability, and extinction of the learned neuroendocrine and immune responses. Together, these findings not only provide an excellent basis to increase our understanding of human biology but may also have far reaching clinical implications. They pave the way for the ultimate aim of employing associative learning protocols as supportive treatment strategies in pharmacological regimens. As a result, medication levels may be reduced, as well as their unwanted side effects, providing a maximized therapeutic outcome to the benefit of the patient.

Keywords
Learning • Neuroendocrine system • Peripheral immune system • Behavioral conditioning • Placebo

Abbreviations

CaN	calcineurin
CNS	central nervous system
CR	conditioned response
CS	conditioned stimulus
CsA	cyclosporine A
CTA	conditioned taste aversion
EEG	electroencephalogram
HPA-axis	hypothalamus-pituitary-adrenal axis
IFN-	interferon
IL-	interleukin-
LPS	lipopolysaccharide
MS	multiple sclerosis
NF-AT	nuclear factor of activated T-cells
NK-cells	natural killer cells
Poly I:C	polyinosinic:polycytidylic acid
SLE	systemic lupus erythematosus
US	unconditioned stimulus

1 Introduction

The phenomenon of learned placebo responses on immune functions is based on two foundations. Firstly, on the bidirectional communication between the brain and the peripheral immune system, which are constantly exchanging information via efferent and afferent pathways (Ader and Cohen 1975; Riether et al. 2008; Schedlowski and Pacheco-López 2010). Secondly, on classical conditioning or associative learning processes, which are often described as the transfer of the response-eliciting property of a biologically significant stimulus (unconditioned stimulus; US) to another stimulus (conditioned stimulus; CS) without that property (Pavlov 1927; Fanselow and Poulos 2005).

In this chapter, we will give a short overview of the communication pathways between the central nervous system (CNS) and the peripheral immune system and we will focus on data that provides evidence for behaviorally conditioned immune responses in experimental animals. We will present findings, which elucidate afferent, efferent, and central mechanisms steering behavioral conditioned immune responses as well as their potential clinical relevance. Finally, we focus on learned neuroendocrine and immune placebo responses in humans and the future challenges in implementing learning protocols in a clinical context in order to supplement pharmacological treatments.

2 Prerequisites of Learned Immune Responses

2.1 Multidirectional Communication Between the CNS, the Neuroendocrine, and the Peripheral Immune System

The functional interaction between behavior and the neuroendocrine and immune systems has been intensively investigated for many decades (Riether et al. 2008; Ader 1976; Besedovsky and del Rey 1996; Blalock and Smith 2007; Janz et al. 1996; Pacheco-López et al. 2004). This research demonstrates that stress exposure affects humoral as well as cellular immune responses and disease outcome (Benschop et al. 1996; Engler et al. 2004; Kelley et al. 1985). The inverse of this is also visible, meaning that activated immune responses were also found to alter mood and behavior (Bernstein 1996; Grigoleit et al. 2012; Maier and Watkins 1998). Regarding the *efferent* mechanisms, neural and humoral pathways link the brain with the peripheral immune system. The sympathetic nervous system seems to be one major efferent neural pathway between the brain and the immune system, since all primary and secondary lymphoid organs are innervated by sympathetic nerve fibers. Leukocytes are expressing alpha and beta-adrenoceptors and catecholamines, such as adrenaline and noradrenaline, which have been shown to affect the circulation, migration, and activity of immunocompetent cells (Felten and Olschowka 1987; Nance and Sanders 2007; Sanders and Kohm 2002; Sanders and Straub 2002). The hypothalamus–pituitary–adrenal axis (HPA), with the release of cortisol, is one of the major humoral pathways via which the CNS is affecting

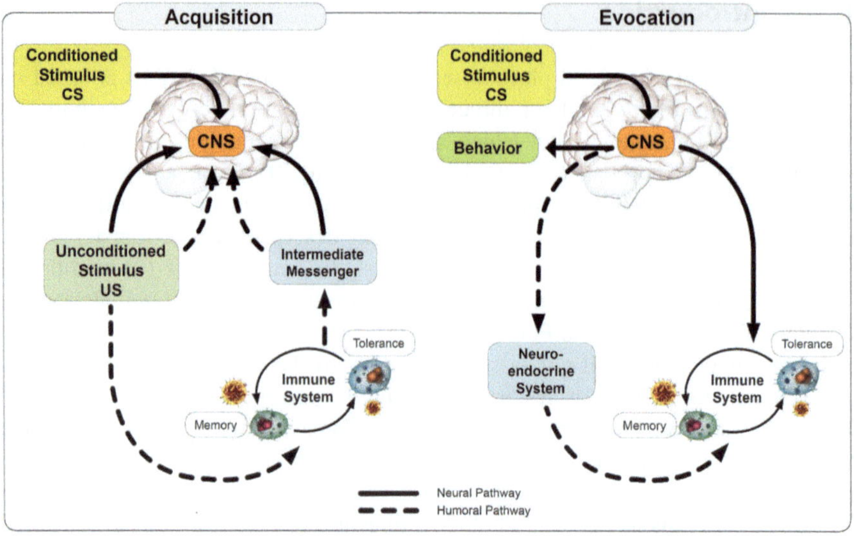

Fig. 1 As with any classical conditioning protocol, the conditioned taste aversion (CTA) paradigm consists of two basic steps: During the acquisition phase, the gustatory stimulus (e.g., saccharin solution), which serves as the conditioned stimulus (CS), is paired with an injection of a drug or substance with immune modulating properties (the unconditioned stimulus/US). The taste is received via neural afferences by the CNS. In parallel, the drug or substance itself (US), or the neural or immunological changes induced by the US, are sensed via neural or humoral afferent pathways by the CNS. The CS and the US are subsequently associated in the brain and this link may be stored like any other learned event. During evocation, when the organism is re-exposed to the CS alone, the gustatory stimulus is now able to re-enlist the stored information via efferent neural and/or humoral pathways, and to induce the immunological changes formally elicited by the US. *CNS* central nervous system, *CS* conditioned stimulus, *US* unconditioned stimulus

immunity (Kelley et al. 1985; Viveros-Paredes et al. 2006). More recently, an efferent vagus nerve-mediated cholinergic signaling path has also been identified, which controls immune functions and proinflammatory responses via the inflammatory reflex (Andersson and Tracey 2012).

With respect to the afferent pathway, it has been suggested that the immune system may act as a "sensory organ," with immune cells being mobile sentinels, informing the brain about the peripheral immune status (Blalock and Smith 2007). As on the efferent arm, immunosensory mechanisms have been categorized into two general types of pathways: a humoral and a neural pathway. Within humoral pathways, neurotransmitters, cytokines as well as prostaglandins, act as putative messengers by either crossing the blood–brain barrier (Banks 2005) or reaching the brain via circumventricular organs (Dantzer et al. 2000; Goehler et al. 2006; Schedlowski et al. 2014) (Fig. 1). The afferent neural pathway, in contrast, requires the translation of immune-borne messengers into neuronal signals. Based on anatomical and functional properties, the vagus nerve, with its relays in brain stem nuclei, is well positioned and suited to detect and inform the brain about changes of the peripheral immune status (Dantzer et al. 2000; Schedlowski et al. 2014; Goehler et al. 2000, 2007; Maier et al. 1998).

2.2 Associative Learning Rules in Behaviorally Conditioned Immune Responses

The phenomenon of behavioral conditioning of immune functions is a fascinating example of a communication network between the brain and the immune system (Ader and Cohen 1975; Riether et al. 2008; Schedlowski and Pacheco-López 2010; Ader 2003). Most learning contexts comprise of one or more initially neutral stimuli (later becoming the conditioned stimulus, CS), the organism, behavioral and physiological responses, and a biologically significant stimulus (unconditioned stimulus, US). Classical, or *Pavlovian,* conditioning can be described as the transfer of the property of a biologically significant stimulus (US) to another stimulus (CS) not possessing that property (Pavlov 1927; Fanselow and Poulos 2005; Carew and Sahley 1986; Domjan 2005; Hawkins et al. 1983). This transfer is thought to occur only if the CS serves as a predictor for the US (Rescorla and Wagner 1972; Pearce 1987; Rescorla 1988), which can in turn modify the immune response on demand at a later time (conditioned response: CR). Thus, classical conditioning can be understood as learning about the temporal or causal relationships between external and internal stimuli, to allow for the appropriate preparatory set of responses before biologically significant events occur.

Three basic steps compose any classical conditioning protocol and are particularly required for learned immune responses. The first step takes place during the acquisition or learning phase, where the US (e.g., antigen, immunomodulatory drug) must either be directly sensed by the CNS or indirectly recognized through changes within the immune response. Secondly, the CNS must now integrate and associate signals caused through the US, as well as the sensory information, provided by the CS. Thirdly, in the *evocation* or memory phase, the re-exposure to the CS must activate those brain areas which integrated the CS/US association, and subsequently modify the immune response via efferent pathways (Pavlov 1927; Bermúdez-Rattoni 2004; Pacheco-López et al. 2006a, 2007; Sewards and Sewards 2001) (Fig. 1). A conditioned taste aversion (CTA) paradigm (Garcia et al. 1955) is commonly employed to study behaviorally conditioned immune changes in experimental animals (Exton et al. 1999, 2000a, 2001). Here, organisms learn to associate a particular taste with a visceral stimulation (Bermúdez-Rattoni 2004). The majority of studies with rodents employ a sweet tasting solution (e.g., saccharin) as a CS, combined with an injection of an immune modulating drug or agent as an US. After one or several pairings of CS and US during the acquisition phase, animals are re-exposed to the CS during the evocation phase. The re-exposure to the CS then induces a behavioral response, which is characterized by an avoidance and/or aversion to the CS and reduced consumption of the sweet tasting solution. More importantly however, the re-exposure to the CS is now eliciting an immune response similar to the effects induced by the drug or agent employed as an US

(Riether et al. 2008; Schedlowski and Pacheco-López 2010; Pacheco-López et al. 2006b; Brittain and Wiener 1985; Hucklebridge 2002; Marković et al. 1993).

3 Mechanisms Mediating the Behaviorally Conditioned Immune Response

Experimental evidence over the last 30 years has demonstrated behaviorally conditioned effects in rodents, both in humoral (Bovbjerg et al. 1987a) and cellular immunity (Exton et al. 2000a; Roudebush and Bryant 1991). This has been observed in processes such as re-enlisting changes in lymphocyte circulation (Exton et al. 2000b) and proliferation (Exton et al. 1998a, b, 2000c; Lysle et al. 1988; Neveu et al. 1986), cytokine production (Exton et al. 1998a, b, 2000c; Coussons-Read et al. 1994), natural killer cell (NK-cell) activity (Coussons-Read et al. 1994; Demissie et al. 1997; Solvason et al. 1991, 1992), as well as endotoxin tolerance (Oberbeck et al. 2003); reviewed in Schedlowski and Pacheco-López (2010).

Based on the pioneering works of Ader and Cohen (Ader and Cohen 1975; Ader 2003), our laboratory has developed a behavioral conditioning model for rats, employing a CTA paradigm using saccharin as a CS and the potent immunosuppressive drug cyclosporine A (CsA) as an US (Exton et al. 2001). CsA is widely used in clinical conditions where a suppression of immune functions, in particular T cell activity, is required. Examples of this vary from transplantation medicine, where the rejection of transplanted organs needs to be prevented, or with chronic inflammatory autoimmune diseases (Kapturczak et al. 2004). CsA exerts its cellular effects through binding to a family of proteins, called immunophilins, thereupon targeting the activity of the Ca^{2+} -activated serine/threonine phosphatase calcineurin (CaN). The immunophilin cyclophilin A, predominantly binds CsA, which in turn enhances the immunophilin's affinity to CaN (reviewed in Kapturczak et al. 2004). An inhibitory complex with calcineurin is subsequently formed, leading to the inhibition of CaN activity. Conclusively, CsA prevents the dephosphorylation of the nuclear factor of activated T-cells (NF-AT) through calcineurin, thereby inhibiting the production of IL-2 and IFN-γ (Batiuk and Halloran 1997).

On the basis of this paradigm, we will briefly summarize the afferent and efferent communication pathways between the brain and the peripheral immune system steering the learned immunosuppression. To date, the *afferent* mechanisms through which the brain detects immunosuppressants such as CsA are poorly understood. In rats, acute peripheral CsA administration induced behavioral changes, such as decreased ambulatory activity in the open field as well as increased defecation 6 h after CsA injection (von Hörsten et al. 1998). These changes are paralleled by increased neuronal activity, which was found in the insular cortex and the amygdala, 1–6 h after intraperitoneal CsA injection. This was analyzed by depth EEG telemetry, c-Fos expression and noradrenaline levels in the amygdala, as determined by micro dialysis (Doenlen et al. 2011; Pacheco-López et al. 2012).

These immediate alterations appeared to be a direct effect of CsA and not indirectly mediated via the vagus nerve, since a vagal deafferentation prior to CsA injection did not prevent the increased neural activity. However, CsA levels could be detected in the cerebellum, the insular cortex and the amygdala, 2–4 hours after CsA administration (Pacheco-López et al. 2012).

Regarding the afferent pathways, we have been able to identify the *central neural substrates* involved in behaviorally conditioned immunosuppression with CsA as an US in rats (Pacheco-López et al. 2005). The conditioned immunosuppression (decreased lymphocyte proliferation as well as reduced IL-2 and IFN-γ production) was differentially affected through brain excitotoxic lesions, demonstrating that the insular cortex is essential for both the acquisition and evocation of the conditioned response. In contrast, the amygdala only appears to mediate the input of visceral information necessary for acquisition, whereas the ventro-medial hypothalamic nucleus seems to participate only in the output pathway to the immune system, which evokes the behaviorally conditioned immune response. Notably, across varying conditioning models and substances used as an US, the insular cortex and the amygdala are essential brain areas for the learned immune response (Schedlowski and Pacheco-López 2010; Pacheco-López et al. 2005).

One major *efferent* neural pathway linking the brain with the immune system is the sympathetic nervous system (Felten and Olschowka 1987; Nance and Sanders 2007; Sanders and Kohm 2002; Sanders and Straub 2002). Within the CTA paradigm, it was repeatedly demonstrated that behaviorally conditioned suppression of cytokine production (IL-2, IFN-γ) and lymphocyte proliferation, is mediated through the splenic nerve, noradrenaline, and adrenoceptor-dependent mechanisms (Exton et al. 1999, 2000c, 2001, 2002). More recently, utilizing the CsA-saccharine conditioning paradigm in rats, the enzymatic activity of CaN has been reduced in splenic lymphocytes via a ß-adrenoceptor dependent mechanism (Pacheco-López et al. 2009; Riether et al. 2011). However, this neuroanatomical pathway with the splenic nerve mediating the learned immunosuppression appears to be just one of many efferent neural routes, which are mobilized during learned immunosuppression (Schedlowski and Pacheco-López 2010; Exton et al. 2000a).

4 Clinical Relevance of Learned Immune Responses in Experimental Animals

A number of studies have addressed the possibility, that conditioned changes in immune functions are able to affect disease outcome. The use of Pavlovian conditioning protocols as a supporting therapy within pharmacological treatment has been put to the test in experimental models of autoimmune or allergic diseases, tumor progression, as well as organ transplantation (Schedlowski and Pacheco-López 2010). Indeed, these protocols, using either cyclophosphamide or CsA as an US, were able to decrease the magnitude of disease and mortality rate in experimental animal models of chronic inflammatory autoimmune diseases such as systemic lupus erythematosus (SLE) (Ader and Cohen 1982; Grota et al. 1987), adjuvant arthritis (Klosterhalfen and Klosterhalfen 1983, 1990) or multiple sclerosis (MS) (Jones et al. 2008). A series of studies addressed behavioral conditioning of asthma-like symptoms, anaphylactic shock (Djurić et al. 1988; Noelpp and Noelpp-Eschenhagen 1951a, b, c; Palermo-Neto and Guimarães 2000), or histamine release (Irie et al. 2001, 2002, 2004; Peeke et al. 1987), indicating that mast cell functions above all, are responsive to learned immune responses (Marone et al. 2003). Delayed-type or contact hypersensitivity responses, a T-lymphocyte-driven immune overreaction produced by a presensitized immune state of the host, could be inhibited by conditioning protocols pairing cyclophosphamide or CsA, respectively, with saccharine flavor (Exton et al. 2000a; Roudebush and Bryant 1991; Bovbjerg et al. 1987b). Behavioral conditioning as supportive therapy has also been studied in the context of tumor progression (Bovbjerg 2003; Hiramoto et al. 1991; Spector 1996). Employing camphor odor as a CS and polyinosinic: polycytidylic acid (poly I:C) as an US, tumor growth was reduced and the survival time was prolonged in conditioned tumor-bearing mice (Ghanta et al. 1985, 1987, 1988, 1995). Conditioning protocols also prolonged the survival time of heterotopically transplanted heart allografts in rats (Exton et al. 1998c; Grochowicz et al. 1991). Moreover, a combination of behaviorally conditioned immunosuppression and administration of a subtherapeutic dose of CsA during the evocation phase, induced synergistic effects preventing rejection in 20–30 % of animals (Exton et al. 1999).

5 Learned Placebo Responses in Immune Functions Within Humans

The increased understanding of neuroimmune mechanisms steering the behaviorally conditioned immune responses, combined with data from experimental animals, formed the basis for studying this interesting phenomenon in humans (Vits et al. 2011) (Table 1). Early observations showed that allergic symptoms can be induced in effected patients despite the absence of allergens (Turnbull 1962). This led to the assumption that learning mechanisms contribute to the pathophysiology of asthma (Turnbull 1962). Experimental evidence comes from studies investigating patients with allergic rhinitis, in which after the association phase, an intranasal saline application was given alongside the CS, resulting in elevated measures of mast cell tryptase in mucosa (Gauci et al. 1994). Similarly, allergic subjects re-exposed to an olfactory cue (CS), formerly paired with a grass allergen challenge, showed increased histamine release (Barrett et al. 2000). In contrast, the antihistaminergic properties of the H_1-receptor antagonist desloratadine, could be behaviorally conditioned in patients suffering from allergic house–dust–mite rhinitis (Goebel et al. 2008). In these patients, the learned placebo response significantly reduced subjective symptoms in the skin prick test as well as the immune response analyzed by basophile activation (Goebel et al. 2008). This result was recently confirmed and demonstrates both reproducible manipulation of expectation and

Table 1 Placebo effects in immune parameters in humans: a systematic review

Conditioned stimulus	Unconditioned stimulus	Conditioned response	Results[a]	Relevant references
Experimental environment	Chemotherapy infusion	Lymphoproliferative response to T-cell mitogens, Con A, and PHA	⇓	Bovbjerg et al. (1990)
Sweet sherbet	Epinephrine	Natural killer cell activity	⇑	Buske-Kirschbaum et al. (1992)
Sweet sherbet	Epinephrine	Natural killer cell activity	~	Kirschbaum et al. (1992)
Taste of cod liver and rose odor	Cyclophosphamide	Clinical improvement of lupus erythematosus	⇓	Olness and Ader (1992)
Experimental environment	Chemotherapy infusion	Natural killer cell and mitogen activity	~	Fredrikson et al. (1993)
Stimulus compound	Epinephrine	Natural killer cell activity/number	⇑	Buske-Kirschbaum et al. (1994)
Saline infusion	Chemotherapy infusion	Natural killer cell activity	⇑	Lekander et al. (1994)
Novel-tasting drink	House dust mite allergen	Tryptase release	⇑	Gauci et al. (1994)
Experimental environment	Chemotherapy infusion	Lymphoproliferative response to T-cell mitogen, Con A	⇑	Lekander et al. (1995)
Novel-tasting drink	Cyclophosphamide	Peripheral leukocyte counts	⇓	Giang et al. (1996)
Novel-tasting drink	Recombinant human IFN-γ	Neopterin and quinolinic acid,	⇑	Longo et al. (1999)
Novel olfactory stimulus	Grass allergen challenge	Histamine release	⇑	Barrett et al. (2000)
Novel-tasting drink	Cyclosporine A Cyclosporine A Cyclosporine A	IL-2 and IFN-γ mRNA expression In vitro release of IL-2 and IFN-γ Lymphocyte proliferation	⇓ ⇓ ⇓	Goebel et al. (2002)
Novel-tasting drink	IFN-β-1a IFN-β-1a	Granulocytes, monocytes, Lymphocytes, IL-6	~ ~	Goebel et al. (2005)
Novel-tasting drink	Desloratadine	Basophil activation	⇓	Goebel et al. (2008)
Placebo ointment	Corticosteroid	Prevalence of relapse of psoriasis	⇓	Ader et al. (2010)
Experimental environment	Chemotherapy infusion	Natural killer cell activity, IFN-γ	⇑	Stockhorst et al. (2000)

(continued)

Table 1 (continued)

Conditioned stimulus	Unconditioned stimulus	Conditioned response	Results[a]	Relevant references
Novel-tasting drink	Cyclosporine A	In vitro release of IL-2 and IFN-γ	⇓	Wirth et al. (2011)
Novel-tasting drink	Lipopolysaccharide	TNF-α, IL-6, IL-10	~	Grigoleit et al. (2012)
Novel-tasting drink	Cyclosporine A	In vitro release of IL-2	⇓	Ober et al. (2012)

[a]~: not significant; ⇑: significant, direction of effect: increase; ⇓: significant, direction of effect: decrease

learning-induced placebo responses (Vits et al. 2013). The effectiveness of the conditioning procedure on another type of allergic reaction (delayed-type hypersensitivity response) was tested in healthy volunteers receiving monthly tuberculin skin tests (Smith and McDaniel 1983). All subjects showed significantly blunted symptom severity as a result of the conditioning process (Smith and McDaniel 1983). However, employing a similar approach, these results could not be replicated (Booth et al. 1995).

These associative learning processes have also been researched in the context of cancer treatment, particularly within chemotherapy (Bovbjerg 2003). Cytotoxic chemotherapy agents generally cause adverse reactions such as nausea and vomiting as well as immunosuppressive side effects. These agents are typically administered in cycles, where each outpatient treatment infusion is followed by a period of recovery prior to the next infusion. Unwanted side effects may be caused through formerly neutral stimuli that are present during infusion of cytostatic medication (US), such as the hospital atmosphere or specific odors. When these formerly neutral stimuli are subsequently encountered alone, they are able to induce side effects as a conditioned response. Several studies examined whether patients who developed pronounced adverse reactions (nausea, vomiting) toward the hospital stimuli, also showed significant changes in immune parameters (Lekander et al. 1994, 1995; Stockhorst et al. 2000; Bovbjerg et al. 1990). Of these studies, three were able to show such an increase, one of which, measured a rise in NK-cell activity and IFN-γ concentration in patients with anticipatory nausea and vomiting in the hospital prior to chemotherapy, compared with assessments conducted at home (i.e., neutral environment) (Stockhorst et al. 2000). Proliferative responses to T-cell mitogens were significantly lower for cell samples taken in the hospital (i.e., after evocation) than for home samples (Bovbjerg et al. 1990) with similar effects reported in ovarian (Lekander et al. 1995) and pediatric patients receiving chemotherapy (Stockhorst et al. 2000). In contrast, Frederikson et al. were unable to show any functional changes (Fredrikson et al. 1993). These incongruent results were found, due to differences between the populations studied, the experimental conditions and the type of cancer. In addition to conditioned nausea (Bovbjerg et al. 1990; Andrykowski 1988; Matteson et al. 2002; Morrow et al. 1991), chemotherapy patients often develop anxiety (DiLorenzo et al. 1995; Jacobsen et al. 1993) and fatigue (Bovbjerg et al. 2005) in response to reminders of chemotherapy as well as images of chemotherapy (Dadds et al. 1997; Redd et al. 1993), raising the possibility that conditioned effects may affect patients during the course of normal life for years even after cessation of pharmacological treatment (Schedlowski and Pacheco-López 2010).

The efficiency of learned immune responses was also tested in patients with MS. When cyclophosphamide infusions were continuously paired with the taste of anise-flavored syrup during the acquisition phase, patients showed a conditioned reduction in peripheral leukocyte numbers when re-exposed to the gustatory stimulus (CS) during evocation (Giang et al. 1996). Furthermore, through pairing subcutaneous IFN-γ injections with a distinctively flavored drink (CS), it was possible to induce elevated levels of neopterin and quinolinic acid serum after re-exposing healthy volunteers to the CS (Longo et al. 1999).

Based on the fact that acute adrenaline administration increases NK-cell numbers and activity (Benschop et al. 1996; Kemeny and Schedlowski 2007; Schedlowski et al. 1996), behaviorally conditioned enhancement in NK-cell activity was observed after evoking a taste–adrenaline association (Buske-Kirschbaum et al. 1992; Buske-Kirschbaum et al. 1994). However, these effects could not be replicated in another study (Kirschbaum et al. 1992).

In a series of experiments, the CsA-saccharin paradigm was extended from experimental animals to healthy humans by pairing CsA four times with a distinctively flavored and colored drinking solution, inducing taste/immune associative learning during acquisition (Goebel et al. 2002). During evocation, subjects were re-exposed 4 times within 48 h to the CS (drink). This re-exposure caused a significant suppression of T-cell function, analyzed by impaired cytokine (IL-2 and IFN-γ) production, reduced cytokine mRNA expression, as well as inhibited T-cell proliferation (Goebel et al. 2002). In a recent study, this learned immunosuppression was repeatedly recalled by exposing the conditioned subjects to the CS again after a pause of 11 days (Wirth et al. 2011). In addition, the immunosuppression could not be induced through mere manipulation of the expectancy of test subjects (Albring et al. 2012). Moreover, plasma noradrenaline concentration, state anxiety and base levels of IL-2 predicted nearly 60 % of the conditioned immunosuppressive response, providing evidence for putative biological and psychological predictors of conditioned placebo responses in general and learned immune responses in particular (Ober et al. 2012).

Recent experimental evidence in humans confirms observations in experimental animals (Niemi et al. 2007) that inducing a learned response in the immune system requires multiple CS-US combinations during acquisition and evocation. Four CS-US pairings during the acquisition trial employing the established CsA-taste paradigm in humans with just one re-exposure to the CS during evocation, did not induce a conditioned suppression of IL-2 production (Albring et al. 2012). Similarly, an injection with lipopolysaccharide (LPS) (Grigoleit et al. 2012) or IFN-β-1a (Goebel et al. 2005) paired with a gustatory stimulus induced a conditioned odor response (Grigoleit et al. 2012) but did not affect the release of pro- and anti-inflammatory cytokines.

Together, these experimental data from human studies can be taken as a "proof of principle," that associative learning protocols, inducing placebo responses in immune functions, may be seriously considered as supportive treatment options to pharmacological therapies (Barshes et al. 2004; Cronin et al. 2000; Doering and Rief 2012). For example, giving cyclophosphamide (US) paired with the taste of cod liver and rose odor (CS) to a patient with SLE, led to clinical improvements when the cyclophosphamide dose was reduced to 50 %, by exchanging every second treatment of the drug, with a placebo infusion in combination with the CS (Olness and Ader 1992). Supporting this, a recent partial reinforcement schedule significantly reduced the amount of corticosteroid needed for the treatment of cutaneous lesions in psoriasis patients (Ader et al. 2010).

6 Learned Placebo Responses in Neuroendocrine Functions Within Humans

Although learned placebo responses in neuroendocrine functions have been successfully demonstrated in experimental animals (Ader 1976; Janz et al. 1996; Pacheco-López et al. 2004; Buske-Kirschbaum et al. 1996), there is only a small number of studies reporting these effects in humans (Buske-Kirschbaum et al. 1992; Goebel et al. 2005; Fehm-Wolfsdorf et al. 1993; Klosterhalfen et al. 2000; Sabbioni et al. 1997; Stockhorst et al. 1999, 2004, 2011a, b) (Table 2). The experimental designs include a variety of procedures, which basically resemble the conditioning protocols that have been established in animal research of learned placebo responses in immune functions. The experimental group is conditioned through the pairing of a CS (e.g., stimulus compound, injection procedure, novel-tasting drink or novel olfactory stimulus) and an US (e.g., administration of adrenaline, insulin, dexamethasone, glucose, IFN-β-1a, sumatriptan, motion sickness), which induces alterations in neuroendocrine responses. After acquisition, subjects of the experimental group are re-exposed to the CS during evocation and alterations of neuroendocrine functions (e.g., concentrations of adrenaline, glucose, cortisol, insulin, norepinephrine, glucagon, vasopressin, ACTH, somatropin) are analyzed reflecting the conditioned response (Table 2).

One of the first observations of learned neuroendocrine placebo responses has been made in schizophrenic patients (reviewed in Lichko 1959), who were treated with insulin shock therapy (US). High doses of intravenous insulin significantly reduced blood glucose levels. This decrease in glucose concentrations (CR) was later on also observed in patients after receiving a saline injection (CS). Today, there are still a number of studies assessing placebo effects in neuroendocrine functions within humans, which have chosen insulin as their US (Fehm-Wolfsdorf et al. 1993; Stockhorst et al. 1999, 2004, 2011a, b).

Through vagal mediation, centrally administered insulin in dogs affected peripheral insulin secretion and blood glucose (Chen et al. 1975; Chowers et al. 1966).

Table 2 Placebo effects in neuroendocrine parameters in humans: a systematic review

Conditioned stimulus	Unconditioned stimulus	Conditioned response	Results[a]	Relevant references
Saline injection	Insulin	Blood glucose	⇓	Lichko (1959)
Sweet sherbet	Epinephrine	Epinephrine	~	Buske-Kirschbaum et al. (1992)
Stimulus compound vs. Injection procedure itself	Insulin	Blood glucose	⇑⇓	Fehm-Wolfsdorf et al. (1993)
Novel-tasting drink	Dexamethasone	Plasma cortisol	⇑	Sabbioni et al. (1997)
Novel olfactory stimulus	Insulin	Epinephrine, norepinephrine,	~	Stockhorst et al. (1999)
	Insulin	glucagon, cortisol, insulin	~	
	Insulin	Blood glucose	⇓	
Novel-tasting drink	Motion sickness	ACTH, Vasopressin	~	Klosterhalfen et al. (2000)
Injection procedure itself	Sumatriptan	Somatropin, plasma cortisol	⇑⇓	Benedetti et al. (2003)
Novel olfactory stimulus	Insulin	Norepinephrine, somatropin	⇑	Stockhorst et al. (2004)
	Insulin	Blood glucose, serum insulin,	~	
	Insulin	glucagon, epinephrine, cortisol	~	
	Glucose	Cortisol, serum insulin	⇑	
	Glucose	Blood glucose, glucagon	~	
	Glucose	Norepinephrine	~	
Novel-tasting drink	IFN-β-1a	Norepinephrine	~	Goebel et al. (2005)
Novel olfactory stimulus	Intranasal insulin	Insulin	⇑	Stockhorst et al. (2011a)
	Intranasal insulin	Blood glucose	~	
	Intranasal insulin	Epinephrine	⇓	
Novel olfactory stimulus	Insulin	Heart rate variability	⇑	Stockhorst et al. (2011b)
	Glucose	Heart rate variability	⇑	

[a]~: not significant; ⇑: significant, direction of effect: increase; ⇓: significant, direction of effect: decrease

Insulin employed as an US therefore, fulfills the requirements of behavioral conditioning (Eikelboom and Stewart 1982; Hopkins and Williams 1997) where the CS and US are both detected and associated by the CNS, and the subsequent CR is elicited through the CNS. This has been demonstrated with a novel olfactory stimulus (CS) which was paired with an injection of insulin, inducing a conditioned decrease of blood glucose levels during evocation, when the CS was given with a placebo injection (Fehm-Wolfsdorf et al. 1993; Stockhorst et al. 1999, 2004, 2011a, b). In contrast, the same CS-US association did not yield statistically significant decreases in blood glucose levels and serum insulin, but a significant increase of norepinephrine and somatropin (Stockhorst et al. 2004, 2011b). Administration of intranasal insulin (US) was also paired with an olfactory CS, where the re-exposure to the CS induced a significant increase of insulin and a significant decrease of epinephrine, but had no effect on blood glucose (Stockhorst et al. 2011a). Another research group administered insulin injections (US) and acquired substantial deviations from baseline glucose level within 11 out of 25 subjects after a placebo injection later on. Conditioned effects were more pronounced in those subjects who were given a "compound CS" (smell, noise, visual stimuli) in addition to the injection procedure itself (Fehm-Wolfsdorf et al. 1993). Furthermore, heart rate variability, reflecting general increases of vagal activity was also increased, when a novel olfactory stimulus (CS) was either paired with insulin (US) or glucose (US) and then later on presented alone (Stockhorst et al. 2011b).

Glucose is the natural stimulus for endogenous insulin secretion and the glucose level is also detected through specific areas within the CNS (Levin et al. 1999; Oomura et al. 1969), which then elicits a counter regulation of either hypo- or hyperglycemia. For this reason, glucose has also been utilized as an US (Stockhorst et al. 2004, 2011b). In one of these studies, intravenously injected glucose (US) was paired with a novel olfactory stimulus (CS). Placebo injections were then given with the CS, which elicited no statistically significant decrease in blood glucose, but a significant increase in cortisol and serum insulin (Stockhorst et al. 2004, 2011b).

Two additional studies reported conditioned changes in plasma cortisol concentrations. Sabbioni et al. measured an increase of plasma cortisol level by re-exposing subjects to a novel tasting drink (CS), which was previously paired with dexamethasone (US) (Sabbioni et al. 1997). A decrease in cortisol as well as an increase of growth hormone was observed, when sumatriptan was utilized as an US during the conditioning procedure (Benedetti et al. 2003).

7 Clinical Implications of Learned Placebo Responses in Neuroendocrine and Immune Functions: Reproducibility, Predictability and Extinction

The phenomenon of learned placebo responses in neuroendocrine and immune functions is a fascinating example for the communication between the brain, the endocrine, and peripheral immune systems. It not only provides an excellent basis to increase our understanding of human biology but also has far reaching clinical implications with the aim of reducing the required dosage of medication, thus limiting unwanted side effects and maximizing the therapeutic outcome for the patient's benefit (Enck et al. 2008, 2013) at the same time. However, before these learning protocols can be considered as a serious additional treatment option in clinical routine, a number of issues need to be resolved, such as the potential reproducibility, predictability, and extinction of learned neuroendocrine and immune responses.

Some of these issues have been addressed within humans, such as analyzing the kinetics of learned immunosuppression, which clearly demonstrate that a single re-exposition to the CS is not sufficient to induce an immunosuppressive response (Albring et al. 2012). This finding confirms previous data of animal studies revealing that once consolidated, the extinction of the taste-CsA engram is prolonged and the more this engram is activated at evocation, the more pronounced is the behaviorally conditioned immunosuppression (Niemi et al. 2007).

As in many, if not in all pharmacological treatments, subjects differ in their ability toward learned placebo effects in immune responses. Thus, the identification of predictors of conditioned placebo responses will be essential for the future application of learning protocols in clinical practice. In a first approach to analyze putative psychological or biological variables associated with learned immunosuppressive placebo responses, we could demonstrate that, plasma noradrenaline, state anxiety, and baseline IL-2 predicted nearly 60 % of the variance in the conditioned decrease in IL-2 concentrations (Ober et al. 2012).

A basic requirement for clinical implementation of learned immunosuppression is that the learned immune response is not restricted to a single event or evocation, but is retrievable and can be recalled on multiple occasions. If conditioned immunosuppression is rapidly extinguished and can only be recalled as a single event, this paradigm can still be utilized as a valuable tool to investigate bidirectional brain–immune interactions and consequently cannot be considered an option for the clinical treatment of patients. In a first attempt to meet the above mentioned basic requirements, we were recently able to demonstrate in both, rodents and humans, that the learned immunosuppression is not restricted to a single event but is retrievable and can be repeatedly recalled (Wirth et al. 2011).

8 Future Direction

Regarding these findings and the potential clinical relevance, it is important to consider that repeated unreinforced re-expositions to the CS will finally lead to an extinction of the conditioned response (Pavlov 1927; Berman and Dudai 2001; Bouton 2002). However, the processes involved in modulating extinction learning in conditioned immunosuppression are for the most part unknown. If the extinction of learned immunosuppression can be modified or even controlled by contextual cues, retention intervals, partial reinforcement, or reconsolidation processes, as has recently been documented for CTA or fear conditioning (Dudai 2012; Monfils et al. 2009; Nader and Hardt 2009; Schiller et al. 2010; Tronson and Taylor

2007), the potential benefits of learned immunosuppression as supportive therapy together with pharmacological regimens will be enormous. In order to better assess and attain these benefits however, further research, focusing on overcoming the extinction process will be of essential significance.

Acknowledgments We would like to thank Jason Wendt for the professional editing service. This work was supported by a grant of the German Research Foundation (FOR 1328; SCHE 341/ 17-1) and Volkswagen-Stiftung (I/83 806).

Conflict of Interest Statement All authors declare that there are no conflicts of interest.

References

Ader R (1976) Conditioned adrenocortical steroid elevations in the rat. J Comp Physiol Psychol 90 (12):1156–1163

Ader R (2003) Conditioned immunomodulation: research needs and directions. Brain Behav Immun 17(Suppl 1):51–57

Ader R, Cohen N (1975) Behaviorally conditioned immunosuppression. Psychosom Med 37 (4):333–340

Ader R, Cohen N (1982) Behaviorally conditioned immunosuppression and murine systemic lupus erythematosus. Science 215(4539):1534–1536

Ader R, Mercurio MG, Walton J, James D, Davis M, Ojha V et al (2010) Conditioned pharmacotherapeutic effects: a preliminary study. Psychosom Med 72(2):192–197

Albring A, Wendt L, Benson S, Witzke O, Kribben A, Engler H et al (2012) Placebo effects on the immune response in humans: the role of learning and expectation. PLoS ONE 7(11):e49477

Andersson U, Tracey KJ (2012) Reflex principles of immunological homeostasis. Annu Rev Immunol 30:313–335

Andrykowski MA (1988) Defining anticipatory nausea and vomiting: differences among cancer chemotherapy patients who report pretreatment nausea. J Behav Med 11(1):59–69

Banks WA (2005) Blood-brain barrier transport of cytokines: a mechanism for neuropathology. Curr Pharm Des 11(8):973–984

Barrett JE, King MG, Pang G (2000) Conditioning rhinitis in allergic humans. Ann N Y Acad Sci 917:853–859

Barshes NR, Goodpastor SE, Goss JA (2004) Pharmacologic immunosuppression. Front Biosci 9:411–420

Batiuk TD, Halloran PF (1997) The downstream consequences of calcineurin inhibition. Transplant Proc 29(1–2):1239–1240

Benedetti F, Pollo A, Lopiano L, Lanotte M, Vighetti S, Rainero I (2003) Conscious expectation and unconscious conditioning in analgesic, motor, and hormonal placebo/nocebo responses. J Neurosci 23(10):4315–4323

Benschop RJ, Jacobs R, Sommer B, Schürmeyer TH, Raab JR, Schmidt RE et al (1996) Modulation of the immunologic response to acute stress in humans by beta-blockade or benzodiazepines. FASEB J 10(4):517–524

Berman DE, Dudai Y (2001) Memory extinction, learning anew, and learning the new: dissociations in the molecular machinery of learning in cortex. Science 291(5512):2417–2419

Bermúdez-Rattoni F (2004) Molecular mechanisms of taste-recognition memory. Nat Rev Neurosci 5(3):209–217

Bernstein IL (1996) Neutral mediation of food aversions and anorexia induced by tumor necrosis factor and tumors. Neurosci Biobehav Rev 20(1):177–181

Besedovsky HO, del Rey A (1996) Immune-neuro-endocrine interactions: facts and hypotheses. Endocr Rev 17(1):64–102

Blalock JE, Smith EM (2007) Conceptual development of the immune system as a sixth sense. Brain Behav Immun 21(1):23–33
Booth RJ, Petrie KJ, Brook RJ (1995) Conditioning allergic skin responses in humans: a controlled trial. Psychosom Med 57(5):492–495
Bouton ME (2002) Context, ambiguity, and unlearning: sources of relapse after behavioral extinction. Biol Psychiatry 52(10):976–986
Bovbjerg DH (2003) Conditioning, cancer, and immune regulation. Brain Behav Immun 17:58–61
Bovbjerg D, Kim YT, Siskind GW, Weksler ME (1987a) Conditioned suppression of plaque-forming cell responses with cyclophosphamide. The role of taste aversion. Ann N Y Acad Sci 496:588–594
Bovbjerg D, Cohen N, Ader R (1987b) Behaviorally conditioned enhancement of delayed-type hypersensitivity in the mouse. Brain Behav Immun 1(1):64–71
Bovbjerg DH, Redd WH, Maier LA, Holland JC, Lesko LM, Niedzwiecki D et al (1990) Anticipatory immune suppression and nausea in women receiving cyclic chemotherapy for ovarian cancer. J Consult Clin Psychol 58(2):153–157
Bovbjerg DH, Montgomery GH, Raptis G (2005) Evidence for classically conditioned fatigue responses in patients receiving chemotherapy treatment for breast cancer. J Behav Med 28 (3):231–237
Brittain RW, Wiener NI (1985) Neural and Pavlovian influences on immunity. Pavlov J Biol Sci 20(4):181–194
Buske-Kirschbaum A, Kirschbaum C, Stierle H, Lehnert H, Hellhammer D (1992) Conditioned increase of natural killer cell activity (NKCA) in humans. Psychosom Med 54(2):123–132
Buske-Kirschbaum A, Kirschbaum C, Stierle H, Jabaij L, Hellhammer D (1994) Conditioned manipulation of natural killer (NK) cells in humans using a discriminative learning protocol. Biol Psychol 38(2–3):143–155
Buske-Kirschbaum A, Grota L, Kirschbaum C, Bienen T, Moynihan J, Ader R et al (1996) Conditioned increase in peripheral blood mononuclear cell (PBMC) number and corticosterone secretion in the rat. Pharmacol Biochem Behav 55(1):27–32
Carew TJ, Sahley CL (1986) Invertebrate learning and memory: from behavior to molecules. Annu Rev Neurosci 9:435–487
Chen M, Woods SC, Porte D Jr (1975) Effect of cerebral intraventricular insulin on pancreatic insulin secretion in the dog. Diabetes 24(10):910–914
Chowers I, Lavy S, Halpern L (1966) Effect of insulin administered intracisternally on the glucose level of the blood and the cerebrospinal fluid in vagotomized dogs. Exp Neurol 14(3):383–389
Coussons-Read ME, Dykstra LA, Lysle DT (1994) Pavlovian conditioning of morphine-induced alterations of immune status: evidence for opioid receptor involvement. J Neuroimmunol 55 (2):135–142
Cronin DC 2nd, Faust TW, Brady L, Conjeevaram H, Jain S, Gupta P et al (2000) Modern immunosuppression. Clin Liver Dis 4(3):619–655
Dadds MR, Bovbjerg DH, Redd WH, Cutmore TR (1997) Imagery in human classical conditioning. Psychol Bull 122(1):89–103
Dantzer R, Konsman JP, Bluthé RM, Kelley KW (2000) Neural and humoral pathways of communication from the immune system to the brain: parallel or convergent? Auton Neurosci 85(1–3):60–65
Demissie S, Ban EE, Rogers CF, Hiramoto NS, Ghanta VK, Hiramoto RN (1997) Afferent signals to the CNS appear not to condition the modulation of interleukin-1 receptors in the hippocampus. Neuroimmunomodulation 4(5–6):298–304
DiLorenzo TA, Jacobsen PB, Bovbjerg DH, Chang H, Hudis CA, Sklarin NT et al (1995) Sources of anticipatory emotional distress in women receiving chemotherapy for breast cancer. Ann Oncol 6(7):705–711
Djurić VJ, Marković BM, Lazarević M, Janković BD (1988) Anaphylactic shock-induced conditioned taste aversion. II. Correlation between taste aversion and indicators of anaphylactic shock. Brain Behav Immun 2(1):24–31

Doenlen R, Krügel U, Wirth T, Riether C, Engler A, Prager G et al (2011) Electrical activity in rat cortico-limbic structures after single or repeated administration of lipopolysaccharide or staphylococcal enterotoxin B. Proc Biol Sci 278(1713):1864–1872

Doering BK, Rief W (2012) Utilizing placebo mechanisms for dose reduction in pharmacotherapy. Trends Pharmacol Sci 33(3):165–172

Domjan M (2005) Pavlovian conditioning: a functional perspective. Annu Rev Psychol 56:179–206

Dudai Y (2012) The restless engram: consolidations never end. Annu Rev Neurosci 35:227–247

Eikelboom R, Stewart J (1982) Conditioning of drug-induced physiological responses. Psychol Rev 89(5):507–528

Enck P, Benedetti F, Schedlowski M (2008) New insights into the placebo and nocebo responses. Neuron 59(2):195–206

Enck P, Bingel U, Schedlowski M, Rief W (2013) The placebo response in medicine: minimize, maximize or personalize? Nat Rev Drug Discov 12(3):191–204

Engler H, Dawils L, Hoves S, Kurth S, Stevenson JR, Schauenstein K et al (2004) Effects of social stress on blood leukocyte distribution: the role of alpha- and beta-adrenergic mechanisms. J Neuroimmunol 156(1–2):153–162

Exton MS, Von Hörsten S, Vöge J, Westermann J, Schult M, Nagel E et al (1998a) Conditioned taste aversion produced by cyclosporine A: concomitant reduction in lymphoid organ weight and splenocyte proliferation. Physiol Behav 63(2):241–247

Exton MS, von Hörsten S, Schult M, Vöge J, Strubel T, Donath S et al (1998b) Behaviorally conditioned immunosuppression using cyclosporine A: central nervous system reduces IL-2 production via splenic innervation. J Neuroimmunol 88(1–2):182–191

Exton MS, Schult M, Donath S, Strubel T, Nagel E, Westermann J et al (1998c) Behavioral conditioning prolongs heart allograft survival in rats. Transplant Proc 30(5):2033

Exton MS, Schult M, Donath S, Strubel T, Bode U, del Rey A et al (1999) Conditioned immunosuppression makes subtherapeutic cyclosporin effective via splenic innervation. Am J Physiol 276(6 Pt 2):R1710–R1717

Exton MS, Elfers A, Jeong WY, Bull DF, Westermann J, Schedlowski M (2000a) Conditioned suppression of contact sensitivity is independent of sympathetic splenic innervation. Am J Physiol Regul Integr Comp Physiol 279(4):1310–1315

Exton MS, von Hörsten S, Strubel T, Donath S, Schedlowski M, Westermann J (2000b) Conditioned alterations of specific blood leukocyte subsets are reconditionable. Neuroimmunomodulation 7(2):106–114

Exton MS, von Auer AK, Buske-Kirschbaum A, Stockhorst U, Göbel U, Schedlowski M (2000c) Pavlovian conditioning of immune function: animal investigation and the challenge of human application. Behav Brain Res 110(1–2):129–141

Exton MS, Herklotz J, Westermann J, Schedlowski M (2001) Conditioning in the rat: an in vivo model to investigate the molecular mechanisms and clinical implications of brain-immune communication. Immunol Rev 184:226–235

Exton MS, Gierse C, Meier B, Mosen M, Xie Y, Frede S et al (2002) Behaviorally conditioned immunosuppression in the rat is regulated via noradrenaline and β-adrenoceptors. J Neuroimmunol 131(1–2):21–30

Fanselow MS, Poulos AM (2005) The neuroscience of mammalian associative learning. Annu Rev Psychol 56:207–234

Fehm-Wolfsdorf G, Gnadler M, Kern W, Klosterhalfen W, Kerner W (1993) Classically conditioned changes of blood glucose level in humans. Physiol Behav 54(1):155–160

Felten SY, Olschowka J (1987) Noradrenergic sympathetic innervation of the spleen: II. Tyrosine hydroxylase (TH)-positive nerve terminals form synapticlike contacts on lymphocytes in the splenic white pulp. J Neurosci Res 18(1):37–48

Fredrikson M, Fürst CJ, Lekander M, Rotstein S, Blomgren H (1993) Trait anxiety and anticipatory immune reactions in women receiving adjuvant chemotherapy for breast cancer. Brain Behav Immun 7(1):79–90

Garcia J, Kimeldorf DJ, Koelling RA (1955) Conditioned aversion to saccharin resulting from exposure to gamma radiation. Science 122(3160):157–158
Gauci M, Husband AJ, Saxarra H, King MG (1994) Pavlovian conditioning of nasal tryptase release in human subjects with allergic rhinitis. Physiol Behav 55(5):823–825
Ghanta VK, Hiramoto RN, Solvason HB, Spector NH (1985) Neural and environmental influences on neoplasia and conditioning of NK activity. J Immunol 135(2):848–852
Ghanta V, Hiramoto RN, Solvason B, Spector NH (1987) Influence of conditioned natural immunity on tumor growth. Ann N Y Acad Sci 496:637–646
Ghanta VK, Miura T, Hiramoto NS, Hiramoto RN (1988) Augmentation of natural immunity and regulation of tumor growth by conditioning. Ann N Y Acad Sci 521:29–42
Ghanta VK, Hiramoto NS, Soong SJ, Hiramoto RN (1995) Conditioning of the secondary cytotoxic T-lymphocyte response to YC8 tumor. Pharmacol Biochem Behav 50(3):399–403
Giang DW, Goodman AD, Schiffer RB, Mattson DH, Petrie M, Cohen N et al (1996) Conditioning of cyclophosphamide-induced leukopenia in humans. J Neuropsychiatry Clin Neurosci 8 (2):194–201
Goebel MU, Trebst AE, Steiner J, Xie YF, Exton MS, Frede S et al (2002) Behavioral conditioning of immunosuppression is possible in humans. FASEB J 16(14):1869–1873
Goebel MU, Hübell D, Kou W, Janssen OE, Katsarava Z, Limmroth V et al (2005) Behavioral conditioning with interferon beta-1a in humans. Physiol Behav 84(5):807–814
Goebel MU, Meykadeh N, Kou W, Schedlowski M, Hengge UR (2008) Behavioral conditioning of antihistamine effects in patients with allergic rhinitis. Psychother Psychosom 77(4):227–234
Goehler LE, Gaykema RP, Hansen MK, Anderson K, Maier SF, Watkins LR (2000) Vagal immune-to-brain communication: a visceral chemosensory pathway. Auton Neurosci 85(1–3):49–59
Goehler LE, Erisir A, Gaykema RPA (2006) Neural-immune interface in the rat area postrema. Neuroscience 140(4):1415–1434
Goehler LE, Lyte M, Gaykema RPA (2007) Infection-induced viscerosensory signals from the gut enhance anxiety: implications for psychoneuroimmunology. Brain Behav Immun 21(6):721–726
Grigoleit J-S, Kullmann JS, Winkelhaus A, Engler H, Wegner A, Hammes F et al (2012) Single-trial conditioning in a human taste-endotoxin paradigm induces conditioned odor aversion but not cytokine responses. Brain Behav Immun 26(2):234–238
Grochowicz PM, Schedlowski M, Husband AJ, King MG, Hibberd AD, Bowen KM (1991) Behavioral conditioning prolongs heart allograft survival in rats. Brain Behav Immun 5 (4):349–356
Grota LJ, Ader R, Cohen N (1987) Taste aversion learning in autoimmune Mrl-lpr/lpr and Mrl +/+mice. Brain Behav Immun 1(3):238–250
Hawkins RD, Abrams TW, Carew TJ, Kandel ER (1983) A cellular mechanism of classical conditioning in Aplysia: activity-dependent amplification of presynaptic facilitation. Science 219(4583):400–405
Hiramoto RN, Hiramoto NS, Rish ME, Soong SJ, Miller DM, Ghanta VK (1991) Role of immune cells in the Pavlovian conditioning of specific resistance to cancer. Int J Neurosci 59(1–3):101–117
Hopkins DF, Williams G (1997) Insulin receptors are widely distributed in human brain and bind human and porcine insulin with equal affinity. Diabet Med 14(12):1044–1050
Hucklebridge F (2002) Behavioral conditioning of the immune system. Int Rev Neurobiol 52:325–351
Irie M, Maeda M, Nagata S (2001) Can conditioned histamine release occur under urethane anesthesia in guinea pigs? Physiol Behav 72(4):567–573
Irie M, Nagata S, Endo Y (2002) Fasting stress exacerbates classical conditioned histamine release in guinea pigs. Life Sci 72(6):689–698
Irie M, Nagata S, Endo Y (2004) Diazepam attenuates conditioned histamine release in guinea pigs. Int J Psychophysiol 51(3):231–238

Jacobsen PB, Bovbjerg DH, Redd WH (1993) Anticipatory anxiety in women receiving chemotherapy for breast cancer. Health Psychol 12(6):469–475

Janz LJ, Green-Johnson J, Murray L, Vriend CY, Nance DM, Greenberg AH et al (1996) Pavlovian conditioning of LPS-induced responses: effects on corticosterone, splenic NE, and IL-2 production. Physiol Behav 59(6):1103–1109

Jones RE, Moes N, Zwickey H, Cunningham C, Gregory WL, Oken B (2008) Treatment of experimental autoimmune encephalomyelitis with alpha lipoic acid and associative conditioning. Brain Behav Immun 22(4):538–543

Kapturczak MH, Meier-Kriesche HU, Kaplan B (2004) Pharmacology of calcineurin antagonists. Transplant Proc 36(2):25–32

Kelley KW, Dantzer R, Mormede P, Salmon H, Aynaud JM (1985) Conditioned taste aversion suppresses induction of delayed-type hypersensitivity immune reactions. Physiol Behav 34 (2):189–193

Kemeny ME, Schedlowski M (2007) Understanding the interaction between psychosocial stress and immune-related diseases: a stepwise progression. Brain Behav Immun 21(8):1009–1018

Kirschbaum C, Jabaaij L, Buske-Kirschbaum A, Hennig J, Blom M, Dorst K et al (1992) Conditioning of drug-induced immunomodulation in human volunteers: a European collaborative study. Br J Clin Psychol 31(Pt 4):459–472

Klosterhalfen W, Klosterhalfen S (1983) Pavlovian conditioning of immunosuppression modifies adjuvant arthritis in rats. Behav Neurosci 97(4):663–666

Klosterhalfen S, Klosterhalfen W (1990) Conditioned cyclosporine effects but not conditioned taste aversion in immunized rats. Behav Neurosci 104(5):716–724

Klosterhalfen S, Rüttgers A, Krumrey E, Otto B, Stockhorst U, Riepl RL et al (2000) Pavlovian conditioning of taste aversion using a motion sickness paradigm. Psychosom Med 62(5):671–677

Lekander M, Furst C, Rotstein S, Blomgren H, Fredrikson M (1994) Does informed adjuvant placebo chemotherapy for breast-cancer elicit immune changes. Oncol Rep 1(4):699–703

Lekander M, Fürst CJ, Rotstein S, Blomgren H, Fredrikson M (1995) Anticipatory immune changes in women treated with chemotherapy for ovarian cancer. Int J Behav Med 2(1):1–12

Levin BE, Dunn-Meynell AA, Routh VH (1999) Brain glucose sensing and body energy homeostasis: role in obesity and diabetes. Am J Physiol 276(5 Pt 2):1223–1231

Lichko AE (1959) On conditioned reflex hypoglycemia in man. Zh Vyssh Nerv Deiat Im I P Pavlova 9:823–829

Longo DL, Duffey PL, Kopp WC, Heyes MP, Alvord WG, Sharfman WH et al (1999) Conditioned immune response to interferon-gamma in humans. Clin Immunol 90(2):173–181

Lysle DT, Cunnick JE, Fowler H, Rabin BS (1988) Pavlovian conditioning of shock-induced suppression of lymphocyte reactivity: acquisition, extinction, and preexposure effects. Life Sci 42(22):2185–2194

Maier SF, Watkins LR (1998) Cytokines for psychologists: implications of bidirectional immune-to-brain communication for understanding behavior, mood, and cognition. Psychol Rev 105 (1):83–107

Maier SF, Goehler LE, Fleshner M, Watkins LR (1998) The role of the vagus nerve in cytokine-to-brain communication. Ann N Y Acad Sci 840:289–300

Marković BM, Dimitrijević M, Janković BD (1993) Immunomodulation by conditioning: recent developments. Int J Neurosci 71(1–4):231–249

Marone G, Granata F, Spadaro G, Genovese A, Triggiani M (2003) The histamine-cytokine network in allergic inflammation. J Allergy Clin Immunol 112(4):83–88

Matteson S, Roscoe J, Hickok J, Morrow GR (2002) The role of behavioral conditioning in the development of nausea. Am J Obstet Gynecol 186(5):239–243

Monfils M-H, Cowansage KK, Klann E, LeDoux JE (2009) Extinction-reconsolidation boundaries: key to persistent attenuation of fear memories. Science 324(5929):951–955

Morrow GR, Lindke J, Black PM (1991) Anticipatory nausea development in cancer patients: replication and extension of a learning model. Br J Psychol 82(1):61–72

Nader K, Hardt O (2009) A single standard for memory: the case for reconsolidation. Nat Rev Neurosci 10(3):224–234
Nance DM, Sanders VM (2007) Autonomic innervation and regulation of the immune system (1987-2007). Brain Behav Immun 21(6):736–745
Neveu PJ, Dantzer R, Le Moal M (1986) Behaviorally conditioned suppression of mitogen-induced lymphoproliferation and antibody production in mice. Neurosci Lett 65(3):293–298
Niemi M-B, Härting M, Kou W, Del Rey A, Besedovsky HO, Schedlowski M et al (2007) Taste-immunosuppression engram: reinforcement and extinction. J Neuroimmunol 188(1–2):74–79
Noelpp B, Noelpp-Eschenhagen I (1951a) Experimental bronchial asthma in guinea pigs, Part I. Methods for objective registration of asthma attacks. Int Arch Allergy Appl Immunol 2 (4):308–320
Noelpp B, Noelpp-Eschenhagen I (1951b) Role of conditioned reflex in bronchial asthma; experimental investigation on the pathogenesis of bronchial asthma. Helv Med Acta 18 (2):142–158
Noelpp B, Noelpp-Eschenhagen I (1951c) Experimental bronchial asthma in guinea pigs. Part II. The role of conditioned reflexes in the pathogenesis of bronchial asthma. Int Arch Allergy Appl Immunol 2(4):321–329
Ober K, Benson S, Vogelsang M, Bylica A, Günther D, Witzke O et al (2012) Plasma noradrenaline and state anxiety levels predict placebo response in learned immunosuppression. Clin Pharmacol Ther 91(2):220–226
Oberbeck R, Kromm A, Exton MS, Schade U, Schedlowski M (2003) Pavlovian conditioning of endotoxin-tolerance in rats. Brain Behav Immun 17(1):20–27
Olness K, Ader R (1992) Conditioning as an adjunct in the pharmacotherapy of lupus erythematosus. J Dev Behav Pediatr 13(2):124–125
Oomura Y, Ono T, Ooyama H, Wayner MJ (1969) Glucose and osmosensitive neurones of the rat hypothalamus. Nature 222(5190):282–284
Pacheco-López G, Niemi M-B, Kou W, Härting M, Del Rey A, Besedovsky HO et al (2004) Behavioural endocrine immune-conditioned response is induced by taste and superantigen pairing. Neuroscience 129(3):555–562
Pacheco-López G, Niemi M-B, Kou W, Härting M, Fandrey J, Schedlowski M (2005) Neural substrates for behaviorally conditioned immunosuppression in the rat. J Neurosci 25(9):2330–2337
Pacheco-López G, Niemi M-B, Engler H, Schedlowski M (2006a) Conditioned enhancement of immune responses. In: Ader R, Felten DL, Cohen N (eds) Psychoneuroimmunology. Academic, San Diego, pp 631–660
Pacheco-López G, Engler H, Niemi M-B, Schedlowski M (2006b) Expectations and associations that heal: immunomodulatory placebo effects and its neurobiology. Brain Behav Immun 20 (5):430–446
Pacheco-López G, Niemi M-B, Engler H, Schedlowski M (2007) Neuro-immune associative learning. In: Bermúdez-Rattoni F (ed) Neural plasticity and memory: from genes to brain imaging. CRC Press, Boca Raton, FL, Chapter 14
Pacheco-López G, Riether C, Doenlen R, Engler H, Niemi M-B, Engler A et al (2009) Calcineurin inhibition in splenocytes induced by pavlovian conditioning. FASEB J 23(4):1161–1167
Pacheco-López G, Doenlen R, Krügel U, Arnold M, Wirth T, Riether C et al (2012) Neurobehavioural activation during peripheral immunosuppression. Int J Neuropsychopharmacol 16:137–149
Palermo-Neto J, Guimarães RK (2000) Pavlovian conditioning of lung anaphylactic response in rats. Life Sci 68(6):611–623
Pavlov IP (1927) Conditioned reflexes. Courier Dover, Mineola, NY
Pearce JM (1987) A model for stimulus generalization in Pavlovian conditioning. Psychol Rev 94 (1):61–73
Peeke HV, Dark K, Ellman G, McCurry C, Salfi M (1987) Prior stress and behaviorally conditioned histamine release. Physiol Behav 39(1):89–93

Redd WH, Dadds MR, Futterman AD, Taylor KL, Bovbjerg DH (1993) Nausea induced by mental images of chemotherapy. Cancer 72(2):629–636

Rescorla RA (1988) Behavioral studies of Pavlovian conditioning. Annu Rev Neurosci 11:329–352

Rescorla RA, Wagner AR (1972) A theory of Pavlovian conditioning: variations in the effectiveness of reinforcement and nonreinforcement. In: Black AH, Prokasy WF (eds) Classical conditioning II. Appleton, New York, NY, pp 64–99

Riether C, Doenlen R, Pacheco-López G, Niemi M-B, Engler A, Engler H et al (2008) Behavioural conditioning of immune functions: how the central nervous system controls peripheral immune responses by evoking associative learning processes. Rev Neurosci 19(1):1–17

Riether C, Kavelaars A, Wirth T, Pacheco-López G, Doenlen R, Willemen H et al (2011) Stimulation of β_2-adrenergic receptors inhibits calcineurin activity in CD4(+) T cells via PKA-AKAP interaction. Brain Behav Immun 25(1):59–66

Roudebush RE, Bryant HU (1991) Conditioned immunosuppression of a murine delayed type hypersensitivity response: dissociation from corticosterone elevation. Brain Behav Immun 5 (3):308–317

Sabbioni ME, Bovbjerg DH, Mathew S, Sikes C, Lasley B, Stokes PE (1997) Classically conditioned changes in plasma cortisol levels induced by dexamethasone in healthy men. FASEB J 11(14):1291–1296

Sanders VM, Kohm AP (2002) Sympathetic nervous system interaction with the immune system. Int Rev Neurobiol 52:17–41

Sanders VM, Straub RH (2002) Norepinephrine, the beta-adrenergic receptor, and immunity. Brain Behav Immun 16(4):290–332

Schedlowski M, Pacheco-López G (2010) The learned immune response: Pavlov and beyond. Brain Behav Immun 24(2):176–185

Schedlowski M, Hosch W, Oberbeck R, Benschop RJ, Jacobs R, Raab HR et al (1996) Catecholamines modulate human NK cell circulation and function via spleen-independent beta 2-adrenergic mechanisms. J Immunol 156(1):93–99

Schedlowski M, Engler H, Grigoleit J-S (2014) Endotoxin-induced experimental systemic inflammation in humans: a model to disentangle immune-to-brain communication. Brain Behav Immun 35:1–8

Schiller D, Monfils M-H, Raio CM, Johnson DC, Ledoux JE, Phelps EA (2010) Preventing the return of fear in humans using reconsolidation update mechanisms. Nature 463(7277):49–53

Sewards TV, Sewards MA (2001) Cortical association areas in the gustatory system. Neurosci Biobehav Rev 25(5):395–407

Smith GR Jr, McDaniel SM (1983) Psychologically mediated effect on the delayed hypersensitivity reaction to tuberculin in humans. Psychosom Med 45(1):65–70

Solvason HB, Ghanta VK, Lorden JF, Soong SJ, Hiramoto RN (1991) A behavioral augmentation of natural immunity: odor specificity supports a Pavlovian conditioning model. Int J Neurosci 61(3–4):277–288

Solvason HB, Ghanta VK, Soong SJ, Rogers CF, Hsueh CM, Hiramoto NS et al (1992) A simple, single, trial-learning paradigm for conditioned increase in natural killer cell activity. Proc Soc Exp Biol Med 199(2):199–203

Spector NH (1996) Neuroimmunomodulation: a brief review. Can conditioning of natural killer cell activity reverse cancer and/or aging? Regul Toxicol Pharmacol 24(1 Pt 2):32–38

Stockhorst U, Gritzmann E, Klopp K, Schottenfeld-Naor Y, Hübinger A, Berresheim HW et al (1999) Classical conditioning of insulin effects in healthy humans. Psychosom Med 61 (4):424–435

Stockhorst U, Spennes-Saleh S, Körholz D, Göbel U, Schneider ME, Steingrüber HJ et al (2000) Anticipatory symptoms and anticipatory immune responses in pediatric cancer patients receiving chemotherapy: features of a classically conditioned response? Brain Behav Immun 14 (3):198–218

Stockhorst U, Mahl N, Krueger M, Huenig A, Schottenfeld-Naor Y, Huebinger A et al (2004) Classical conditioning and conditionability of insulin and glucose effects in healthy humans. Physiol Behav 81(3):375–388
Stockhorst U, de Fries D, Steingrueber H-J, Scherbaum WA (2011a) Unconditioned and conditioned effects of intranasally administered insulin vs placebo in healthy men: a randomised controlled trial. Diabetologia 54(6):1502–1506
Stockhorst U, Huenig A, Ziegler D, Scherbaum WA (2011b) Unconditioned and conditioned effects of intravenous insulin and glucose on heart rate variability in healthy men. Physiol Behav 103(1):31–38
Tronson NC, Taylor JR (2007) Molecular mechanisms of memory reconsolidation. Nat Rev Neurosci 8(4):262–275
Turnbull JW (1962) Asthma conceived as a learned response. J Psychosom Res 6:59–70
Vits S, Cesko E, Enck P, Hillen U, Schadendorf D, Schedlowski M (2011) Behavioural conditioning as the mediator of placebo responses in the immune system. Philos Trans R Soc Lond B Biol Sci 366(1572):1799–1807
Vits S, Cesko E, Benson S, Rueckert A, Hillen U, Scahdendorf D et al (2013) Cognitive factors mediate placebo responses in patients with house dust mite allergy. PLoS ONE 8(11):e79576
Viveros-Paredes JM, Puebla-Pérez AM, Gutiérrez-Coronado O, Sandoval-Ramírez L, Villaseñor-García MM (2006) Dysregulation of the Th1/Th2 cytokine profile is associated with immunosuppression induced by hypothalamic-pituitary-adrenal axis activation in mice. Int Immunopharmacol 6(5):774–781
von Hörsten S, Exton MS, Vöge J, Schult M, Nagel E, Schmidt RE et al (1998) Cyclosporine A affects open field behavior in DA rats. Pharmacol Biochem Behav 60(1):71–76
Wirth T, Ober K, Prager G, Vogelsang M, Benson S, Witzke O et al (2011) Repeated recall of learned immunosuppression: evidence from rats and men. Brain Behav Immun 25(7):1444–1451

Placebo Responses on Cardiovascular, Gastrointestinal, and Respiratory Organ Functions

Karin Meissner

Contents

1 Introduction 184
2 Common Paradigms for the Investigation of Autonomic Placebo Responses 185
3 Neurobiological Basis of Placebo Responses on Autonomic Organ Functions 186
3.1 The Autonomic Nervous System (ANS) 186
3.2 The Central Autonomic Network (CAN) 186
4 Effects of Expectation and Conditioning on Blood Pressure 187
5 Placebo Responses on Coronary Arteries 188
6 Expectation Effects on Heart Rate 189
7 Placebo Responses on Gastrointestinal Motility 190
8 Placebo Responses on Nausea and Gastric Activity 191
9 Placebo Responses on Lung Function and Respiratory Activity 193
10 Reduction of Stress and Sympathetic Activity During Placebo Analgesia 194
11 Sympathetic Placebo Responses to Caffeine-Related Stimuli 195
12 Summary 195
13 Specificity of Peripheral Placebo Responses 197
References 198

Abstract

It is widely acknowledged that placebo responses are accompanied by physiological changes in the central nervous system, but little is known about placebo responses on end organ functions. The present chapter aims to fill this gap by reviewing the literature on peripheral placebo responses. Overall, there is a wide range of placebo and nocebo responses on various organ functions of the

K. Meissner (✉)
Institute of Medical Psychology, Ludwig-Maximilians-University Munich, Goethestrasse 31, 80336 Munich, Germany
e-mail: karin.meissner@med.lmu.de

F. Benedetti et al. (eds.), *Placebo*, Handbook of Experimental Pharmacology 225,
DOI 10.1007/978-3-662-44519-8_11

cardiovascular, the gastrointestinal system, and the respiratory system. Most of these studies used expectation paradigms to elicit placebo and nocebo responses. Expectations can affect heart rate, blood pressure, coronary diameter, gastric motility, bowel motility, and lung function. Classical conditioning can induce placebo respiratory depression after prior exposure to opioid drugs, and habitual coffee drinkers show physiological arousal in response to coffee-associated stimuli. Similar to findings in placebo pain research, peripheral placebo responses can be target specific. The autonomic nervous system is a likely candidate to mediate peripheral placebo responses. Further studies are necessary to identify the brain mechanisms and pathways involved in peripheral placebo responses.

Keywords
Placebo • Nocebo • Verbal suggestion • Expectation • Conditioning • Autonomic nervous system • Central autonomic network

1 Introduction

It has long been suggested that placebo interventions primarily affect subjective symptoms. However, mounting evidence from different neurobiological approaches indicates that placebo responses are also measurable on a physiological level. For example, several studies have shown that pain relief after administration of a placebo pill described as a potent painkiller is partially due to the release of endogenous opioids in the brain's descending pain-modulatory system. Furthermore, the motor improvement in Parkinsonian patients after placebo administration is associated with the release of dopamine in the striatum as well as decreased pathological activity of neurons in the subthalamic nucleus (de la Fuente-Fernandez et al. 2001; Benedetti et al. 2005). Obviously, the physiological changes during placebo analgesia are not limited to the brain: Eippert et al. (2009) induced placebo analgesia in healthy volunteers and showed that the transmission of pain signals from the forearm to the brain was blocked already at the level of the spinal cord.

Hence, placebo responses are neurobiological phenomena and comprise meaningful changes in the central nervous system. In this endeavor, less attention has been paid to the question of whether placebo responses can also affect bodily processes outside the central nervous system. Data from clinical trials indicate that placebo responses may well affect organ functions regulated by the autonomic nervous system, for example, heart rate and local blood flow (Meissner et al. 2007, 2011). The present chapter therefore gives an overview of the current knowledge about placebo responses on peripheral organ functions and their autonomic correlates. Beforehand, common research paradigms to study the placebo response are explained, and a neurobiological framework for placebo responses on peripheral organ functions is provided.

2 Common Paradigms for the Investigation of Autonomic Placebo Responses

The two most prominent paradigms to induce placebo responses in experimental studies are verbal suggestions and classical conditioning. In the first case, the placebo is administered with verbal suggestions on the alleged effects of the intervention. The verbal suggestions usually describe the expected effects of the intervention very specifically and thus induce specific expectations in the recipient, which then trigger a placebo response (when the outcome is positive), or a nocebo response (when the expected outcome is negative; see Enck et al. 2013; Meissner et al. 2011). While placebo responses on subjective outcomes frequently mirror the content of the verbal suggestions, this is not always true for placebo responses on peripheral organ functions, as will be shown later in this chapter.

The second approach to generate placebo responses is classical conditioning. Thereby, an organism learns to associate certain perceptual stimuli (e.g., a colored drink with a novel taste) with the bodily response to the drug. This learning procedure is repeated several times until the organism responds to a placebo drug and the novel stimuli as if the original drug had been administered. For example, when the drug cyclophosphamide is used for pharmacological conditioning, the learned placebo response equals the immunosuppressive effects of cyclophosphamide (Vits et al. 2011). Interestingly, the biological pathways that mediate the effects of pharmacological conditioning are drug specific (Amanzio and Benedetti 1999; Benedetti et al. 2011).

In addition to paradigms that combine placebo administration with verbal suggestions, the so-called "open–hidden paradigm" allows to assess the role of expectation for the treatment success independently from using placebo interventions. This paradigm delivers active treatment under open and hidden conditions. Increased efficacy of open interventions compared to hidden ones has been shown for many conditions, including migraine, anxiety, and Parkinson's disease (Colloca et al. 2004; Benedetti et al. 2004; Kam-Hansen et al. 2014). For example, two groups of patients with postoperative pain received the same amount of analgesics via infusion 2 h after surgery. Only the patients in the "open" group, however, were informed about the time point of drug administration. In contrast, the "hidden" group did not know when the drug was infused. As a result, the "open" group displayed faster pain relief (Benedetti et al. 2004).

Verbal suggestions, classical conditioning, and the open–hidden paradigm have been used to investigate placebo responses on peripheral organ functions. Before summarizing the most important findings from these studies, a neurobiological framework for placebo responses on peripheral organ functions is proposed. This framework clearly shows that the ANS is a likely candidate to mediate placebo responses on peripheral organ functions.

3 Neurobiological Basis of Placebo Responses on Autonomic Organ Functions

3.1 The Autonomic Nervous System (ANS)

The autonomic nervous system (ANS) provides a highly specific communication between the organs and the brain (Jänig 2006). Each organ of the body is connected to the central nervous system through efferent and afferent autonomic pathways. To maintain optimal physiological balance in the body, i.e., homeostasis, the central nervous system—on the one hand—must receive afferent inputs that report the condition of all tissues of the body (Craig 2002). This is implemented by the afferent "sensory" branches of the ANS, which convey information about the mechanical, thermal, chemical, metabolic, and hormonal state of bowels, skin, muscles, and joints from the body to the brain. According to current understanding, these pathways also play a mandatory role for visceral influences on perception, cognition, emotion, and behavior (Critchley and Harrison 2013).

On the other hand, the efferent "motor" branches of the ANS mediate physiological changes in a variety of systems including the respiratory, cardiac, vasomotor, digestive, and endocrine systems (Morrison 2001). Most target organs are innervated by both the sympathetic and the parasympathetic branches of the ANS. Their axons leave the central nervous system via cranial nerves or ventral roots. Sympathetic motor neurons synapse upon specialized neurons in the paravertebral and prevertebral ganglia, which in turn innervate the smooth muscles and glands of all organ systems (Shields 1993).

3.2 The Central Autonomic Network (CAN)

The premotor neurons of the ANS are located in the brainstem. They are under the inhibitory and excitatory control of a number of cortical and subcortical structures, i.e., the "central autonomic network" (CAN) (Beissner et al. 2013). In particular, the CAN comprises the anterior and midcingulate cortices, the insula, the dorsolateral prefrontal cortex, the amygdala, the hippocampal formation, and the hypothalamus (but is not limited to these brain regions; see Beissner et al. 2013). Thereby, the CAN integrates information from higher-order centers with incoming afferent information and thus allows for an optimal adaptation of body systems to internal and external demands (Thayer and Lane 2000; Critchley et al. 2011). Notably, several structures of the CAN, such as the dorsolateral prefrontal cortex and the anterior cingulate cortex, have been shown to play a key role for the initiation and mediation of placebo responses on pain (Zubieta et al. 2005; Krummenacher et al. 2010; Benedetti et al. 2006).

While parasympathetic responses have long been recognized to display a high degree of organ specificity, sympathetic responses are frequently equaled with the "fight-or-flight" response. However, the acute stress response is just one of many options how the CAN can exert control over sympathetic outflow (Morrison 2001).

Indeed, both parasympathetic and sympathetic efferents allow for a precise and organ-specific regulation of peripheral organ functions (Hagemann et al. 2003; Oppenheimer 2007). The specificity of the efferent pathways of the CAN is nicely illustrated by studies showing different autonomic patterns for discrete emotions (for review, see Kreibig 2010).

In the following, the experimental data on peripheral placebo responses will be reviewed. As will be shown, both expectation and conditioning can elicit placebo responses on various organ functions of the cardiovascular, the gastrointestinal, and the respiratory system.

4 Effects of Expectation and Conditioning on Blood Pressure

Elevated blood pressure, i.e., hypertension, is a major problem for human health. Accumulating evidence indicates that several types of hypertension are initiated and maintained by an increased sympathetic tone (Guyenet 2006). Blood pressure decreases over time are frequently present in the placebo groups of clinical trials on hypertension. However, when controlling for possible confounding factors (e.g., regression to the mean, spontaneous fluctuations, or habituation), the placebo response on hypertension seems to be negligible (Hrobjartsson and Gotzsche 2010).

Nevertheless, several experimental studies reported that expectations can influence blood pressure. For example, Agras and colleagues (1982) studied the contribution of verbal suggestions for the reduction of blood pressure in hypertensive patients. Thirty patients were randomly allocated to two groups that both underwent three sessions of relaxation training. The "immediate group" was told that relaxation training would produce immediate lowering of blood pressure, while the "delayed group" received information that the positive effects on blood pressure would be delayed, and that even a slight rise in blood pressure during the first relaxation sessions may occur. Results indicated a significant decrease in systolic blood pressure in the immediate group, as compared to the delayed group, while diastolic blood pressure did not differ between groups.

Amigo and colleagues (1993) examined the immediate effects of verbal suggestions on blood pressure and heart rate. Hypertensive and normotensive subjects were assigned to one of four groups, which received instructions that blood pressure would either decrease, or increase or would not change, or they received no instruction at all. After a first measurement of blood pressure and heart rate, patients were given a brief explanation, why it was necessary to repeat the measurement after a few minutes. According to group assignment they were also told whether the blood pressure would be expected to rise, fall, or remain the same. Normotensive as well as hypertensive patients showed changes of systolic blood pressure in the suggested direction, while diastolic blood pressure and heart rate remained unaffected.

Meissner and Ziep (2011) investigated the specificity of the verbally induced placebo response on blood pressure in healthy participants. Participants were randomly assigned to either a placebo treatment, or a homeopathic "verum" treatment, or no treatment. Placebo and homeopathic treatments were administered in a double-blinded design, together with information on the alleged effects of the homeopathic remedy on blood pressure (i.e., a decrease of blood pressure). Blood pressure, heart rate, and gastric activity as well as autonomic indices (sympathetic and parasympathetic components of heart rate variability, electrodermal activity) were continuously measured prior to and following placebo administration. Results showed a placebo response on systolic blood pressure in comparison to the untreated controls. No group-specific changes of any of the other measures occurred, indicating a rather specific placebo response on systolic blood pressure.

In an additional study, we asked healthy participants to inhale a placebo spray together with the verbal suggestion that it contained an effective pharmacological drug to either increase or decrease blood pressure, or that the spray was an ineffective placebo (Zimmermann-Viehoff et al. 2013). Blood pressure, heart rate, stroke volume, peripheral resistance, heart rate variability, and skin conductance levels were continuously assessed. We did not find a placebo or nocebo response on blood pressure, even though manipulation checks indicated the successful induction of treatment expectations in the two "active" placebo groups. This sheds an interesting light on the role of expectations for peripheral placebo responses: conscious expectations alone may not be sufficient to elicit a bodily placebo response.

Only recently, we started to investigate conditioned placebo effects on blood pressure (unpublished data). Healthy volunteers received either caffeine capsules or placebo capsules for two consecutive days together with a novel stimulus in a double-blind design. The novel stimulus was a green-colored strawberry milk aromatized with lavender oil, which has been successfully used for the pharmacological conditioning of immune responses (Wirth et al. 2011). On day 3, all subjects received placebo capsules and the special milk. Caffeine reliably increased diastolic blood pressure on days 1 and 2, but there was no conditioned placebo response on diastolic blood pressure on day 3 (i.e., the testing day). Possibly, an increased number of learning sessions is necessary to elicit a conditioned placebo response on blood pressure.

5 Placebo Responses on Coronary Arteries

In randomized controlled trials on ischemic heart disease, the patients in the placebo groups often show clinical improvement over time (Olshansky 2007). Whether these improvements are due to placebo responses or to placebo-unrelated factors, such as regression to the mean or spontaneous improvement, needs to be investigated (Meissner et al. 2007). However, from a theoretical point of view, placebo responses on coronary arteries may work via two mechanisms: reduction of pain associated with angina pectoris, and improvement of coronary blood flow and, thus, myocardial perfusion.

We recently examined the hypothesis that placebo interventions in patients with chest pain may directly alter the blood flow in the coronary arteries (Ronel et al. 2011). After the completion of routine diagnostic catheterization, patients with biomarker-negative chest pain were randomized to either a verbal suggestion group or a control group and received an intracoronary saline injection via the catheter. Only in the verbal suggestion group, however, patients were informed by the cardiologist that a pharmacological drug to improve coronary perfusion was going to be administered. Patients and physicians were blinded with regard to the study medication. Angiograms were performed immediately before and after the intervention. In addition, blood pressure and heart rate were repeatedly assessed, and patients were asked for acute chest pain and perceived stress just before and after the coronary injection. Remarkably, the coronary diameter was significantly affected by the verbal suggestions. Contrary to our expectation, however, we observed a coronary vasoconstriction in the verbal suggestion group as compared to controls. This vasoconstriction was accompanied by a reduction of chest pain. This study was the first to show a direct placebo response on the coronary arteries. The pain reduction in the verbal suggestion group indicated a positive expectation toward the "drug." The coronary vasoconstriction could not be explained by stress, since heart rate, blood pressure, and perceived stress did not change differentially between groups. Possibly, the verbal suggestions reduced the sympathetic outflow to the coronary arteries, thereby reducing oxygen demand, and thus coronary perfusion (Ronel et al. 2011).

6 Expectation Effects on Heart Rate

Benedetti and colleagues (2003a) studied the effects of open versus hidden infusions of the beta-blocker propranolol on heart rate changes in healthy volunteers. In the open condition, the injection was performed in full view of the subjects, who were told that the drug would decrease their heart rate and blood pressure. In the hidden condition, patients were not aware when the drug was infused. As expected, the open condition was more effective compared to the hidden condition. When propranolol was replaced by atropine (a muscarinic acetylcholine receptor antagonist), the open condition group—in accordance with the information they had received—showed larger heart rate increases. Hence, the heart rate response to pharmacologic drugs was modulated by verbally induced expectations.

Benedetti and colleagues (2004) and Lanotte and colleagues (2005) compared heart rate responses to either open or hidden stimulation of the subthalamic region in Parkinsonian patients. In an earlier study the clinical benefit of deep brain stimulation was larger in the open compared to the hidden condition, indicating an effect of expectation (Benedetti et al. 2003a). Now the group focused on effect sizes and response thresholds of changes in heart rate and heart rate variability. A difference between the open and the hidden stimulation emerged only, when the ventralmost pole of the subthalamic region was stimulated. This part of the subthalamic region relates to associative and limbic functions. This suggests that expectation may increase the excitability of limbic structures, which could enhance not only autonomic responses but also motor improvement in Parkinsonian patients.

7 Placebo Responses on Gastrointestinal Motility

Gastric contractions are regulated by the gastric pacemaker, which is under the control of the ANS. Increased gastric motility in the fasting state is associated with decreases in dominant frequency (Daniel 2001; Parkman et al. 1999), and the well-documented postprandial dip in frequency shortly after eating also is of vagal origin (Kaneko et al. 1995). In contrast, the inhibition of cholinergic activity with atropine or vagotomy increases the pacemaker frequency in dogs and humans (Sarna and Daniel 1974; Stoddard et al. 1981; Parkman et al. 1999). Thus, gastric contractions get longer when vagal outflow to the stomach is enhanced and shorter when vagal outflow is reduced.

We recently studied the effects of different verbal suggestions on gastric myoelectrical activity (Meissner 2009). In a within-subject design, healthy volunteers received a placebo pill on three separate days in randomized order together with the verbal suggestion that the pill would stimulate, relax, or not affect gastric activity. In truth, they received a placebo pill on every testing day. In addition to gastric myoelectrical activity, we assessed skin conductance levels, heart rate, and heart rate variability prior to and following the interventions, and participants were asked for perceived changes in gastric activity. The evaluation of changes in gastric myoelectrical activity indicated that in the stimulating condition the duration of stomach contractions increased, while there was a trend in the opposite direction in the relaxation condition (Fig. 1). Remarkably, participants, who reported an increase of gastric activity, also showed larger gastric placebo responses. Notably, neither heart rate nor autonomic measures (skin conductance levels and heart rate variability) showed condition-specific changes. Possibly, the placebo intervention selectively enhanced vagal outflow to the gastric pacemaker (Meissner 2009).

Similar to gastric motility, bowel motility is under close control of the ANS (Mayer 2011). One trial investigated whether reduced bowel activity after surgery can be affected by verbally induced expectations (Disbrow et al. 1993). Forty patients who underwent surgery were allocated to either a verbal suggestion group or a control group. The suggestion group received the information that regular bowel movements would return soon after surgery, while the control group received no such information prior to surgery. Both groups received the same standard treatment. As expected, the first emptying of the bowels occurred earlier in the suggestion group compared to controls. Expectation may have an influence on bowel motility.

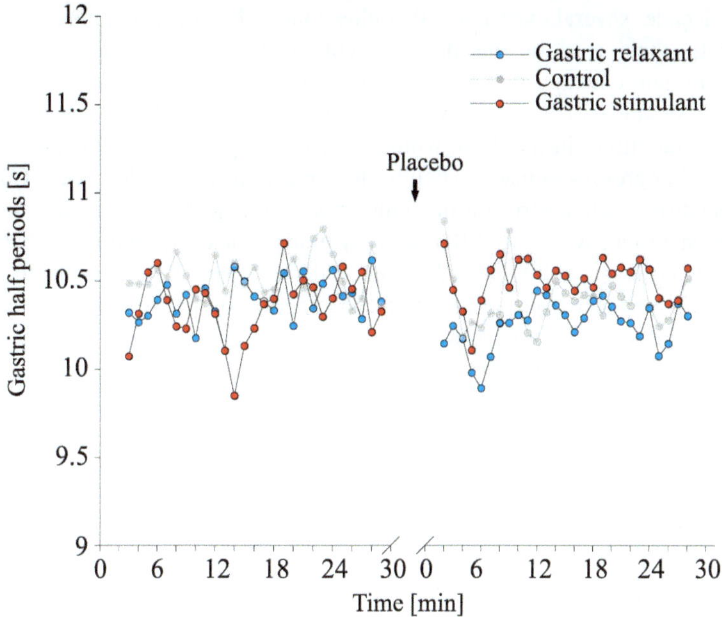

Fig. 1 Minute-by-minute changes in gastric contraction periods (i.e., half periods) before and after placebo interventions to either stimulate, or relax, or not affect gastric motility [according to Meissner (2009)]

8 Placebo Responses on Nausea and Gastric Activity

Nausea is a disabling condition. It frequently occurs as a side effect of medical treatments (e.g., chemotherapy and surgery), pregnancy, and motion. Drugs help to prevent emesis but do not completely alleviate nausea (Jordan et al. 2007). Nausea is closely associated with gastric activity. When an individual develops nausea, the normal rhythm of the gastric pacemaker of about three cycles per minute is disrupted and replaced by a faster rhythm between four and nine cycles per minute (i.e., "tachygastria") (Stern et al. 1985, 2007). Tachygastria interrupts the normal contractile activity of the stomach, and therefore causes delayed gastric emptying, increased oral–cecal transit times, and finally vomiting (Muth 2006). Nausea is a key symptom of motion sickness, which is due to the conflicting inputs arising from the visual and vestibular systems. The perceptual mismatch results in a cascade of psychological and physiological reactions, including nausea, dizziness, urge to vomit, tachygastria, and enhanced sympathetic activity (Meissner et al. 2009).

Besides pain, nausea is one of the few medical conditions for which placebo responses can be reliably found in clinical populations (Hrobjartsson and Gotzsche 2010). Early case studies showed that verbal suggestions could reliably reverse the cessation of normal gastric activity induced by an emetic (Wolf 1950). During the last decade, several experimental studies manipulated expectations in order to investigate their role in the development of nausea and motion sickness. Williamson and colleagues (2004) used an optokinetic drum to induce symptoms of motion sickness in two groups of healthy participants: While one group was informed that sitting in the drum would induce strong symptoms of motion sickness, the other group was made to believe to perceive positive side effects. Contrary to expectation, tachygastria during motion sickness was lower in the negative expectation group, while no differences in motion sickness symptoms occurred. Levine and colleagues (2006) used placebo pills with verbal suggestions to induce either positive expectations, or negative expectations, or no expectation to develop motion sickness in response to a nauseating stimulus. Again, only the negative-expectation group showed improvements, namely less symptoms of nausea, motion sickness, and tachygastria.

Klosterhalfen and colleagues (2009) investigated the role of negative expectation and classical conditioning for symptom worsening during exposure to a nauseating stimulus. Both factors turned out to contribute to the impairment of motion sickness (i.e., the nocebo response). Interestingly, conditioned nocebo responses were stronger in women, while men responded stronger to negative expectations. The same group also investigated expectation-induced placebo responses on motion sickness and tachygastria. While simple placebo instructions in combination with a placebo ginger solution did not induce a placebo response, verbal suggestions, when combined with a preconditioning procedure, reliably reduced the subjective symptoms of motion sickness, but not tachygastria (Weimer et al. 2012; Horing et al. 2013). We recently started to study the effects of placebo acupuncture point stimulation on experimentally induced nausea. In pain-related conditions, placebo acupuncture is associated with a larger placebo response than oral placebos, probably due to enhanced expectation toward invasive procedures (Meissner et al. 2013; Kaptchuk et al. 2006; Meissner and Linde 2013). First results indicate that the mock stimulation of a non-acupuncture point elicited a placebo response on nausea, while tachygastria was non-significantly reduced (unpublished data).

Mounting evidence for the clinical relevance of expectation and conditioning for nausea and vomiting comes from the field of chemotherapy research. A meta-analysis of 17 clinical trials investigated the relationship between expectations for post-chemotherapy nausea and subsequent nausea. Indeed, chemotherapy patients, who had higher expectations for nausea, were more likely to experience post-chemotherapy nausea (Colagiuri and Zachariae 2010). Classical conditioning plays a role for posttreatment nausea as well (Bovbjerg 2006) and also affects anticipatory nausea and vomiting (ANV) (Roscoe et al. 2011; Stockhorst et al. 2007). ANV is learned during chemotherapy infusions, when novel sights, smells, and sounds become associated with the experience of nausea and vomiting. After repeated chemotherapy cycles these novel stimuli acquire the ability to trigger such responses even before the next chemotherapeutic drug is infused. The role of autonomic changes for both anticipatory and post-chemotherapy nausea needs investigation.

9 Placebo Responses on Lung Function and Respiratory Activity

The ANS plays an essential role in the regulation of the smooth muscles in the respiratory tract (Canning and Fischer 2001). Activation of vagal efferents induces bronchoconstriction, whereas sympathetic activation (probably by circulating catecholamines) dilates the airways (Jänig 2006).

By 1992, more than 20 studies had examined the acute bronchial responses to verbal suggestions (c.f. Isenberg et al. 1992). In these studies, asthmatic patients were typically given an alleged bronchoconstrictor for inhalation, which in truth was a pure saline solution. On average, one-third of patients responded with a significant bronchoconstriction. This nocebo response could be distinguished from changes due to repeated inhalation of saline by appropriate control groups and did not depend on the type of measurement. Also non-asthmatic subjects responded to nocebo suggestions with subtle but significant bronchoconstriction. Some studies showed that suggestions of bronchoconstriction could even inhibit drug-induced bronchodilation (McFadden et al. 1969; Luparello et al. 1970; Strupp et al. 1974). Furthermore, verbally induced bronchoconstriction was abolished by anticholinergic agents, suggesting that the nocebo response on the airways is mediated by enhanced vagal activation of the lung (Luparello et al. 1968; Butler and Steptoe 1986). Two studies provided evidence that placebo interventions can reduce the acute bronchoconstriction that was induced by either nocebo suggestions (Butler and Steptoe 1986) or metacholine (Kemeny et al. 2007).

However, most of these studies used small sample sizes and did not systematically examine various somatic, environmental, and demographic factors that could have influenced the results (Isenberg et al. 1992). A recent study used a rigorously controlled design to investigate the reliability and consistency of acute placebo responses on lung function in asthma (Wechsler et al. 2011; online protocol). Placebo interventions and verbal suggestions induced a clear placebo response on subjective asthma symptoms. Surprisingly, however, no objective placebo response on lung function was elicited (Wechsler et al. 2011). These results indicate that the more positive results of previous studies need replication with rigorous trial designs.

Two studies investigated whether conditioned placebo analgesia is accompanied by placebo respiratory depression as a typical side effect of opioid drugs (Benedetti et al. 1998, 1999a). Postoperative patients received buprenorphine for 2 days after surgery, while respiratory depression was objectively assessed. On the third day, the patients received placebo instead of buprenorphine in a single-blind design. Remarkably, patients with a large respiratory response to the opioid drug on the prior days showed a likewise respiratory depressant response to placebo administration. This conditioned placebo respiratory depression disappeared after administration of the opioid-antagonist naloxone, indicating mediation by the endogenous opioid system. Interestingly, the placebo response on respiration was independent from the placebo response on pain, suggesting the involvement of different subpopulations of opioid receptors.

10 Reduction of Stress and Sympathetic Activity During Placebo Analgesia

Various areas of the CAN are components of pain-modulatory areas. The insula, for example, receives visceral as well as nociceptive inputs. Moreover, the anterior cingulate cortex, heavily connected with the CAN, plays a central role in the affective and motivational components of pain (Benarroch 2006). Therefore, it is not surprising that painful stimuli can initiate autonomic responses.

Pain sensations usually increase sympathetic activity (Rainville et al. 2005; Rhudy and Meagher 2003). Several studies investigated the question of whether the placebo response on pain is accompanied by a reduction of this pain-induced sympathetic activation. Pollo and colleagues (2003) measured heart rate as well as the sympathetic and parasympathetic components of heart rate variability during expectation-induced placebo analgesia in healthy volunteers. Placebo responses were induced by using a multiple open–hidden paradigm with different drugs (i.e., saline, naloxone, atropine, and propranolol). In this design, the placebo response is estimated by the difference between cardiac responses to the same substance in the open and the hidden condition. The placebo response on pain was associated with lower increases of heart rate as well as the sympathetic (low-frequency-) component of heart rate variability. The opiate antagonist naloxone abolished not only placebo analgesia but also the concomitant reduction of heart rate and sympathetic activity. The beta-blocker propranolol did not affect placebo analgesia but inhibited the pain-induced increase of heart rate and sympathetic activity independently from expectation. Muscarinic blockade with atropine had no effect, neither on placebo analgesia nor on the cardiac measures. Taken together, these results suggest that the decrease in heart rate, which accompanies placebo analgesia, is due to a reduction of beta-adrenergic sympathetic activity. Furthermore, the placebo analgesic response appears to be independent from changes in heart rate and sympathetic activity, but not vice versa (see also Benedetti et al. 2006). The cardiac placebo response could thus be secondary to pain reduction, or could be part of the neurobiological placebo analgesic response.

Flaten and colleagues investigated the relationship between placebo analgesia and changes in subjective stress and sympathetic cardiac activity (Aslaksen and Flaten 2008). Regression analyses showed that the placebo response on pain was predicted by reductions of subjective stress levels, while stress ratings in turn were predicted by reductions of sympathetic cardiac activity and improved mood. A follow-up study showed that stress ratings dropped down immediately after placebo administration, that is, before pain was induced (Aslaksen et al. 2011). The authors concluded that the reduction of sympathetic activity observed during placebo analgesia is part of the placebo response on pain rather than a consequence of pain reduction.

11 Sympathetic Placebo Responses to Caffeine-Related Stimuli

A series of studies investigated peripheral placebo responses in response to coffee-related stimuli. Coffee is the most common drug to increase subjective and physiological arousal. Due to elevated sympathetic activity caffeine increases blood pressure, skin conductance, startle responses, plasma epinephrine, free fatty acids, as well as feelings of alertness and energy (Benowitz et al. 1995; Flaten 2013).

Flaten and Blumenthal (1999) investigated whether the increase of subjective and autonomic arousal in response to coffee can be classically conditioned. For this, they tested whether the caffeine-specific stimuli of decaffeinated coffee would elicit subjective and objective arousal in habitual coffee drinkers. Indeed, subjective arousal as well as skin conductance responses and startle reflexes increased after decaffeinated coffee in comparison to orange juice. These placebo responses, however, might also reflect the different expectations toward the coffee-like drink and the orange juice. In a follow-up study Mikalsen and colleagues (2001) therefore also manipulated the levels of expectation toward the drinks by informing subjects either correctly or falsely about the caffeine content of each drink. Interestingly, expectation did not affect the arousal levels. These findings suggest that conditioning rather than expectation mediated the placebo-induced increases of subjective and physiological arousal in habitual coffee drinkers (see also Schneider et al. 2006; Walach et al. 2002; Walach and Schneider 2009; Lotshaw et al. 1996; Flaten et al. 2003).

12 Summary

Table 1 summarizes the findings from the literature on peripheral placebo and nocebo responses. Most of these studies used expectation paradigms to elicit a placebo response. In the cardiovascular system, expectations increased or decreased blood pressure and heart rate according to the intention of the suggestion. Verbal suggestions to improve blood flow reduced coronary artery diameter and chest pain in patients undergoing heart catheterization; the coronary vasoconstriction was possibly due to reduced sympathetic outflow to the heart. In the gastrointestinal system, expectations of enhanced gastric activity increased the duration of gastric contraction periods, and positive expectations reduced the time to first bowel emptying after surgery. In the respiratory system, expectations of a negative effect on lung function induced bronchoconstriction, which could be reversed by positive verbal suggestions. In a recent study, positive expectations reduced asthma symptoms but did not improve lung function. In the field of placebo analgesia, expectations for pain relief reduced cardiac sympathetic activity. Expectations for nausea relief reduced nausea, but not associated tachygastria. However, high expectations of nausea paradoxically reduced both nausea and tachygastria. Finally, a few studies investigated conditioned placebo responses on peripheral organ functions. Conditioned placebo respiratory depression was demonstrated in

Table 1 Summary of placebo and nocebo responses on peripheral organ functions and subjective symptoms in expectation and conditioning paradigms

Placebo response	Expectation paradigms												Conditioning paradigms		
	Blood pressure		Coronary diameter	Heart rate		Gastric motility	Bowel motility	Lung function		Pain	Nausea		Pain	Arousal	Nausea
	↑	↓	↑	↑	↓	↑	↑	↑	↓	↓	↑	↓	↓	↑	↑
Subjective responses															
Pain			↓							↓			↓		
Nausea											↓, ↑	↓			↑
Asthma symptoms								↓	↑						
Stress			↔							↓				↑	
Peripheral responses															
Cardiovascular system															
Systolic blood pressure	↑	↓	↔			↔									
Diastolic blood pressure	↔	↔	↔			↔									
Heart rate	↔	↔	↔	↑	↓	↔				↓					
Heart rate variability (parasympathetic)		↔				↔				↔					
Heart rate variability (sympathetic)		↔				↔				↓					
Coronary diameter			↓												
Gastrointestinal system															
Gastric contraction periods						↑									
Tachygastria						↔					↓	↔			
Bowel emptying							↑								
Respiratory system															
Bronchoconstriction		↔						↔, ↓	↑						
Respiratory activity													↓		
Electrodermal system															
Skin conductance level		↔				↔								↓	

postoperative patients after exposure to opioid drugs, and habitual coffee drinkers developed increased stress and skin conductance levels in response to coffee-associated stimuli.

13 Specificity of Peripheral Placebo Responses

Some of the studies reported above suggested that peripheral placebo responses can display a high degree of target specificity. The idea of spatially directed placebo responses fits nicely with current knowledge about viscerotopic representations of both afferent and efferent autonomic functions in the CAN (see above). However, it should be noted that the degree of specificity probably depends on site-specific suggestions during placebo administration. If neither verbal nor nonverbal cues of the placebo intervention suggest that a site-specific response will occur, the placebo response may well be nonspecific (Watson et al. 2006, 2007).

It is important to note that target specificity of placebo responses is well recognized in other fields of placebo research: For example, placebo analgesia induced by verbal suggestions and local administration of a placebo cream occurred only in the placebo-treated part of the body (Montgomery and Kirsch 1996; Benedetti et al. 1999b). These site-specific responses were blocked by the opiate antagonist naloxone, suggesting a somatotopic activation of the endogenous opioid pathways (Benedetti et al. 1999b). Indeed, a recent brain imaging study demonstrated site-specific activation of the descending pain-modulatory system during pain stimulation at different body sites (Ritter et al. 2014). It would be challenging to likewise compare the neurobiological pathways of placebo interventions targeting different autonomic organs, such as the heart and the stomach.

The issue of specificity raises the question of how specific responses on peripheral organ functions are induced. It has been shown that the anticipation of future events plays a central role in the formation of placebo responses on pain (e.g., Wager et al. 2004). The anticipation of end organ changes may likewise play a role for eliciting a specific placebo response. There is ample evidence for anticipatory response patterning in the central autonomic network. For example, heart rate and blood pressure can increase just before the onset of physical activity—a phenomenon called "central command" (Williamson 2010). Expectation-induced increases of heart rate and blood pressure may thus be explained by the activation of "central command" circuits. Likewise, placebo interventions that increase gastric activity may activate neural circuits for the "cephalic phase response," which is induced by anticipation of a meal and prepares the stomach for the ingestion and digestion of food (Power and Schulkin 2008). The use of neural circuitry for purposes other than their original function is a fundamental organizational principle of the brain (Anderson 2010). This concept of "reuse" could also explain why peripheral placebo responses induced by expectation appear to be subject to restrictions with regard to the type of change that can be induced. In more than one study in this review the placebo responses did not exactly correspond to the intentions of the verbal suggestions. Rather, opposite or no responses were induced. This may indicate that no appropriate neural circuit was available or activated, to mediate the specific response. In contrast, autonomic placebo responses induced by classical conditioning appear to be more flexible in this regard, probably due to the fact that associative learning can take place without consciousness (Benedetti et al. 2003b).

Conclusions and Future Directions

The studies presented in this review clearly indicate that cardiovascular, gastrointestinal, and pulmonary functions can be affected by placebo interventions. Like in other fields of placebo research most of these studies investigated short-term placebo responses. The potential of placebo interventions to induce acute changes on peripheral organ functions, however, represents a basic prerequisite for sustained placebo responses to occur. Furthermore, also short-term placebo responses can be clinically important. In acute bronchoconstriction, for example, the acute placebo response might significantly add to a lifesaving treatment response.

Further studies are necessary to elucidate the brain centers and the efferent pathways involved in the initiation and mediation of peripheral placebo responses. In general, approaches to increase the strength of placebo interventions could help to evaluate the full potentials and limitations of placebo responses on peripheral organ functions. Such approaches may comprise the combined use of both expectation and conditioning paradigms as well as the manipulation of cognitive, emotional, and bodily factors that maximize treatment expectations.

Acknowledgments I gratefully acknowledge Hans Distel (†) and Anja Haile for their helpful comments on the manuscript. This work has been supported by the Theophrastus Foundation (Germany) and the Schweizer-Arau-Foundation (Germany).

References

Agras WS, Horne M, Taylor CB (1982) Expectation and the blood-pressure-lowering effects of relaxation. Psychosom Med 44(4):389–395

Amanzio M, Benedetti F (1999) Neuropharmacological dissection of placebo analgesia: expectation-activated opioid systems versus conditioning-activated specific subsystems. J Neurosci 19(1):484–494

Amigo I, Cuesta V, Fernandez A, Gonzalez A (1993) The effect of verbal instructions on blood-pressure measurement. J Hypertens 11(3):293–296. doi:10.1097/00004872-199303000-00010

Anderson ML (2010) Neural reuse: a fundamental organizational principle of the brain. Behav Brain Sci 33(4):245–266. doi:10.1017/S0140525X10000853, discussion 266-313

Aslaksen PM, Flaten MA (2008) The roles of physiological and subjective stress in the effectiveness of a placebo on experimentally induced pain. Psychosom Med 70(7):811–818. doi:10.1097/PSY.0b013e31818105ed

Aslaksen PM, Bystad M, Vambheim SM, Flaten MA (2011) Gender differences in placebo analgesia: event-related potentials and emotional modulation. Psychosom Med 73(2):193–199. doi:10.1097/PSY.0b013e3182080d73

Beissner F, Meissner K, Bar KJ, Napadow V (2013) The autonomic brain: an activation likelihood estimation meta-analysis for central processing of autonomic function. J Neurosci 33 (25):10503–10511. doi:10.1523/JNEUROSCI.1103-13.2013
Benarroch EE (2006) Pain-autonomic interactions. Neurol Sci 27(Suppl 2):S130–S133. doi:10.1007/s10072-006-0587-x
Benedetti F, Amanzio M, Baldi S, Casadio C, Cavallo A, Mancuso M, Ruffini E, Oliaro A, Maggi G (1998) The specific effects of prior opioid exposure on placebo analgesia and placebo respiratory depression. Pain 75(2–3):313–319
Benedetti F, Amanzio M, Baldi S, Casadio C, Maggi G (1999a) Inducing placebo respiratory depressant responses in humans via opioid receptors. Eur J Neurosci 11(2):625–631
Benedetti F, Arduino C, Amanzio M (1999b) Somatotopic activation of opioid systems by target-directed expectations of analgesia. J Neurosci 19(9):3639–3648
Benedetti F, Maggi G, Lopiano L, Lanotte M, Rainero I, Vighetti S, Pollo A (2003a) Open versus hidden medical treatments: the patient's knowledge about a therapy affects the therapy outcome. Prevention Treatment 6(1):1a
Benedetti F, Pollo A, Lopiano L, Lanotte M, Vighetti S, Rainero I (2003b) Conscious expectation and unconscious conditioning in analgesic, motor, and hormonal placebo/nocebo responses. J Neurosci 23(10):4315–4323
Benedetti F, Colloca L, Lanotte M, Bergamasco B, Torre E, Lopiano L (2004) Autonomic and emotional responses to open and hidden stimulations of the human subthalamic region. Brain Res Bull 63(3):203–211. doi:10.1016/j.brainresbull.2004.01.010
Benedetti F, Mayberg HS, Wager TD, Stohler CS, Zubieta JK (2005) Neurobiological mechanisms of the placebo effect. J Neurosci 25(45):10390–10402. doi:10.1523/JNEUROSCI.3458-05.2005
Benedetti F, Arduino C, Costa S, Vighetti S, Tarenzi L, Rainero I, Asteggiano G (2006) Loss of expectation-related mechanisms in Alzheimer's disease makes analgesic therapies less effective. Pain 121(1–2):133–144. doi:10.1016/j.pain.2005.12.016
Benedetti F, Amanzio M, Rosato R, Blanchard C (2011) Nonopioid placebo analgesia is mediated by CB1 cannabinoid receptors. Nat Med 17(10):1228–1230. doi:10.1038/nm.2435
Benowitz NL, Jacob P 3rd, Mayan H, Denaro C (1995) Sympathomimetic effects of paraxanthine and caffeine in humans. Clin Pharmacol Ther 58(6):684–691. doi:10.1016/0009-9236(95)90025-X
Bovbjerg DH (2006) The continuing problem of post chemotherapy nausea and vomiting: contributions of classical conditioning. Auton Neurosci 129(1–2):92–98. doi:10.1016/j.autneu.2006.07.016
Butler C, Steptoe A (1986) Placebo responses: an experimental study of psychophysiological processes in asthmatic volunteers. Br J Clin Psychol 25:173–183
Canning BJ, Fischer A (2001) Neural regulation of airway smooth muscle tone. Respir Physiol 125 (1–2):113–127. doi:10.1016/s0034-5687(00)00208-5
Colagiuri B, Zachariae R (2010) Patient expectancy and post-chemotherapy nausea: a meta-analysis. Ann Behav Med 40(1):3–14. doi:10.1007/s12160-010-9186-4
Colloca L, Lopiano L, Lanotte M, Benedetti F (2004) Overt versus covert treatment for pain, anxiety, and Parkinson's disease. Lancet Neurol 3(11):679–684. doi:10.1016/S1474-4422(04)00908-1
Craig AD (2002) How do you feel? Interoception: the sense of the physiological condition of the body. Nat Rev Neurosci 3(8):655–666. doi:10.1038/nrn894
Critchley HD, Harrison NA (2013) Visceral influences on brain and behavior. Neuron 77(4):624–638. doi:10.1016/j.neuron.2013.02.008
Critchley HD, Nagai Y, Gray MA, Mathias CJ (2011) Dissecting axes of autonomic control in humans: insights from neuroimaging. Auton Neurosci 161(1–2):34–42. doi:10.1016/j.autneu.2010.09.005

Daniel EE (2001) Physiology and pathophysiology of the interstitial cell of Cajal: from bench to bedside. III. Interaction of interstitial cells of Cajal with neuromediators: an interim assessment. Am J Physiol Gastrointest Liver Physiol 281(6):G1329–G1332
de la Fuente-Fernandez R, Ruth TJ, Sossi V, Schulzer M, Calne DB, Stoessl AJ (2001) Expectation and dopamine release: mechanism of the placebo effect in Parkinson's disease. Science 293 (5532):1164–1166. doi:10.1126/science.1060937
Disbrow EA, Bennett HL, Owings JT (1993) Effect of preoperative suggestion on postoperative gastrointestinal motility. West J Med 158(5):488–492
Eippert F, Finsterbusch J, Bingel U, Buchel C (2009) Direct evidence for spinal cord involvement in placebo analgesia. Science 326(5951):404. doi:10.1126/science.1180142
Enck P, Bingel U, Schedlowski M, Rief W (2013) The placebo response in medicine: minimize, maximize or personalize? Nat Rev Drug Discov 12(3):191–204. doi:10.1038/nrd3923
Flaten MA (2013) Placebo responses, antagonistic responses, and homeostasis. In: Colloca L, Flaten MA, Meissner K (eds) Placebo and pain: from bench to bedside. Academic, Boston
Flaten MA, Blumenthal TD (1999) Caffeine-associated stimuli elicit conditioned responses: an experimental model of the placebo effect. Psychopharmacology (Berl) 145(1):105–112
Flaten MA, Aasli O, Blumenthal TD (2003) Expectations and placebo responses to caffeine-associated stimuli. Psychopharmacology (Berl) 169(2):198–204. doi:10.1007/s00213-003-1497-8
Guyenet PG (2006) The sympathetic control of blood pressure. Nat Rev Neurosci 7(5):335–346. doi:10.1038/nrn1902
Hagemann D, Waldstein SR, Thayer JF (2003) Central and autonomic nervous system integration in emotion. Brain Cogn 52(1):79–87. doi:10.1016/S0278-2626(03)00011-3
Horing B, Weimer K, Schrade D, Muth ER, Scisco JL, Enck P, Klosterhalfen S (2013) Reduction of motion sickness with an enhanced placebo instruction: an experimental study with healthy participants. Psychosom Med 75(5):497–504. doi:10.1097/PSY.0b013e3182915ee7
Hrobjartsson A, Gotzsche PC (2010) Placebo interventions for all clinical conditions. Cochrane Database Syst Rev 1, CD003974. doi:10.1002/14651858.CD003974.pub3
Isenberg SA, Lehrer PM, Hochron S (1992) The effects of suggestion and emotional arousal on pulmonary function in asthma: a review and a hypothesis regarding vagal mediation. Psychosom Med 54(2):192–216
Jänig W (2006) The integrative action of the autonomic nervous system. Neurobiology of homeostasis. Cambridge University Press, Cambridge
Jordan K, Schmoll HJ, Aapro MS (2007) Comparative activity of antiemetic drugs. Crit Rev Oncol Hematol 61(2):162–175. doi:10.1016/j.critrevonc.2006.08.003
Kam-Hansen S, Jakubowski M, Kelley JM, Kirsch I, Hoaglin DC, Kaptchuk TJ, Burstein R (2014) Altered placebo and drug labeling changes the outcome of episodic migraine attacks. Sci Transl Med 6(218):218ra215. doi:10.1126/scitranslmed.3006175
Kaneko H, Sakakibara M, Mitsuma T, Morise K (1995) Possibility of postprandial electrogastrography for evaluating vagal/nonvagal cholinergic activity in humans, through simultaneous analysis of postprandial heart rate variability and serum immunoreactive hormone levels. Am J Gastroenterol 90(4):603–609
Kaptchuk TJ, Stason WB, Davis RB, Legedza AR, Schnyer RN, Kerr CE, Stone DA, Nam BH, Kirsch I, Goldman RH (2006) Sham device v inert pill: randomised controlled trial of two placebo treatments. BMJ 332(7538):391–397. doi:10.1136/bmj.38726.603310.55
Kemeny ME, Rosenwasser LJ, Panettieri RA, Rose RM, Berg-Smith SM, Kline JN (2007) Placebo response in asthma: a robust and objective phenomenon. J Allergy Clin Immunol 119(6):1375–1381. doi:10.1016/j.jaci.2007.03.016
Klosterhalfen S, Kellermann S, Braun S, Kowalski A, Schrauth M, Zipfel S, Enck P (2009) Gender and the nocebo response following conditioning and expectancy. J Psychosom Res 66(4):323–328. doi:10.1016/j.jpsychores.2008.09.019
Kreibig SD (2010) Autonomic nervous system activity in emotion: a review. Biol Psychol 84 (3):394–421. doi:10.1016/j.biopsycho.2010.03.010

Krummenacher P, Candia V, Folkers G, Schedlowski M, Schonbachler G (2010) Prefrontal cortex modulates placebo analgesia. Pain 148(3):368–374. doi:10.1016/j.pain.2009.09.033
Lanotte M, Lopiano L, Torre E, Bergamasco B, Colloca L, Benedetti F (2005) Expectation enhances autonomic responses to stimulation of the human subthalamic limbic region. Brain Behav Immun 19(6):500–509. doi:10.1016/j.bbi.2005.06.004
Levine ME, Stern RM, Koch KL (2006) The effects of manipulating expectations through placebo and nocebo administration on gastric tachyarrhythmia and motion-induced nausea. Psychosom Med 68(3):478–486. doi:10.1097/01.psy.0000221377.52036.50
Lotshaw SC, Bradley JR, Brooks LR (1996) Illustrating caffeine's pharmacological and expectancy effects utilizing a balanced placebo design. J Drug Educ 26(1):13–24. doi:10.2190/Uucl-E5v6-Xc25-5mc6
Luparello T, Lyons HA, Bleecker ER, McFadden ER Jr (1968) Influences of suggestion on airway reactivity in asthmatic subjects. Psychosom Med 30(6):819–825
Luparello TJ, Leist N, Lourie CH, Sweet P (1970) The interaction of psychologic stimuli and pharmacologic agents on airway reactivity in asthmatic subjects. Psychosom Med 32(5):509–513
Mayer EA (2011) Gut feelings: the emerging biology of gut-brain communication. Nat Rev Neurosci 12(8):453–466. doi:10.1038/Nrn3071
McFadden ER Jr, Luparello T, Lyons HA, Bleecker E (1969) The mechanism of action of suggestion in the induction of acute asthma attacks. Psychosom Med 31(2):134–143
Meissner K (2009) Effects of placebo interventions on gastric motility and general autonomic activity. J Psychosom Res 66(5):391–398. doi:10.1016/j.jpsychores.2008.09.004
Meissner K, Linde K (2013) What are the best placebo interventions for the treatment of pain? In: Colloca L, Flaten MA, Meissner K (eds) Placebo and pain: from bench to bedside. Academic, Boston
Meissner K, Ziep D (2011) Organ-specificity of placebo effects on blood pressure. Auton Neurosci 164(1–2):62–66. doi:10.1016/j.autneu.2011.06.006
Meissner K, Distel H, Mitzdorf U (2007) Evidence for placebo effects on physical but not on biochemical outcome parameters: a review of clinical trials. BMC Med 5:3. doi:10.1186/1741-7015-5-3
Meissner K, Enck P, Muth ER, Kellermann S, Klosterhalfen S (2009) Cortisol levels predict motion sickness tolerance in women but not in men. Physiol Behav 97(1):102–106. doi:10.1016/j.physbeh.2009.02.007
Meissner K, Kohls N, Colloca L (2011) Introduction to placebo effects in medicine: mechanisms and clinical implications. Philos Trans R Soc Lond B Biol Sci 366(1572):1783–1789. doi:10.1098/rstb.2010.0414
Meissner K, Fassler M, Rucker G, Kleijnen J, Hrobjartsson A, Schneider A, Antes G, Linde K (2013) Differential effectiveness of placebo treatments: a systematic review of migraine prophylaxis. JAMA Intern Med 173(21):1941–1951. doi:10.1001/jamainternmed.2013.10391
Mikalsen A, Bertelsen B, Flaten MA (2001) Effects of caffeine, caffeine-associated stimuli, and caffeine-related information on physiological and psychological arousal. Psychopharmacology (Berl) 157(4):373–380. doi:10.1007/s002130100841
Montgomery G, Kirsch I (1996) Mechanisms of placebo pain reduction: an empirical investigation. Psychol Sci 7(3):174–176. doi:10.1111/j.1467-9280.1996.tb00352.x
Morrison SF (2001) Differential control of sympathetic outflow. Am J Physiol Regul Integr Comp Physiol 281(3):R683–R698. doi:10.1046/j.1440-1681.2001.03402.x
Muth ER (2006) Motion and space sickness: intestinal and autonomic correlates. Auton Neurosci 129(1–2):58–66. doi:10.1016/j.autneu.2006.07.020
Olshansky B (2007) Placebo and nocebo in cardiovascular health: implications for healthcare, research, and the doctor-patient relationship. J Am Coll Cardiol 49(4):415–421. doi:10.1016/j.jacc.2006.09.036
Oppenheimer S (2007) Cortical control of the heart. Cleve Clin J Med 74(Suppl 1):S27–S29. doi:10.3949/ccjm.74.Suppl_1.S27

Parkman HP, Trate DM, Knight LC, Brown KL, Maurer AH, Fisher RS (1999) Cholinergic effects on human gastric motility. Gut 45(3):346–354
Pollo A, Vighetti S, Rainero I, Benedetti F (2003) Placebo analgesia and the heart. Pain 102(1–2):125–133
Power ML, Schulkin J (2008) Anticipatory physiological regulation in feeding biology: cephalic phase responses. Appetite 50(2–3):194–206. doi:10.1016/j.appet.2007.10.006
Rainville P, Bao QV, Chretien P (2005) Pain-related emotions modulate experimental pain perception and autonomic responses. Pain 118(3):306–318. doi:10.1016/j.pain.2005.08.022
Rhudy JL, Meagher MW (2003) Individual differences in the emotional reaction to shock determine whether hypoalgesia is observed. Pain Med 4(3):244–256
Ritter C, Hebart MN, Wolbers T, Bingel U (2014) Representation of spatial information in key areas of the descending pain modulatory system. J Neurosci 34(13):4634–4639. doi:10.1523/JNEUROSCI.4342-13.2014
Ronel J, Mehilli J, Ladwig KH, Blattler H, Oversohl N, Byrne RA, Bauer A et al (2011) Effects of verbal suggestion on coronary arteries: results of a randomized controlled experimental investigation during coronary angiography. Am Heart J 162(3):507–511. doi:10.1016/j.ahj.2011.06.016
Roscoe JA, Morrow GR, Aapro MS, Molassiotis A, Olver I (2011) Anticipatory nausea and vomiting. Support Care Cancer 19(10):1533–1538. doi:10.1007/s00520-010-0980-0
Sarna SK, Daniel EE (1974) Threshold curves and refractoriness properties of gastric relaxation oscillators. Am J Physiol 226(4):749–755
Schneider R, Gruner M, Heiland A, Keller M, Kujanova Z, Peper M, Riegl M, Schmidt S, Volz P, Walach H (2006) Effects of expectation and caffeine on arousal, well-being, and reaction time. Int J Behav Med 13(4):330–339. doi:10.1207/s15327558ijbm1304_8
Shields RW Jr (1993) Functional anatomy of the autonomic nervous system. J Clin Neurophysiol 10(1):2–13. doi:10.1097/00004691-199301000-00002
Stern RM, Koch KL, Leibowitz HW, Lindblad IM, Shupert CL, Stewart WR (1985) Tachygastria and motion sickness. Aviat Space Environ Med 56(11):1074–1077
Stern RM, Koch KL, Levine ME, Muth ER (2007) Gastrointestinal response. In: Cacioppo JT, Tassinary LG, Berntson GG (eds) Handbook of psychophysiology. Cambridge University Press, Cambridge, pp 211–230
Stockhorst U, Enck P, Klosterhalfen S (2007) Role of classical conditioning in learning gastrointestinal symptoms. World J Gastroenterol 13(25):3430–3437
Stoddard CJ, Smallwood RH, Duthie HL (1981) Electrical arrhythmias in the human stomach. Gut 22(9):705–712
Strupp HH, Levenson RW, Manuck SB, Snell JD, Hinrichsen JJ, Boyd S (1974) Effects of suggestion on total respiratory resistance in mild asthmatics. J Psychosom Res 18(5):337–346. doi:10.1016/0022-3999(74)90053-1
Thayer JF, Lane RD (2000) A model of neurovisceral integration in emotion regulation and dysregulation. J Affect Disord 61(3):201–216
Vits S, Cesko E, Enck P, Hillen U, Schadendorf D, Schedlowski M (2011) Behavioural conditioning as the mediator of placebo responses in the immune system. Philos Trans R Soc Lond B Biol Sci 366(1572):1799–1807. doi:10.1098/rstb.2010.0392
Wager TD, Rilling JK, Smith EE, Sokolik A, Casey KL, Davidson RJ, Kosslyn SM, Rose RM, Cohen JD (2004) Placebo-induced changes in FMRI in the anticipation and experience of pain. Science 303(5661):1162–1167. doi:10.1126/science.1093065
Walach H, Schneider R (2009) Does the presence of a pharmacological substance alter the placebo effect? – results of two experimental studies using the placebo-caffeine paradigm. Hum Psychopharmacol 24(7):549–558. doi:10.1002/hup.1054
Walach H, Schmidt S, Dirhold T, Nosch S (2002) The effects of a caffeine placebo and suggestion on blood pressure, heart rate, well-being and cognitive performance. Int J Psychophysiol 43(3):247–260

Watson A, El-Deredy W, Bentley DE, Vogt BA, Jones AK (2006) Categories of placebo response in the absence of site-specific expectation of analgesia. Pain 126(1–3):115–122. doi:10.1016/j.pain.2006.06.021

Watson A, El-Deredy W, Vogt BA, Jones AK (2007) Placebo analgesia is not due to compliance or habituation: EEG and behavioural evidence. Neuroreport 18(8):771–775. doi:10.1097/WNR.0b013e3280c1e2a8

Wechsler ME, Kelley JM, Boyd IO, Dutile S, Marigowda G, Kirsch I, Israel E, Kaptchuk TJ (2011) Active albuterol or placebo, sham acupuncture, or no intervention in asthma. N Engl J Med 365(2):119–126. doi:10.1056/NEJMoa1103319

Weimer K, Schulte J, Maichle A, Muth ER, Scisco JL, Horing B, Enck P, Klosterhalfen S (2012) Effects of ginger and expectations on symptoms of nausea in a balanced placebo design. PLoS ONE 7(11):e49031. doi:10.1371/journal.pone.0049031

Williamson J (2010) The relevance of Central Command for the neural cardiovascular control of exercise. Exp Physiol. doi:10.1113/expphysiol.2009.051870

Williamson MJ, Thomas MJ, Stern RM (2004) The contribution of expectations to motion sickness symptoms and gastric activity. J Psychosom Res 56(6):721–726. doi:10.1016/S0022-3999(03)00130-2

Wirth T, Ober K, Prager G, Vogelsang M, Benson S, Witzke O, Kribben A, Engler H, Schedlowski M (2011) Repeated recall of learned immunosuppression: evidence from rats and men. Brain Behav Immun 25(7):1444–1451. doi:10.1016/j.bbi.2011.05.011

Wolf S (1950) Effects of suggestion and conditioning on the action of chemical agents in human subjects; the pharmacology of placebos. J Clin Invest 29(1):100–109. doi:10.1172/JCI102225

Zimmermann-Viehoff F, Meissner K, Koch J, Weber CS, Richter S, Deter HC (2013) Autonomic effects of suggestive placebo interventions to increase or decrease blood pressure: a randomized controlled trial in healthy subjects. J Psychosom Res 75(1):32–35. doi:10.1016/j.jpsychores.2013.03.011

Zubieta JK, Bueller JA, Jackson LR, Scott DJ, Xu Y, Koeppe RA, Nichols TE, Stohler CS (2005) Placebo effects mediated by endogenous opioid activity on mu-opioid receptors. J Neurosci 25 (34):7754–7762. doi:10.1523/JNEUROSCI.0439-05.2005

Placebo and Nocebo Effects in Itch and Pain

Andrea W.M. Evers, Danielle J.P. Bartels, and Antoinette I.M. van Laarhoven

Contents

1 Itch and Pain 206
2 Placebo Effects in Itch and Pain 207
3 Contagious Itch 207
4 Placebo and Nocebo Effects in Chronic Itch 208
5 Experimental Studies on Placebo and Nocebo Effects on Itch 209
5.1 Role of Verbal Suggestion in Itch 209
5.2 Role of Conditioning and Verbal Suggestion on Itch 210
6 Individual Predictors of Placebo and Nocebo Effects on Itch 211
7 Clinical Relevance and Implications for Innovative Treatment Strategies 211
References 212

Abstract

Physical complaints, such as pain, can be effectively altered by placebo and nocebo effects due to induction of positive or negative expectations. While verbal suggestion and conditioning are recognized as playing a key role in placebo and nocebo effects on pain, these mechanisms have barely been investigated with regard to other somatosensory sensations, such as itch. Results on contagious itch in both animals and humans suggest that itch sensations might be even more susceptible for placebo and nocebo effects than pain. Research on placebo and nocebo effects on pain and itch can further deliver insight into the common and specific mechanisms underlying these effects in different physical complaints. Work of our research group on verbal suggestions inducing nocebo

A.W.M. Evers (✉) • D.J.P. Bartels • A.I.M. van Laarhoven
Health, Medical and Neuropsychology Unit, Institute of Psychology, Leiden University, PO Box 9555, 2300 Leiden, The Netherlands

Department of Medical Psychology, Radboud university medical center, Nijmegen, The Netherlands
e-mail: a.evers@fsw.leidenuniv.nl

F. Benedetti et al. (eds.), *Placebo*, Handbook of Experimental Pharmacology 225,
DOI 10.1007/978-3-662-44519-8_12

effects demonstrated an important role of verbal suggestions with regard to itch, with stronger nocebo effects on itch than a comparable procedure for pain. Recent work also demonstrated that placebo and nocebo effects on itch sensations were most effectively induced by procedures that consist of both conditioning and verbal suggestion principles. This work adds to previous prospective studies showing that expectation mechanisms, such as preservative worrying about negative consequences, are relatively consistent predictors of future disease outcomes, including itch, in chronic somatic conditions. Future studies should focus on the specific psychoneurobiological mechanisms of placebo and nocebo effects in various physical sensations, to get insight into the common and specific effects and to contribute to the long-term and clinically relevant use of placebo effects in clinical practice.

Keywords
Nocebo • Placebo • Expectations • Itch • Pain • Pruritus

1 Itch and Pain

Itch is the most frequent physical symptom of patients with chronic skin diseases. Itch is defined as a sensation that provokes the desire to scratch (Ikoma et al. 2005, 2006; Verhoeven et al. 2007, 2008; Yosipovitch et al. 2003). Like pain, itch is a somatosensory sensation that can be of considerable burden to patients with long-term and chronic symptoms. A substantial number of the patients with chronic skin diseases regularly suffer from itch and enduring scratching behavior (Verhoeven et al. 2007, 2008; Yosipovitch et al. 2003).

Although the specific neurophysiological processing of itch is not yet known, the physiological mechanisms of acute itch show many similarities to acute pain. Itch is a somatosensory experience that can also be considered as nociceptive, like pain, since it serves as a defense mechanism against threat. Parallel to pain, inflammatory mediators, such as bradykinin, histamine, and prostaglandins, also have an acute sensitizing effect on peripheral nociceptors. In addition, both itch and pain can be accompanied by comparable neuroendocrine and autonomous responses as a reaction to the nociceptive input. Moreover, the cortical representation of brain areas activated by both sensations also shows a broad overlap, for example, with similar activation patterns of, for example, the somatosensory, anterior cingulate, prefrontal, and premotor cortex, being involved in both acute itch and pain. Finally, in line with chronic pain, there is evidence that central and peripheral sensitization mechanisms also play a role in the increased sensitivity to itch of patients with chronic itch (see also van Laarhoven et al. 2011; Yosipovitch et al. 2003).

Like pain, itch significantly lowers patients' quality of life, which in turn unfavorably affects the outcome of dermatological therapies and patients' skin status (Verhoeven et al. 2007, 2008; Yosipovitch et al. 2003). Patients with chronic skin diseases frequently complain about the large impact of itch on daily life, including the belief that "itch is worse than pain." The high individual variability in treatment outcomes and interactions between neurobiological and psychological factors in itch suggest that psychological factors play a significant role in itch sensations of patients (Evers et al. 2009; Verhoeven et al. 2008; Yosipovitch et al. 2003). One of these factors are learning processes that induce specific expectations with regard to treatment outcomes, usually described as placebo and nocebo effects.

2 Placebo Effects in Itch and Pain

Placebos are known to affect a broad variety of cognitive-behavioral and physiological responses and play a role in various symptoms and conditions, including pain (Atlas and Wager 2012; Benedetti 2008; Benedetti et al. 2003; Colloca et al. 2013; Price et al. 2008). Placebo effects can also contribute to interindividual variability in treatment outcomes for a range of symptoms (Atlas and Wager 2012; Benedetti 2008; Benedetti et al. 2003; Colloca et al. 2013; Price et al. 2008). Particularly, verbal suggestion and conditioning have been identified as main processes responsible for placebo and nocebo effects on pain (Colloca et al. 2008, 2010; Jensen et al. 2012; Klinger et al. 2007; Martin-Pichora et al. 2011; Stewart-Williams and Podd 2004). Since pain and itch show many physiological and psychological similarities (van Laarhoven et al. 2007, 2010, 2011, 2012, 2013; Verhoeven et al. 2008; Yosipovitch et al. 2003), expectations may also influence itch. However, in contrast to pain, much less research has been conducted on placebo and nocebo effects on other somatosensory sensations, such as itch, on both an experimental and clinical level. Investigating the role of placebo and nocebo effects in different somatosensory sensations, such as itch and pain, may provide insight into common and specific mechanisms underlying these effects. This understanding of the mechanisms might help improve therapeutic interventions for patients suffering from clinical conditions associated with chronic itch or pain.

3 Contagious Itch

A specific phenomenon in itch, i.e., contagious itch, suggests a substantial role for placebo and nocebo effects on itch. Itch in particular seems to be highly susceptible to suggestion, as demonstrated by the phenomenon of contagious itch (Feneran et al. 2013; Holle et al. 2012; Lloyd et al. 2011; Niemeier and Gieler 2000; Papoiu et al. 2011). The first preliminary evidence for contagious itch was demonstrated by a study showing that people scratch more during a lecture about itch than during a neutral lecture (Niemeier and Gieler 2000). In this study, the frequency of scratching responses could be aggravated during a lecture about itch in which individuals were shown itch-related pictures, such as fleas, mites, scratch marks, allergic reactions, in contrast to a neutral lecture (Niemeier and Gieler 2000). More recently, this phenomenon has also been demonstrated in animal studies of monkeys watching videos of other monkeys scratching (Feneran et al. 2013). In addition, neuroimaging studies suggest that the itch brain activation patterns during contagious itch in humans are comparable to the patterns when applying somatosensory itch stimuli (Holle et al. 2012). A similar mechanism might be responsible for the findings that patients with atopic dermatitis and other chronic itch conditions have the tendency to react more strongly to itch inductions, for example, by verbal suggestions (Papoiu et al. 2011; Scholz and Hermanns 1994; van Laarhoven et al. 2007, 2013). These studies might support the particularly important role of expectancy mechanisms and related placebo and nocebo effects in somatosensory sensations of itch.

4 Placebo and Nocebo Effects in Chronic Itch

There is some evidence that placebo and nocebo effects in itch, and related scratching behavior, are not only relevant during acute itch induction, but also play a role in long-term processes, such as in patients with chronic itch. It is, for example, well-known that long-term use of negative expectancy mechanisms, like preservative worrying, can significantly affect the disease outcome in patients with chronic itch (Verhoeven et al. 2009). In addition, patients with chronic itch conditions have been reported to have a lowered threshold for sensory stimuli in general and particularly for itch. This latter has been shown in laboratory settings when exposed to different mechanical, electrical, or chemical stimuli inducing itch, although evidence is somewhat inconsistent in this area (Ikoma et al. 2005, 2006; van Laarhoven et al. 2007, 2013). This altered pattern might be due to changes in information processing, such as an attentional or interpretational bias to general or itch-specific stimuli, or catastrophizing expectations about aversive stimuli. These patterns might finally result into a generalized changed perception of somatosensory stimuli or sensitization processes (Curatolo et al. 2006; Smith et al. 2008; van Laarhoven et al. 2007, 2013). Preliminary evidence suggests that these processes particularly occur with regard to the dominant symptom of patients, e.g., itch in chronic itch and pain in chronic pain (van Laarhoven et al. 2007, 2013). Generally, patients with chronic physical symptoms of pain and itch might react differently to sensory stimuli, by sensitizing more easily to long-term exposure to the pain and itch stimuli due to the possible threat value, while healthy subjects tend to habituate to such stimuli (Ikoma et al. 2005, 2006; van Laarhoven et al. 2007, 2013). Long-term sensitization processes in patients might also result from learning processes during their chronic condition to early attend to potentially aversive symptoms due to the possible threat value of these symptoms. This might result in the interpretation of possible ambiguous stimuli as sign of the symptom of condition, which might induce negative expectations and aggravate further sensitization.

Next to experimental studies, there is also increasing evidence from prospective studies for the role of expectancy mechanisms (and consequently placebo and nocebo effects) in itch sensations for patient with chronic itch. The most important indicator related to expectancy mechanisms is the cognitive status of preservative worrying. About 40 % of patients with a chronic skin condition show tendencies of preservative worrying that are related to more disease severity and worse treatment outcomes (Verhoeven et al. 2008). In our own research following psoriasis patients during the natural course of their disease, worrying was significantly related to future disease severity and levels of itch 1 month later during a half-year measurement period (Verhoeven et al. 2009). This evidence is in line with research in other chronic diseases, such as rheumatoid arthritis, showing that worrying is one of the most consistent predictors of future disease outcome and physical and psychological disability, including itch and pain symptoms (Evers et al. 2003, 2014; Verhoeven et al. 2008, 2009).

5 Experimental Studies on Placebo and Nocebo Effects on Itch

5.1 Role of Verbal Suggestion in Itch

Based on the large amount of indirect evidence from experimental and clinical studies, nocebo effects on itch were for the first time systematically investigated by experimentally inducing nocebo effects on itch and pain by verbal suggestions in our research group (van Laarhoven et al. 2011). Healthy female subjects were randomly assigned to various experimental conditions and received condition-corresponding verbal information about the stimuli to be applied subsequently. The suggestions were designed to produce either high or low expectations for itch or pain. All subjects received the same somatosensory quantitative sensory testing (QST) stimuli by monofilament stimulation, electrical stimulation, and histamine iontophoresis. Results showed that nocebo effects could be induced on itch by manipulating expectations regarding the different somatosensory stimuli applied by giving verbal suggestions (van Laarhoven et al. 2011). As expected, subjects who received verbal suggestions to induce high pain expectations reported significantly higher levels of pain than subjects who received verbal suggestions inducing low expectations for pain. More importantly, we also showed that nocebo effects could be induced on other physical sensations than pain, namely on itch by verbal suggestions (van Laarhoven et al. 2011). This study further showed that the perception of different ambiguous stimuli can be influenced by negative suggestions, in such a way that negative expectations can adversely influence the intensity of itch or pain experienced. When comparing nocebo effects, the most striking finding was that the results seemed to be generally more pronounced for itch than for pain (van Laarhoven et al. 2011) during almost all QST applications, suggesting that the perception of itch may be more susceptible to suggestion than the perception of pain. Although the specific reasons for this finding are unknown at this time, future research should focus on possible differences between itch and pain in psychoneurobiological response patterns, such as behavioral scratching responses in itch in contrast to retreating responses in pain, different affective and motivational components, such as anticipatory anxiety, or specific neurological pathways from various cortical activation patterns (Scott et al. 2007, 2008; van Laarhoven et al. 2007, 2013; Verhoeven et al. 2008; Yosipovitch et al. 2003) that might play a role in the possibly larger role of expectations in itch than in pain.

5.2 Role of Conditioning and Verbal Suggestion on Itch

Although knowledge of the mechanisms underlying placebo and nocebo effects has advanced in recent years, there is still some debate whether placebo and nocebo effects mainly result from conditioning or expectations induced by (verbal) suggestion (Colloca et al. 2008, 2010; Jensen et al. 2012; Klinger et al. 2007; Martin-Pichora et al. 2011; Stewart-Williams and Podd 2004). Consequently, we were interested to study whether nocebo effects on itch can be enhanced when verbal suggestions are given in combination with conditioning. In pain, placebo and nocebo effects have shown to be much stronger when expectations are induced by the combination of suggestions and conditioning (Colloca et al. 2008, 2010; Klinger et al. 2007; Stewart-Williams and Podd 2004). In another study of our research group (Bartels et al. 2014), the role of both conditioning and verbal suggestion procedures in placebo and nocebo effects on itch were studied. Expectations regarding itch stimuli were induced in healthy subjects by verbal suggestion, conditioning, or a combination of both procedures. Healthy subjects were told that the purpose of the study was to determine sensitivity to itch stimuli. Expectations of lowered, neutral, and heightened levels of itch were raised in subjects by telling them that different cues (colored lights on the computer screen) indicated that the itch stimulus intensity would be altered. The conditioning procedure consisted of the repeated pairing of the color cues with different itch stimulus intensities, in order to induce low, medium, and high itch expectations regarding the stimuli. In this study, a new electrical QST method (DS5 Isolated Bipolar Constant Current Stimulator) was used to induce itch, which proved successful in eliciting reproducible itch sensations of moderate to high intensity in healthy subjects. Results showed that placebo and nocebo effects can be induced on itch (Bartels et al. 2014). In line with research on pain, the combination of conditioning and verbal suggestion induced the largest effects rather than either procedure alone. In view of the comparable evidence in pain, these results show that these procedures also play a role in placebo and nocebo effects in other somatosensory sensations, such as itch (Colloca et al. 2008, 2010; Klinger et al. 2007; Stewart-Williams and Podd 2004). Moreover, larger and longer-lasting placebo and nocebo effects were mostly induced by combining these procedures when compared to these procedures alone (Colloca et al. 2008, 2010; Klinger et al. 2007; Stewart-Williams and Podd 2004). The results of our study also indicate that the combination of conditioning and verbal suggestion is most promising in inducing placebo and nocebo effects on various somatosensory sensations, such as itch and pain.

6 Individual Predictors of Placebo and Nocebo Effects on Itch

Individual characteristics are known to be associated with interindividual variation in placebo and nocebo responding with regard to pain and other symptoms and conditions (Geers et al. 2005, 2007; Morton et al. 2009; Pecina et al. 2013), and are likely to also play a role in placebo and nocebo effects on itch. Particularly because of the high variability in placebo and nocebo responding, research aims to clarify the role of individual characteristics, such as expectation-related personality characteristics of optimism and pessimism, in placebo and nocebo effects on itch and pain (Geers et al. 2005, 2007; Morton et al. 2009; Niemeier and Gieler 2000). One might suggest that individual psychological characteristics of positive or negative expectation tendencies (e.g., neuroticism or catastrophizing) may influence placebo or nocebo responding. In line with previous studies, we also found some evidence in our study (Bartels et al. 2014) that individual characteristics of worrying, psychological distress, extraversion, neuroticism, and hope were significantly correlated with the nocebo effect (Geers et al. 2005, 2007; Morton et al. 2009; Pecina et al. 2013). However, almost no significant correlations were found between these characteristics and the placebo effect. This suggests divergent mechanisms underlying placebo and nocebo effects, such as psychoneurobiological differences, e.g., by the activation of dopamine neurotransmission in placebo effects, or anxiety or stress-related processes of cortisol levels in nocebo effects (Scott et al. 2007, 2008).

7 Clinical Relevance and Implications for Innovative Treatment Strategies

The major clinical challenge remains whether experimental laboratory findings on placebo and nocebo effects on induced itch of short duration in healthy subjects can be generalized to patients in a clinical setting. There is some evidence showing that placebo effects are stronger when contextual factors, e.g., by imitating a clinical setting, are included (Benedetti 2008; Colloca et al. 2013). In addition, the evidence from natural settings, such as studies from contagious itch in humans and animals, suggests a strong clinical relevance of these effects in clinical practice (Feneran et al. 2013; Holle et al. 2012; Lloyd et al. 2011; Niemeier and Gieler 2000; Papoiu et al. 2011). Next to specific placebo and nocebo expectations patterns in chronic itch, it is important to further study whether patients with chronic itch or pain might possibly also be more sensitive to expectation learning processes. Expectations regarding pain or itch sensations might be easier induced and consequently elicit stronger patterns of placebo or nocebo responses in patients compared to healthy subjects (van Laarhoven et al. 2007, 2013). In the long term, this knowledge may help improve therapeutic interventions by enhancing favorable expectations and reducing unfavorable expectations in patients suffering from chronic itch conditions. Moreover, the role of individual characteristics in placebo and nocebo responsiveness might finally be used to personalize interventions and to optimize treatment outcomes. For the diagnostics and treatment of patients with chronic itch, screening instruments and effective psychological treatments, particularly cognitive behavioral approaches, are becoming increasingly available (Evers et al. 2009; Verhoeven et al. 2008). Findings suggest that these treatments are not only effective in reducing itch and scratching behavior, but that they also result in improvements of the skin condition and in a decrease of the health care use of patients with chronic skin diseases in the longer term (Evers et al. 2009; Verhoeven et al. 2008; Yosipovitch et al. 2003). However, treatment outcomes might be substantially improved by making optimally use of placebo effects with regard to itch in clinical practice. For example, adding therapeutic expectation induction trainings regarding outcomes to the regular treatment options may optimize the regular treatment effects or—in case of contraindication for the regular treatments—even partly replace regular pharmacological and topical treatment regimes. Finally, common and specific mechanisms underlying placebo and nocebo effects on itch in relation to pain should be investigated further at a perceptive and neurobiological level to further elucidate the mechanisms underlying placebo and nocebo effects on different somatosensory sensations.

Acknowledgments There are no financial or other arrangements that might lead to a conflict of interest.

References

Atlas LY, Wager TD (2012) How expectations shape pain. Neurosci Lett 520(2):140–148

Bartels DJP, van Laarhoven AIM, Haerkamp EA, Wilder-Smith OH, Donderds ART, van Middendorp H, van de Kerkhof PCM, Evers AWM (2014) Role of conditioning and verbal suggestion in placebo and nocebo effects on itch. PLoS ONE 9:e91727

Benedetti F (2008) Mechanisms of placebo and placebo-related effects across diseases. Annu Rev Pharmacol 48:33–60

Benedetti F, Pollo A, Lopiano L, Lanotte M, Vighetti S, Rainero I (2003) Conscious expectation and unconscious conditioning in analgesic, motor, and hormonal placebo/nocebo responses. J Neurosci 23:4315–4323

Colloca L, Sigaudo M, Benedetti F (2008) The role of learning in nocebo and placebo effects. Pain 136(1–2):211–218

Colloca L, Petrovic P, Wager TD, Ingvar M, Benedetti F (2010) How the number of learning trials affects placebo and nocebo responses. Pain 151(2):430–439

Colloca L, Klinger R, Flor H, Bingel U (2013) Placebo analgesia: psychological and neurobiological mechanisms. Pain 154(4):511–514

Curatolo M, Arendt-Nielsen L, Petersen-Felix S (2006) Central hypersensitivity in chronic pain: mechanisms and clinical implications. Phys Med Rehabil Clin N Am 17:287–302

Evers AWM, Kraaimaat FW, Geenen R, Jacobs JWG, Bijlsma JWJ (2003) Pain coping and social support as predictors of long-term functional disability and pain in early rheumatoid arthritis. Behav Res Ther 41:1295–1310

Evers AWM, Duller P, de Jong EMGJ, Otero ME, Verhaak CM, van der Valk PGM, van de Kerkhof PCM, Kraaimaat FW (2009) Effectiveness of a brief multidisciplinary itch-coping programme in adults with atopic dermatitis. Acta Derm-Venereol 89:57–63

Evers AWM, Verhoeven EWM, Middendorp H, Sweep FGJ, Kraaimaat FW, Donders ART, Eijsbouts AE, van Laarhoven AIM, de Brouwer S, Wirken L, Radstake TRDJ, van Riel PLCM (2014) Daily stressors, stress vulnerability, immune and HPA-axis activity, and short-term fluctuations in rheumatoid arthritis activity. Ann Rheum Dis 73(9):1683–1688

Feneran AN, O'Donnell R, Press A, Yosipovitch G, Cline M, Dugan G, Papoiu ADP, Nattkemper LA, Chan YM, Shively CA (2013) Monkey see, monkey do: contagious itch in nonhuman primates. Acta Derm Venereol 93:27–29

Geers AL, Helfer SG, Kosbab K, Weiland PE, Landry SJ (2005) Reconsidering the role of personality in placebo effects: dispositional optimism, situational expectations, and the placebo response. J Psychosom Res 58(2):121–127

Geers AL, Kosbab K, Helfer SG, Weiland PE, Wellman JA (2007) Further evidence for individual differences in placebo responding: an interactionist perspective. J Psychosom Res 62 (5):563–570

Holle H, Warne K, Seth AK, Critchley HD, Ward J (2012) Neural basis of contagious itch and why some people are more prone to it. Proc Natl Acad Sci USA 109(48):19816–19821

Ikoma A, Handwerker H, Miyachi Y, Schmelz M (2005) Electrically evoked itch in humans. Pain 113(1):148–154

Ikoma A, Steinhoff M, Stander S, Yosipovitch G, Schmelz M (2006) The neurobiology of itch. Nat Rev Neurosci 7(7):535–547

Jensen KB, Kaptchuk TJ, Kirsch I, Raicek J, Lindstrom KM, Berna C, Gollub RL, Ingvar M, Kong J (2012) Nonconscious activation of placebo and nocebo pain responses. Proc Natl Acad Sci USA 109(39):15959–15964

Klinger R, Soost S, Flor H, Worm M (2007) Classical conditioning and expectancy in placebo hypoalgesia: a randomized controlled study in patients with atopic dermatitis and persons with healthy skin. Pain 128(1–2):31–39

Lloyd DM, Hal E, Hall S, McGlone FP (2011) Can itch-related visual stimuli alone provoke a scratch response in healthy individuals? Br J Dermatol 164(6):1299–1303

Martin-Pichora AL, Mankovsky-Arnold TD, Katz J (2011) Implicit versus explicit associative learning and experimentally induced placebo hypoalgesia. J Pain Res 4:67–77

Morton DL, Watson A, El-Deredy W, Jones AK (2009) Reproducibility of placebo analgesia: effect of dispositional optimism. Pain 146(1–2):194–198

Niemeier V, Gieler U (2000) Observations during itch-inducing lecture. Dermatol Psychosom 1:15–18

Papoiu ADP, Wang H, Coghill RC, Chan Y-H, Yosipovitch G (2011) Contagious itch in humans. A study of visual "transmission" of itch in atopic dermatitis and healthy subjects. Br J Dermatol 164(6):1299–1303

Pecina M, Azhar H, Love TM, Lu T, Fredrickson BL, Stohler CS, Zubieta JK (2013) Personality trait predictors of placebo analgesia and neurobiological correlates. Neuropsychopharmacology 38(4):639–646

Price DD, Finniss DG, Benedetti F (2008) A comprehensive review of the placebo effect: recent advances and current thought. Annu Rev Psychol 59:565–590

Scholz O, Hermanns N (1994) Illness behavior and cognitions influence the perception of itching of patients suffering from atopic dermatitis. Z Klin Psychol 23:127–135

Scott DJ, Stohler CS, Egnatuk CM, Wang H, Koeppe RA, Zubieta JK (2007) Individual differences in reward responding explain placebo-induced expectations and effects. Neuron 55(2):325–336

Scott DJ, Stohler CS, Egnatuk CM, Wang H, Koeppe RA, Zubieta JK (2008) Placebo and nocebo effects are defined by opposite opioid and dopaminergic responses. Arch Gen Psychiatry 65 (2):220–231

Smith BW, Tooley EM, Montague EQ, Robinson AE, Cosper CJ, Mullins PG (2008) Habituation and sensitization to heat and cold pain in women with fibromyalgia and healthy controls. Pain 140:420–428

Stewart-Williams S, Podd J (2004) The placebo effect: dissolving the expectancy versus conditioning debate. Psychol Bull 130(2):324–340

van Laarhoven AI, Kraaimaat FW, Wilder-Smith OH, van de Kerkhof PC, Cats H, van Riel PL, Evers AW (2007) Generalized and symptom-specific sensitization of chronic itch and pain. J Eur Acad Dermatol Venereol 21:1187–1192

van Laarhoven AIM, Kraaimaat FW, Wilder-Smith OH, Kerkhof PCM, Evers AWM (2010) Heterotopic pruritic conditioning and itch - analogous to DNIC in pain? Pain 149:332–337

van Laarhoven AI, Vogelaar ML, Wilder-Smith OH, van Riel PL, van de Kerkhof PC, Kraaimaat FW, Evers AW (2011) Induction of nocebo and placebo effects on itch and pain by verbal suggestions. Pain 152(7):1486–1494

van Laarhoven AI, Walker AL, Wilder-Smith OH, Kroeze S, van Riel PL, van de Kerkhof PC, Kraaimaat FW, Evers AW (2012) Role of induced negative and positive emotions in sensitivity to itch and pain in women. Br J Dermatol 167(2):262–269

Van Laarhoven AIM, Kraaimaat FW, Wilder-Smith OH, van Riel PLCM, van de Kerkhof PCM, Evers AWM (2013) Sensitivity to itch and pain in patients with psoriasis and rheumatoid arthritis. Exp Dermatol 22:530–534

Verhoeven EW, Kraaimaat FW, van de Kerkhof PC, van Weel C, Duller P, van der Valk PG, van den Hoogen HJ, Bor JH, Schers HJ, Evers AW (2007) Prevalence of physical symptoms of itch, pain and fatigue in patients with skin diseases in general practice. Br J Dermatol 156:1346–1349

Verhoeven EWM, de Klerk S, Kraaimaat FW, van de Kerkhof PCM, de Jong EMGJ, Evers AWM (2008) Biopsychosocial mechanisms of chronic itch in patients with skin diseases: a review. Acta Derm-Venereol 88:211–218

Verhoeven EWM, Kraaimaat FW, de Jong EMGJ, Schalkwijk J, van de Kerkhof PCM, Evers AWM (2009) Individual differences in the effect of daily stressors on psoriasis: a prospective study. Br J Dermatol 161:295–299

Yosipovitch G, Greaves MW, Schmelz M (2003) Itch. Lancet 361(9358):690–694

Part III

Clinical Practice and Clinical Trials

Clinical and Ethical Implications of Placebo Effects: Enhancing Patients' Benefits from Pain Treatment

Regine Klinger and Herta Flor

Contents

1 Introduction 218
2 Theoretical Concepts for the Clinical Application of Placebo Effects 219
2.1 Psychological Mechanisms of Placebo Analgesia 219
2.1.1 Shaping Expectancy and Inducing Learning via "Classical Conditioning" 219
2.1.2 Shaping Expectancy via "Instruction" 220
2.1.3 Shaping Expectancy and Inducing Learning via "Social Learning" 220
2.2 Interaction Between Conditioning, Social Learning, and Expectancy in Placebo Analgesia 221
2.3 The Neurobiological Basis of Placebo Analgesia 221
3 A Model of Analgesic Placebo Effects 222
3.1 Learning History 222
3.1.1 Preexisting Experiences and Preexisting Expectations 222
3.2 Current Context 224
3.2.1 Current Expectation 224
4 Transfer of the Model to the Clinical Application of Analgesic Placebo Effects 225
4.1 Learning History: Analgesics Can Reactivate Previous Experiences and Expectations 225
4.2 Current Context in Which the Analgesic Is Given 225
4.2.1 Current Information About Pain Medication 225
4.2.2 Current Specific Context and Cues 226
4.2.3 Desire for Pain Relief 228
4.2.4 Certainty About Pain Relief 228
4.2.5 Current Expectations About the Course of Pain Reduction 229
4.2.6 Current Selective Attention 229

R. Klinger (✉)
Outpatient Clinic of Behavior Therapy, Department of Psychology, University of Hamburg, Hamburg, Germany
e-mail: rklinger@uni-hamburg.de

H. Flor
Department of Cognitive and Clinical Neuroscience, Central Institute of Mental Health, Heidelberg University, Mannheim, Germany
e-mail: herta.flor@zi-mannheim.de

F. Benedetti et al. (eds.), *Placebo*, Handbook of Experimental Pharmacology 225,
DOI 10.1007/978-3-662-44519-8_13

4.2.7 Current Pain Experience: Expected or Unexpected Pain Event? 229
4.2.8 Overall Analgesic Effect: Addition of Pharmacological and Placebo Components 229
5 Ethical Aspects of Placebos and Patient Information 230
References 232

Abstract

Expectancy and learning are the core psychological mechanisms of placebo analgesia. They interact with further psychological processes such as emotions and motivations (e.g., anxiety, desire for relief), somatic focus, or cognitions (e.g., attitudes toward the treatment). The development of placebo responsiveness and the actual placebo response in a person is the result of the complex interaction between factors traced back to the individual learning history related to analgesic drugs or treatments and factors of the current context referring to the analgesic or placebo treatment. The aim of this chapter is to depict these complex interactions in a new model of analgesic placebo effects. It joins aspects of the learning history (preexisting experiences and preexisting expectations) of a patient with aspects of the current context (current expectation as a result of external and internal situation in which a pain medication/treatment/placebo is taken, e.g., current information about pain medication, current specific context/cues, desire for pain relief, certainty about upcoming pain relief, current expectation about pain reducing course, current selective attention, increased pain experience, or decreased pain experience). In order to exploit placebo efficacy for an analgesic treatment it is worthwhile to assess in which direction each of these factors exerts its influence in order to maximize placebo effects for a specific patient. By applying placebo mechanisms in this differentiated way, the efficacy of pain treatment can be deliberately boosted.

Keywords

Placebo analgesia • Psychological and neurobiological mechanisms of placebo analgesia • Clinical implications • Model of analgesic placebo effects

1 Introduction

The good evidence for the efficacy of analgesic placebo effects raises the question how to exploit them for clinical pain treatment. Until now, the fact that patients can reduce their pain via placebo effects has rarely been explicitly used in the treatment of acute and chronic pain. In this chapter, we will discuss how the research on placebo effects can be utilized in clinical pain treatment and how the efficacy of pain treatment can be boosted by the exploitation of placebo effects. We believe that specific knowledge about the psychological and neurobiological mechanisms underlying placebo effects aids both patients and the therapists in maximizing their use. Most studies on placebo effects were conducted in healthy participants raising the question if analgesic placebo effects act in a similar manner in patients. This question needs to be addressed in more detail in clinical research.

2 Theoretical Concepts for the Clinical Application of Placebo Effects

As noted above, a better understanding of the psychobiological mechanisms of placebo analgesia is needed to enhance the transfer of placebo effects into clinical applications. The research conducted in recent years has mainly focused on the neurobiological correlates of the placebo effect and the active mechanisms of shaping expectancy via "instruction," "classical conditioning," and "social learning." At the same time, little attention has been paid to possible specific interactions between existing attitudes toward medication in general (e.g., positive/negative attitudes) or prior experience with analgesic treatments and the efficacy of placebo interventions. In most research, the placebo manipulation itself (inducing specific expectations) is an independent variable, but not the already existing pattern of attitudes and learning experiences. In this section we describe the psychological and psychobiological underpinnings of placebo effects and integrate them into a model of processes of analgesic placebo effects, which can be used as a starting point for clinical interventions.

2.1 Psychological Mechanisms of Placebo Analgesia

2.1.1 Shaping Expectancy and Inducing Learning via "Classical Conditioning"

Based on the model of classical conditioning, the placebo effect is viewed as a learnt response, which is triggered by the placebo stimulus ("classically conditioned stimulus"). According to the traditional stimulus substitution model (Ader 1993; Price et al. 2008) the repeated association of an initially neutral stimulus (appearance, color, flavor of the drug) with the unconditioned stimulus (US; pharmacological effect of the drug) leads to a conditioned response (CR; placebo effect). The placebo stimulus therefore becomes a conditioned stimulus (CS; inert "vehicle" of a drug, e.g., appearance, color, and flavor of a tablet). It leads to a placebo response or CR, which is similar to the original pharmacological effect of the corresponding active drug (UR). After this association has been established, the inert agent alone (the placebo) can trigger the effect. Thus, treatments (e.g., analgesic therapies) can have a positive effect based on their associations with previously experienced successful treatments. The nocebo effect can be viewed similarly using the model of classical conditioning. Here the UR refers to the adverse side effects of a drug. They become associated with the agent, in this case the "active (negative) component (US)," which in turn produces the nocebo effect as a CR. It is important to note that the classical conditioning processes shape overt expectancies about positive or negative effects of treatments but also form associative experiences that can be out of the patient's awareness but can still have an effect on his or her behavior, for example, by adding a positive or negative emotional response to certain contexts or cues that were associated with drug effects. Although most research has been conducted on pharmacological analgesic interventions, it is important to realize that placebo effects are also active in other pain treatments, for example, physiotherapy or psychological interventions.

2.1.2 Shaping Expectancy via "Instruction"

According to expectancy theory, the placebo effect is produced by instructions and the anticipatory expectations or response expectancies they induce (Kirsch et al. 2004; Price et al. 1999; Kirsch 1997). Similarly, a negative expectancy, e.g., the expectation that a drug will produce adverse side effects, can produce a nocebo effect. From this point of view, placebo/nocebo effects are a subcategory of expectancy effects and their strength and certainty directly modulate the placebo response. A range of mediating mechanisms and concepts has been proposed to explain why expectancies should trigger a placebo effect. On the one hand, higher control beliefs are postulated, which reduce anxiety and stress (Weisenberg et al. 1996). On the other hand, an altered (selective) attention to pain reduction can be assumed, where negative components are disregarded (Turner et al. 1994). Conversely, negative expectancies can reduce control beliefs, thereby increasing anxiety and stress as well as the selective attention paid to negative components.

2.1.3 Shaping Expectancy and Inducing Learning via "Social Learning"

Pain is influenced by social interactions and can be modulated through the observation of others (Craig 1987). The observation of a painful experience of another person can cause a more painful sensation in the observer when he or she experiences the same situation like in the observation. Colloca and Benedetti (2009) showed that the participants, who were observing an analgesic effect in another person when a special light occurred, also displayed analgesia when they were exposed to the same light (analgesic placebo). This social or observational learning thus also plays an important role in placebo effects. It generates a substantial placebo analgesia in the observer, which is positively correlated with the grade of empathy for the observed person (Colloca and Benedetti 2009). Social learning produces placebo effects of the same magnitude as classical conditioning. Recent research has shown that also nocebo hyperalgesia also undergoes social and observational learning (Świder and Bąbel 2013; Vögtle et al. 2013). Observational learning of placebo and nocebo effects may be of special relevance in inpatient clinical settings where patients observe the interaction of other patients with the healthcare personnel and their responses to pain. They are also relevant in outpatient clinical settings, e.g., when patients are sitting in the dentist's waiting room and can hear other patients crying out in pain in the doctor's office.

2.2 Interaction Between Conditioning, Social Learning, and Expectancy in Placebo Analgesia

Conditioning, social learning, and expectancy processes cannot be expected to occur independently. These core psychological mechanisms interact closely with each other in the placebo effect (Klinger et al. 2007). During conditioning, connections between events are learned, the CS provides information about the US, and previous experiences are represented. If placebos are viewed as conditioned stimuli, they can trigger these previous experiences and response patterns. However, it is also important that the relationship between the events is learnt. This suggests that cognitive processes are involved also in the conditioning process and that conditioning processes can increase the expectation for a positive or negative effect. Specifically, a previous positive experience with a drug (US, pain relief) leads to an association between characteristics of the drug (e.g., appearance, flavor, smell) and the response (UR, pain relief): the "surrounding stimuli" of the drug become the CS ("placebo stimulus") and supply the information that a similar effect (pain relief) may be expected the next time the drug is taken. The placebo on its own can then trigger a response (CR, placebo effect). If the information content of the CS ("placebo"), for which a subject has as an expectancy, is further amplified (by the suggestion of a positive drug effect), the CR (placebo effect) can be further increased. In social learning, conditioning and expectancy processes are also thought to interact in their effect on the observer.

2.3 The Neurobiological Basis of Placebo Analgesia

The described placebo and nocebo mechanisms are associated with specific physiological responses in the central nervous system and the periphery. From a neurobiological point of view, placebo analgesia is closely connected with the endogenous opioid system. Levine et al. (1978) showed that placebo analgesia is in fact a highly complex psychobiological process leading to the release of endogenous opioids. Naloxone, an opioid receptor antagonist, reversed the analgesic placebo effect on postoperative pain following a dental extraction (Levine et al. 1978).

Placebos activate brain circuits that are also involved in opioid analgesia. For example, Petrovic et al. (2002) applied heat pain stimuli to healthy subjects and either administered the opioid remifentanil or a placebo (i.v. NaCl labeled as a "potent analgesic"). Under both conditions, areas of the rostral anterior cingulate cortex (rACC) were activated. In addition, there was an increased connectivity between the rACC and the periaqueductal gray. The dorsolateral prefrontal cortex was also found to be activated during placebo analgesia. The strength of its activation, which was especially high during the phase of anticipating pain, was correlated with the subsequent reduction in pain by a placebo (Eippert et al. 2009b; Wager et al. 2004; Watson et al. 2009). Thus, placebos seem to activate the endogenous descending pain-modulating system (Basbaum and Fields 1978), which inhibits afferent nociceptive information (Millan 2002). This leads to a reduction in perceived pain. If pain stimuli are applied in the placebo condition, the majority of the functional imaging studies indicate a reduced activity in brain areas involved in the processing of pain (Bingel et al. 2006; Petrovic et al. 2002; Wager et al. 2004). Functional imaging studies of the spinal cord show that this effect occurs already at the level of the spinal cord (Eippert et al. 2009b).

Positron emission tomography (PET) studies also confirmed that the endogenous opioid system plays a key role in placebo analgesia. Using [11C] Carfentanil PET, Wager et al. (2007) and Zubieta et al. (2005) showed an amplification of opioid-induced neurotransmission in cingulo-frontal areas and subcortical relay stations during placebo analgesia. As noted above, placebo analgesia can be canceled or considerably reduced by administering the opioid antagonist naloxone (Benedetti et al. 1999; Eippert et al. 2009a, b; Levine et al. 1978; Wall 1999), indicating the important role of the endogenous opioid system; however, Benedetti et al. (1999) showed that both opioid-independent mechanisms are also involved in placebo analgesia. The endocannabinoid system was shown to also play an important role in nonopioid placebo effects (Benedetti et al. 2011). Interesting systems for future studies include, for example, the dopaminergic or the serotonergic effects.

3 A Model of Analgesic Placebo Effects

Beyond expectancy and learning, the described core psychological mechanisms interact with further psychological processes such as emotions and motivations (e.g., anxiety, desire for relief), somatic focus, or cognitions (e.g., attitudes toward the treatment) (Colloca and Benedetti 2007; Geers et al. 2006; Finniss and Benedetti 2005; Price et al. 1999, 2008; Vase et al. 2003; Lyby et al. 2012). The development of placebo responsiveness and the actual placebo response in a person is the result of the complex interaction between factors traced back to the individual learning history related to analgesic drugs or treatments and factors of the current context referring to the analgesic or placebo treatment. Figure 1 depicts these complex interactions, which are discussed in detail below. In order to exploit placebo efficacy for an analgesic treatment it is worthwhile to assess in which direction each of these factors exerts its influence in order to maximize placebo effects for a specific patient.

3.1 Learning History

3.1.1 Preexisting Experiences and Preexisting Expectations

As noted above, in most placebo research, the placebo manipulation itself (inducing specific expectations) is an independent variable, but not the already existing pattern of attitudes. However, it is obvious that a positive attitude compared to an existing negative attitude toward pain medication will result in a stronger placebo effect after the appropriate conditioning and expectancy manipulation. Thus it is

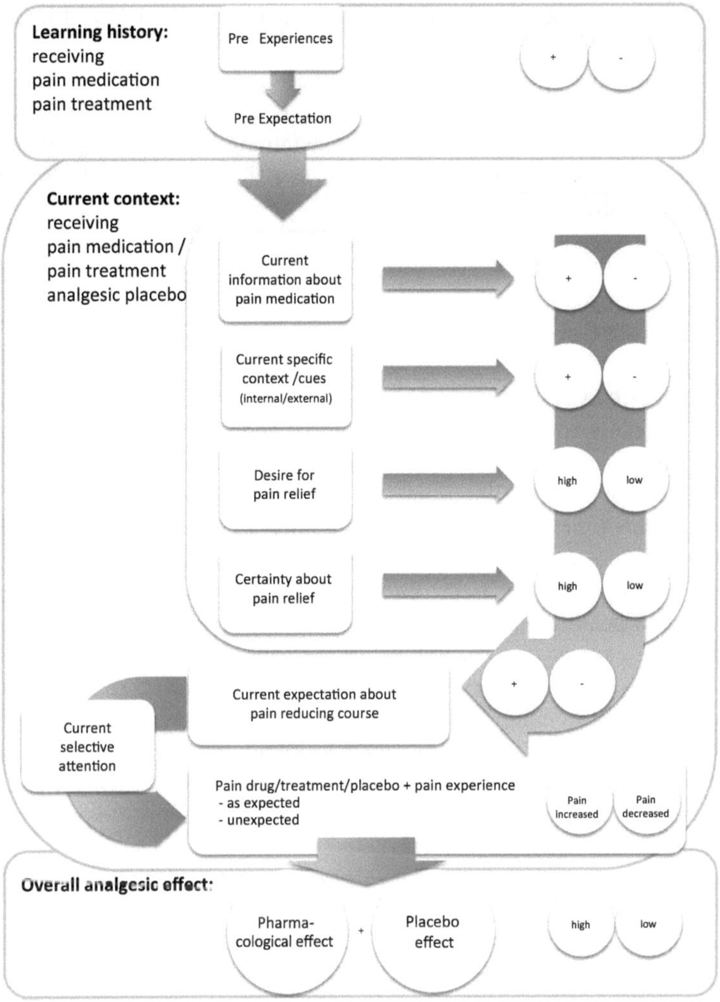

Fig. 1 A model of analgesic placebo effects

important to assess prior learning with analgesic treatment when treating persons with pain.

A history of intake of analgesic medication can be seen as a recurrent association between the pharmacological action and visual or other sensory or emotional aspects of an analgesic. People learn how analgesics act, they learn to presume their effects, and they build up attitudes toward them. Their *preexisting experiences* with analgesics form their *preexisting expectations* about the effects of pain medication. This constitutes the background for receiving current medication or placebos and needs to be considered as an important influence on the efficacy of the current intervention.

The current intake of pain medication or placebos that goes along with current expectancy about the placebo effect might also be seen as a conditioned stimulus that reactivates earlier stimulus associations stemming from learning history.

Thus, not only prior experience but also associations evoked by the current context act upon the analgesic response.

3.2 Current Context

3.2.1 Current Expectation

The current context comprises both the external and internal situation in which a pain medication/treatment/placebo is taken. Initially, the *current information* about pain treatment plays an important role. Is the information positive (e.g., explanations about the positive action of the drug) or negative (e.g., predominant explanation of side effects of the drug/treatment)? Furthermore, the *current external context* is important. Does the current external context contain cues that are associated with healing or cure and do these cues create confidence in the drug/treatment? The *current internal context* of the patient plays an important role as well. Is the patient in an emotionally positive (e.g., confident, relaxed, balanced, concentrated) or in an emotionally negative (e.g., depressed, anxious, angry, distracted) mood? Is the patient's *desire for pain relief* pronounced (e.g., the patient absolutely needs pain relief) or is it rather weak? Moreover, the patient's *certainty about upcoming pain relief* should be explored. How certain is the patient that the pain drug/treatment will reduce his/her pain?

In summary, these factors shape the *current expectations about the course of the analgesic treatment*. This current expectation about the course of analgesic treatment determines the *current selective attention* of a patient. The patient pays attention to either the positive or the negative effects of the pain medication, treatment, or analgesic placebo—thus the pain medication or treatment or at least the analgesic placebo is associated with either *increased pain experience* or *decreased pain experience*. This current pain experience can be in line with the expected pain (reinforcement) or can be different from this expectation (uncertainty, disappointment). Depending on the direction, the current pain experience is increased in the case of an unexpected pain experience as uncertainty and disappointment of the patient will channel his or her attention on pain-increasing features; the current pain experience is decreased in the case of an expected pain experience because reinforcement and assurance of the patient will channel his or her attention on pain-decreasing features (placebo effect). In the case of the intake of an analgesic this current context variable can either increase or decrease the total amount of the drug efficacy by enhancing or reducing the additional placebo component, which adds to the pharmacological component of the drug.

4 Transfer of the Model to the Clinical Application of Analgesic Placebo Effects

The model opens a wide range of placebo applications in the realm of pain treatment in clinical settings while considering ethical standards. This applies to any analgesic and defines the "placebo" component as an additive to the pharmacological component of the analgesic. However, it applies also to other medical and psychological treatments used in pain management. Knowing the principles and mechanisms behind the placebo effect allows for a wide range of interventions (also cf. Finniss and Benedetti 2005), which must be based on a defined concept.

4.1 Learning History: Analgesics Can Reactivate Previous Experiences and Expectations

The model of classical conditioning indicates that previous experiences with pain and pain reducing interventions are remembered as a learned response. Every new experience occurs on the basis of this learning history and is influenced by it (Colloca et al. 2008a, b). If, for example, a new analgesic drug is prescribed and the patient takes it, then its additional, additive placebo effect will greatly depend on previous experiences with analgesic drugs. Expectations that are produced by instructions can also become conditioned stimuli, which can reactivate previously learned associations.

For the routine clinical situation this can have both positive and negative consequences and meanings. Reactivating positive associations could channel the experiences with a new analgesic in a positive direction. For example, the physician or therapist who is delivering a new medication can ask about earlier experiences with analgesics and can emphasize earlier positive associations. However, asking about prior experiences can also reveal negative associations that need to be counteracted for the new substance in the dialogue with the patient (e.g., "Your previous medication had a different spectrum of activity, so it cannot be compared with the new medication you are taking now").

4.2 Current Context in Which the Analgesic Is Given

4.2.1 Current Information About Pain Medication

The power of instructions is one of the important results that placebo research has revealed. When an analgesic is given, the current information about its effects shapes the current expectation about its efficacy (c.f. Fig. 1). One important point for the prescribing physician is to emphasize positive drug effects and to avoid overemphasizing side effects. Due to limited contact hours in clinical settings, there is a high probability that the focus of an interaction between the patient and the therapist is on informing patients about side effects rather than the provision of information on positive drug effects. It is therefore important to explain the positive drug effects as well as the mechanisms of drug action. Personal interaction rather than only written material is especially helpful (Kaptchuk et al. 2008) and supports the patient in accepting the medication and profiting from it.

Building Realistic Positive Expectancy About Analgesic Treatment

The expectation of a positive effect enhances and strengthens drug efficacy. A patient's expectation toward an analgesic is produced to a considerable extent by the information provided about the product. The positive effect of a drug or intervention used in pain management should be explained to the patient as fully as possible. This allows the additive placebo effect of analgesics and other therapeutic measures to be fully exploited. What is important here is that the provided information should be guided as closely as possible by the expected range of effects of the analgesic, in order to maintain credibility and avoid disappointment should the envisioned success fail to materialize (cf. Klinger et al. 2007). Although few conclusive facts are known as yet about the interaction between existing attitudes of patients toward the drug and the placebo effect, it seems reasonable to assume that this will have a key influence. If pain therapists are aware of these attitudes, they will be able to use existing positive attitudes as a starting point and build on them. It should be noted in this context that aspects of the therapist's own personality, acting as a provider of this information, must also be taken in to account when interacting with the patient.

Avoiding Negative Expectations About Analgesics

Negative expectancies about the effect of an analgesic can reduce its efficacy (nocebo effect) (Colloca and Benedetti 2007; Price et al. 2008). In this context, the provision of information on the analgesic treatment in question is once again clinically relevant. In order to avoid nocebo effects, one-sided negative and frightening information should be minimized when administering analgesics. For example, the information "We can try this pain killer, but I don't think it will do any good" will have more chances of being effective if you communicate it as "This pain killer does not help all patients with your condition, but those in whom it works profit a great deal." Apart from this channeling of expectancy, it is very important to determine the patient's preexisting attitude to the drug. Potentially negative attitudes can be addressed and corrected. Whether certain anxieties in the communication of information can also have a positive effect on the mode of action of a drug remains to be investigated. However, it is conceivable that the presence of side effects will in fact enhance the attribution of a positive effect, because it increases the credibility of the substance.

4.2.2 Current Specific Context and Cues

Enhancing and Strengthening the Analgesic Efficacy of a Drug

The open administration of an analgesic, where it is in full view and perceived by the patient, produces better results than its hidden administration (cf. Colloca et al. 2004; Benedetti et al. 2003). The easier it is to perceive the administration of a drug, for example, by seeing, feeling, smelling, and/or tasting it, the better the placebo effect can be exploited. The basis for this is the principle of classical conditioning (association between external stimuli of a medication and its effect). This learning process can also refer to associations between internal (psychological and/or psychophysiological) stimuli and the analgesic effect. For example, a psychologically poor preoperative condition (e.g., anxiety, depression) in patients undergoing surgery leads to a higher postoperative consumption of analgesics (Taenzer et al. 1986), because the negative precondition will reduce placebo and can enhance nocebo effects. Both internal and external context variables can be easily controlled for the exploitation of the analgesic placebo effect in everyday clinical practice.

External Current Specific Context and Cues

In the light of the importance of the immediate context of the use of an analgesic, it is advisable to direct the patient's attention toward the drug, the infusion, or the injection in everyday inpatient practice in order to assign a positive value to the context of pain management and link it to the effect of the drug. The problem in postoperative pain management is that the analgesic medication mostly disappears in the stimulus-saturated patient's room. The pillbox comprises many tablets and in most cases the patient cannot identify the specific analgesics. Often the night nurse allocates the medication and neither the day nurse nor the ward physician knows which analgesics the patient receives. Thus, postoperative pain management approximates in many cases a hidden medication condition and both the nurse and the physician are virtually blinded. To enhance open medication and thus the effects of analgesics, it is important to highlight the current analgesics, for example, by labeling them or enhancing their value by a special design which enables the patient to focus attention on them. Also information about the medication, for example, through written material, could emphasize the analgesic effects. This can also be achieved with patient-controlled analgesia, which is positive, because it enhances the sense of control of the patient. Here the provision of the analgesic can also be emphasized by visual and/or auditory cues. Moreover, the therapist's attire, the appearance of the office or the sick room, as well as the interaction and communication of therapist play an important role in establishing confidence in the medication that is being used (Kaptchuk et al. 2008; Chung et al. 2012).

Internal Current Specific Context and Cues

The internal context of a patient is an important factor that influences the efficacy of drug action. For example, if a patient is in a pain catastrophizing, anxious, or depressed mood, the analgesic effect will be decreased (Pavlin et al. 2005; Khan et al. 2011; De Cosmo et al. 2008; Ip et al. 2009). Moreover, a patient who is exposed to feelings of helplessness and surrenders to the pain experience feels more pain than a patient who experiences control over the pain (Weisenberg et al. 1996). Thus, the pain therapist should support the patient in decreasing catastrophizing and should convince the patient that he or she can control the pain situation with both analgesics and by strengthening control beliefs.

Minimizing Withdrawal Symptoms in Analgesic Detoxification: Hidden Reduction of the Analgesics Makes Withdrawal Easier

The coupling with context variables does not only apply to the analgesic efficacy of drugs, but also to their negative effects (nocebo) and withdrawal symptoms. Just as the hidden administration of an analgesic can switch off its additive placebo component (Colloca et al. 2004), the hidden reduction of medication can ameliorate adverse effects of the medication reduction or withdrawal symptoms. As already noted by Fordyce (1976) who suggested the use of a pain cocktail (the medication dissolved in juice) at fixed time points to reduce negative learning effects associated with medication reduction, withdrawal from medication can be made less negative to the patient if the patients is informed that the reduction will be performed so that it is unnoticed by the patient. This eliminates all stimuli in the context of drug intake that previously predicted its use and led to anticipatory conditioned responses. This can, for instance, be implemented by administering the drug in a beverage (associative cue) and reducing the amount of the drug while a constant amount of the beverage is maintained. The beverage takes on the function of a placebo agent and can maintain the psychological effect of the drug beyond the physical withdrawal. The same mechanisms can also be used to extend the effect of a drug by inserting placebo trials between verum trials. This may be useful when drugs have strong side effects that can be lowered by interspersing placebos.

4.2.3 Desire for Pain Relief

Besides expectation, desire for pain relief plays a central role in the efficacy of the placebo effect (Price et al. 2008; Price and Fields 1997). To examine the contribution of desire for pain relief, the comparison between patients and healthy controls is interesting. Patients with pain disorders have a higher desire to find possibilities that reduce their pain than healthy people. Patients depend on medication for pain relief and their desire for help is therefore high. Thus, they could be more tuned to their bodily sensations and might, as a consequence, expect more immediate relief from medications. Klinger et al. (2007) compared a sample of patients with atopic dermatitis to healthy controls with respect to their analgesic placebo response. For the patients, in contrast to the healthy controls, the verbal instructions alone were not sufficient to maintain placebo analgesia over time. In these patients the induction of expectancies that are not followed by the experience of analgesia produced a loss of the placebo analgesia in a second trial and this effect could be interpreted as disappointment. For clinical use it is important to bear in mind that pain therapists can disappoint their patients when they promise highly effective medication and provide them with ineffective analgesics. Therefore, overstatements or false promises of placebo efficacy and analgesia should be avoided. This topic requires further investigation.

4.2.4 Certainty About Pain Relief

There is evidence that certainty about pain relief following the intake of an analgesic increases the magnitude of its placebo component. Elsenbruch et al. (2012) showed that the probability of attributed certainty to receive a pain reducing medication was positively correlated with pain relief. For the clinician it is a challenging task to balance the information about the analgesic effects without overestimating them and without giving information that is too uncertain about the outcome.

4.2.5 Current Expectations About the Course of Pain Reduction

The current expectation about the course of pain reduction by an analgesic or a supposed analgesic is the result of these complex interactions between the learning history and the current context of the intake of pain medication. It can be positive or negative. In case of a still negative expectation, the healthcare provider should attempt to discuss the patient's concerns about the medication. The direction of the expectation modulates the subsequent selective attention.

4.2.6 Current Selective Attention

As noted, pain can be modulated by attentional processes (Tracey and Mantyh 2007; Aldrich et al. 2000; Crombez et al. 2005; Van Damme et al. 2010). A positive expectation, i.e., expected pain reduction toward the pain medication, leads to a selective attention for pain reducing effects. In case of a negative expectation, the focus is on negative aspects thus making pain more prominent and the analgesic effect is reduced. In clinical practice it is important to support the patient to focus on the pain reducing aspects after receiving pain medication, for example, through information. This could also entail the use of pain diaries that focus on being pain free rather than on the amount of pain that is experienced (Flor 2012) to turn the patient's attention on indicators of the pleasant state of having less pain and to enhance placebo effects. This is especially interesting with respect to chronic pain because of its association with a shift of attention to indicators of pain and with alterations in the processing of rewards and goals (Scott et al. 2007), where pain relief supersedes other goals and reinforcers.

4.2.7 Current Pain Experience: Expected or Unexpected Pain Event?

Pain is evaluated in terms of prior expectancy and thus based on differences between expected and experienced pain. A confirmation of expectation will lead to decreased pain (placebo effect) or to increased pain in the case of non-confirmation (reduction of the placebo effect) (Nakamura et al. 2012). When a person expects a pain-free intervention and experiences pain then this pain will be rated higher because it was unexpected.

4.2.8 Overall Analgesic Effect: Addition of Pharmacological and Placebo Components

The overall analgesic effect comprises the pharmacological and the additional placebo component (Colloca and Benedetti 2005). The variance of this effect primarily results from the variance in the placebo component. Therefore, the effectiveness of an analgesic can be enhanced by exploiting the placebo component. Conversely, the effect of an analgesic could be decreased when its internal placebo effect is suppressed. For example, when an analgesic is applied in a confidential atmosphere and the patient has received sufficient and predominantly positive information about its effectiveness, the probability that its placebo component will be enhanced is high and this supports the overall analgesic effect and vice versa.

5 Ethical Aspects of Placebos and Patient Information

From an ethical point of view it is mandatory that patients are informed and that they consent when verum medication is substituted by placebos in a clinical context. But also the maximization of naturally occurring placebo effects can profit from educating the patient about placebo and nocebo effects. In order to use the placebo effect in clinical practice in a way that is ethically acceptable, it is important to explain the mechanism through which the effect operates in a transparent way. Such an educational provision of information ought to explain the placebo effect based on the models of classical conditioning and expectancy, but also its neurobiological foundations. The ability of patients to understand such neurophysiological and neurobiological connections tends to be underestimated; yet such an education has in itself a significant positive effect on the perception of pain (Moseley et al. 2004). It can increase the conviction that the analgesic placebo effect can be usefully applied. Patients can be included within their own competencies. For example, if patients have understood the principles of classical conditioning as applied to the placebo effect, they themselves will be able to shape the context of taking the medication to optimize the administration of the drug (e.g., taking medication consciously rather than on the side, increasing the effect combining it with positive coping skills such as relaxation exercises). Similarly, they can specifically examine their own expectations toward the drug and possibly seek out additional information in order to improve their attitude to the drug. Table 1 gives examples how one can provide the patient with this placebo-relevant information.

Conclusions and Outlook

The evidence basis of placebo analgesia argues clearly for its effectiveness. It is of great clinical interest to use this phenomenon for clinical pain treatment. This presumes that the placebo effect can be reliably applied within ethical borders. This chapter pointed out the basic psychological mechanisms and the psychobiological underpinnings, which represent the core functioning of analgesic placebo effects and which can be reliably replicated according to current evidence.

Furthermore, this chapter focused on the interaction of placebos with other central influencing factors by presenting an integrated model for placebo analgesia. Based on this model, starting points for interventions that use placebo analgesia were shown. The main applications are within the open medication practice. To exploit the additional placebo component of a pain medication is a very important intervention in the clinical area and improves pain treatment.

Table 1 Clinical application of placebo effects: starting points for open medication

	Transparent explanation of placebo mechanisms	Examples
1.	General information with neurobiological implications	Today we know that every pain medication has a pharmacological and a psychological component, which is the placebo effect. The placebo effect is due to learning and instruction. When a placebo effect is created, the intervention that causes it can on its own create almost the same effect as the actual treatment Neurobiological studies have shown that placebos affect the same structures in the brain as the actual pain treatment. In the case of placebo analgesia the opioid system is involved and endorphins, which act as the body's own pain medication, are released. You can learn to produce this effect yourself by maximizing the placebo effect
2.a	Explanation of classical conditioning mechanisms	The placebo effect is produced and maintained via the coupling of the pain-relieving action of a drug or another treatment with the context in which it is provided or with certain cues that signal its presence. Once there is a sufficient association, the context or cue such as the sight of the shape or color of a pill or the box in which the pill is kept can by itself elicit an effect which is comparable to the pain-relieving effect of the treatment
2.b	Explanation of implications (classical conditioning)	Create your personal positive context of healing and take your pain medication consciously in this context For example – Associate the look, smell, taste, or feeling of the drug with its positive effects including positive thoughts, pleasant surroundings, relaxing strategies – Be aware of the positive components of the pain medication – Pain medication reactivates previous experiences. Recall positive memories to counteract negative pain-related reminders
3.a	Explanation of expectancy via instruction	Instructions and expectancy play an important role in placebo (positive) and nocebo (negative) treatment effects. The way you will be instructed about the pain medication will influence its effectiveness. When you will be told that it will decrease your pain, you expect a pain reduction and this expectancy will additionally help to decrease your pain. When you are only told about the side effects, you might concentrate on them and the pain reducing effect will be less

(continued)

Table 1 (continued)

	Transparent explanation of placebo mechanisms	Examples
3.b	Explanation of implications (expectancy via instruction)	As your physician/doctor/therapist I will realistically emphasize the positive aspects of the presrcibed pain treatment and I will explain the realistic effects of it without overestimating its side effects: For example – Your pain treatment is very effective in reducing your back pain; however, a small amount of pain could remain – It is important that you feel informed about the action of your pain medication. Information based on facts will enable you to focus on the positive, pain reducing aspects of the medication; this will bring out the placebo efficacy, probably because of a selective, specific perception of the positive aspects – Take into consideration that the drug will not act at once, due to its releasing factors it will take (tell the specific duration: . . .) time – Concentrate on the treatments' pain reducing components In case of previous positive experiences with this medication: – You have had a positive experience with this kind of medication in the past, so this is a good prognosis for this type of pain medication In case of previous negative experiences: – You have had a negative experience with this kind of medication in the past; you will now receive a pain treatment related to a different class of medication; this will act through different pathways and should therefore be more beneficial

Further studies should show this additional effect in contexts other than pain medication treatments, e.g., physical therapy and psychological pain therapy.

One of the future tasks of the public health system must be to educate medical and psychological staff about the properties and underlying mechanisms of placebos so that they can optimize the placebo component of their active treatments.

Acknowledgments This research was funded by grants by the Deutsche Forschungsgemeinschaft (FOR 1328/1) to R.K. (Kl 1350/3-1) and H.F. (Fl 156/33-1).

References

Ader R (1993) Conditioned responses in pharmacotherapy research. Psychol Med 23(2):297–299

Aldrich S, Eccleston C, Crombez G (2000) Worrying about chronic pain: vigilance to threat and misdirected problem solving. Behav Res Ther 38(5):457–470

Basbaum AI, Fields HL (1978) Endogenous pain control mechanisms: review and hypothesis. Ann Neurol 4(5):451–462. doi:10.1002/ana.410040511
Benedetti F, Arduino C, Amanzio M (1999) Somatotopic activation of opioid systems by target-directed expectations of analgesia. J Neurosci 19(9):3639–3648
Benedetti F, Maggi G, Lopiano L, Lanotte M, Rainerao I, Vighetti S (2003) Open versus hidden medical treatments: the patient's knowledge about therapy affects the therapy outcome. Prevention Treatment. doi:10.1037/1522-3736.6.1.61a
Benedetti F, Amanzio M, Rosato R, Blanchard C (2011) Nonopioid placebo analgesia is mediated by CB1 cannabinoid receptors. Nat Med 17(10):1228–1230. doi:10.1038/nm.2435
Bingel U, Lorenz J, Schoell E, Weiller C, Büchel C (2006) Mechanisms of placebo analgesia: rACC recruitment of a subcortical antinociceptive network. Pain 120(1–2):8–15. doi:10.1016/j.pain.2005.08.027
Chung H, Lee H, Chang DS, Kim HS, Lee H, Park HJ et al (2012) Doctor's attire influences perceived empathy in the patient-doctor relationship. Patient Educ Couns 89(3):387–391. doi:10.1016/j.pec.2012.02.017
Colloca L, Benedetti F (2005) Placebos and painkillers: is mind as real as matter? Nat Rev Neurosci 6(7):545–552. doi:10.1038/nrn1705
Colloca L, Benedetti F (2007) Nocebo hyperalgesia: how anxiety is turned into pain. Curr Opin Anaesthesiol 20(5):435–439. doi:10.1097/ACO.0b013e3282b972fb
Colloca L, Benedetti F (2009) Placebo analgesia induced by social observational learning. Pain 144(1–2):28–34. doi:10.1016/j.pain.2009.01.033
Colloca L, Lopiano L, Lanotte M, Benedetti F (2004) Overt versus covert treatment for pain, anxiety, and Parkinson's disease. Lancet Neurol 3(11):679–684. doi:10.1016/S1474-4422(04)00908-1
Colloca L, Sigaudo M, Benedetti F (2008a) The role of learning in nocebo and placebo effects. Pain 136(1–2):211–218. doi:10.1016/j.pain.2008.02.006
Colloca L, Tinazzi M, Recchia S, Le Pera D, Fiaschi A, Benedetti F et al (2008b) Learning potentiates neurophysiological and behavioral placebo analgesic responses. Pain 139(2):306–314. doi:10.1016/j.pain.2008.04.021
Craig KD (1987) Consequences of caring: pain in the human context. Can Psychol 28:311–321
Crombez G, Van Damme S, Eccleston C (2005) Hypervigilance to pain: an experimental and clinical analysis. Pain 116(1–2):4–7. doi:10.1016/j.pain.2005.03.035
De Cosmo G, Congedo E, Lai C, Primieri P, Dottarelli A, Aceto P (2008) Preoperative psychologic and demographic predictors of pain perception and tramadol consumption using intravenous patient-controlled analgesia. Clin J Pain 24(5):399–405. doi:10.1097/AJP.0b013e3181671a08
Eippert F, Bingel U, Schoell ED, Yacubian J, Klinger R, Lorenz J et al (2009a) Activation of the opioidergic descending pain control system underlies placebo analgesia. Neuron 63(4):533–543. doi:10.1016/j.neuron.2009.07.014
Eippert F, Finsterbusch J, Bingel U, Büchel C (2009b) Direct evidence for spinal cord involvement in placebo analgesia. Science 326(5951):404. doi:10.1126/science.1180142
Elsenbruch S, Kotsis V, Benson S, Rosenberger C, Reidick D, Schedlowski M et al (2012) Neural mechanisms mediating the effects of expectation in visceral placebo analgesia: an fMRI study in healthy placebo responders and nonresponders. Pain 153(2):382–390. doi:10.1016/j.pain.2011.10.036
Finniss DG, Benedetti F (2005) Mechanisms of the placebo response and their impact on clinical trials and clinical practice. Pain 114(1–2):3–6. doi:10.1016/j.pain.2004.12.012
Flor H (2012) New developments in the understanding and management of persistent pain. Curr Opin Psychiatry 25(2):109–113. doi:10.1097/YCO.0b013e3283503510
Fordyce WE (1976) Behavioral methods for chronic pain and illness. C.V. Mosby, St. Louis
Geers AL, Helfer SG, Weiland PE, Kosbab K (2006) Expectations and placebo response: a laboratory investigation into the role of somatic focus. J Behav Med 29(2):171–178. doi:10.1007/s10865-005-9040-5

Ip HY, Abrishami A, Peng PW, Wong J, Chung F (2009) Predictors of postoperative pain and analgesic consumption: a qualitative systematic review. Anesthesiology 111(3):657–677. doi:10.1097/ALN.0b013e3181aae87a
Kaptchuk TJ, Kelley JM, Conboy LA, Davis RB, Kerr CE, Jacobson EE et al (2008) Components of placebo effect: randomised controlled trial in patients with irritable bowel syndrome. BMJ 336(7651):999–1003. doi:10.1136/bmj.39524.439618.25
Khan RS, Ahmed K, Blakeway E, Skapinakis P, Nihoyannopoulos L, Macleod K et al (2011) Catastrophizing: a predictive factor for postoperative pain. Am J Surg 201(1):122–131. doi:10.1016/j.amjsurg.2010.02.007
Kirsch I (1997) Specifying nonspecifics: psychological mechanisms of placebo effects. In: Harrington A (ed) The placebo effect: an interdisciplinary exploration. Cambridge University Press, Cambridge, MA, pp 166–186
Kirsch I, Lynn SJ, Vigorito M, Miller RR (2004) The role of cognition in classical and operant conditioning. J Clin Psychol 60(4):369–392. doi:10.1002/jclp.10251
Klinger R, Soost S, Flor H, Worm M (2007) Classical conditioning and expectancy in placebo hypoalgesia: a randomized controlled study in patients with atopic dermatitis and persons with healthy skin. Pain 128(1–2):31–39. doi:10.1016/j.pain.2006.08.025
Levine JD, Gordon NC, Fields HL (1978) The mechanism of placebo analgesia. Lancet 2 (8091):654–657
Lyby PS, Forsberg JT, Asli O, Flaten MA (2012) Induced fear reduces the effectiveness of a placebo intervention on pain. Pain 153(5):1114–1121. doi:10.1016/j.pain.2012.02.042
Millan MJ (2002) Descending control of pain. Prog Neurobiol 66(6):355–474
Moseley GL, Nicholas MK, Hodges PW (2004) A randomized controlled trial of intensive neurophysiology education in chronic low back pain. Clin J Pain 20(5):324–330
Nakamura Y, Donaldson GW, Kuhn R, Bradshaw DH, Jacobson RC, Chapman CR (2012) Investigating dose-dependent effects of placebo analgesia: a psychophysiological approach. Pain 153(1):227–237. doi:10.1016/j.pain.2011.10.024
Pavlin DJ, Sullivan MJ, Freund PR, Roesen K (2005) Catastrophizing: a risk factor for postsurgical pain. Clin J Pain 21(1):83–90
Petrovic P, Kalso E, Petersson KM, Ingvar M (2002) Placebo and opioid analgesia– imaging a shared neuronal network. Science 295(5560):1737–1740. doi:10.1126/science.1067176
Price DD (1997) The contribution of desire and expectation to placebo analgesia: implications for new research strategies. In: Harrington A (ed) The placebo effect. An interdisciplinary exploration. Harvard University Press, Cambridge
Price DD, Milling LS, Kirsch I, Duff A, Montgomery GH, Nicholls SS (1999) An analysis of factors that contribute to the magnitude of placebo analgesia in an experimental paradigm. Pain 83(2):147–156
Price DD, Finniss DG, Benedetti F (2008) A comprehensive review of the placebo effect: recent advances and current thought. Annu Rev Psychol 59:565–590. doi:10.1146/annurev.psych.59.113006.095941
Scott DJ, Stohler CS, Egnatuk CM, Wang H, Koeppe RA, Zubieta JK (2007) Individual differences in reward responding explain placebo-induced expectations and effects. Neuron 55(2):325–336. doi:10.1016/j.neuron.2007.06.028
Świder K, Bąbel P (2013) The effect of the sex of a model on nocebo hyperalgesia induced by social observational learning. Pain 154(8):1312–1317. doi:10.1016/j.pain.2013.04.001S0304-3959(13)00155-3 [pii]
Taenzer P, Melzack R, Jeans ME (1986) Influence of psychological factors on postoperative pain, mood and analgesic requirements. Pain 24(3):331–342
Tracey I, Mantyh PW (2007) The cerebral signature for pain perception and its modulation. Neuron 55(3):377–391. doi:10.1016/j.neuron.2007.07.012
Turner JA, Deyo RA, Loeser JD, Von Korff M, Fordyce WE (1994) The importance of placebo effects in pain treatment and research. JAMA 271(20):1609–1614

Van Damme S, Legrain V, Vogt J, Crombez G (2010) Keeping pain in mind: a motivational account of attention to pain. Neurosci Biobehav Rev 34(2):204–213. doi:10.1016/j.neubiorev.2009.01.005

Vase L, Robinson ME, Verne GN, Price DD (2003) The contributions of suggestion, desire, and expectation to placebo effects in irritable bowel syndrome patients. An empirical investigation. Pain 105(1–2):17–25

Vögtle E, Barke A, Kröner-Herwig B (2013) Nocebo hyperalgesia induced by social observational learning. Pain 154(8):1427–1433. doi:10.1016/j.pain.2013.04.041S0304-3959(13)00196-6 [pii]

Wager TD, Rilling JK, Smith EE, Sokolik A, Casey KL, Davidson RJ et al (2004) Placebo-induced changes in FMRI in the anticipation and experience of pain. Science 303(5661):1162–1167. doi:10.1126/science.1093065

Wager TD, Scott DJ, Zubieta JK (2007) Placebo effects on human mu-opioid activity during pain. Proc Natl Acad Sci USA 104(26):11056–11061. doi:10.1073/pnas.0702413104

Wall PD (1999) Textbook of pain. Churchill Livingston, Edinburgh

Watson A, El-Deredy W, Iannetti GD, Lloyd D, Tracey I, Vogt BA et al (2009) Placebo conditioning and placebo analgesia modulate a common brain network during pain anticipation and perception. Pain 145(1–2):24–30. doi:10.1016/j.pain.2009.04.003

Weisenberg M, Schwarzwald J, Tepper I (1996) The influence of warning signal timing and cognitive preparation on the aversiveness of cold-pressor pain. Pain 64(2):379–385

Zubieta JK, Bueller JA, Jackson LR, Scott DJ, Xu Y, Koeppe RA et al (2005) Placebo effects mediated by endogenous opioid activity on mu-opioid receptors. J Neurosci 25 (34):7754–7762. doi:10.1523/JNEUROSCI.0439-05.2005

Van Damme S, Legrain V, Vogt J, Crombez G (2010) Keeping pain in mind: a motivational account of attention to pain. Neurosci Biobehav Rev 34(2):204–213. doi:10.1016/j.neubiorev.2009.01.005

Vase L, Robinson ME, Verne GN, Price DD (2003) The contributions of suggestion, desire, and expectation to placebo effects in irritable bowel syndrome patients. An empirical investigation. Pain 105(1–2):17–25

[illegible]

Traditional and Innovative Experimental and Clinical Trial Designs and Their Advantages and Pitfalls

Katja Weimer and Paul Enck

Contents

1 Introduction 238
2 Experimental Study Designs to Explore the Placebo Response 240
2.1 Manipulating Timing 240
2.1.1 Open/Hidden Treatment Paradigm 240
2.1.2 Delayed Response Paradigm 242
2.2 Manipulating Information 243
2.2.1 The Balanced Placebo Design 243
2.2.2 The Balanced Crossover Design 244
2.2.3 Modifying the Chances to Receive Drug or Placebo 245
2.2.4 Inverse Enrichment 246
2.3 Habituation, Sensitization, Learning 247
2.3.1 "No-Treatment" Controls in the Laboratory 247
2.3.2 Providing Models (Social Learning) 248
2.3.3 Providing Reinforcement (Instrumental Learning) 249
2.4 Predicting Placebo Responders 249
2.5 Avoiding Ethical Conflicts 250
2.6 One Size Fits All? The "Free Choice" Paradigm 251
3 Clinical Designs to Explore the Placebo Effect 252
3.1 Identifying Placebo Responders 252
3.1.1 Crossover Designs 252
3.1.2 Placebo and Drug Run-Ins 253
3.1.3 Randomized Run-in/Withdrawal 254
3.1.4 Repetitive Drug Application Phases 255
3.2 Controlling the Natural Course of Disease 256

K. Weimer • P. Enck (✉)
Department of Psychosomatic Medicine, University Hospital Tübingen, Tübingen, Germany
e-mail: paul.enck@uni-tuebingen.de

F. Benedetti et al. (eds.), *Placebo*, Handbook of Experimental Pharmacology 225,
DOI 10.1007/978-3-662-44519-8_14

3.2.1 Waiting Lists, Treatment as Usual 256
3.2.2 The "Zelen Design" or the "Cohort Multiple Randomized Controlled Trial" 257
3.2.3 Registry Trials 258
3.3 Improving Assay Sensitivity 258
3.3.1 Effective Blinding 259
3.3.2 Active Placebos 259
3.4 Improving Trial Acceptability 260
3.4.1 Unbalanced Randomization 260
3.4.2 Step-Wedge Design 261
3.4.3 Preference Design 262
3.4.4 Cluster Randomization 263
3.5 Developing Individualized Medicine 264
3.5.1 Previous Drug History 264
3.5.2 "Placebo by Proxy" 264
3.5.3 Patient Registry 265
3.6 Dismissing Placebos in RCTs 266
3.6.1 Head-to-Head Trials and CER 266
3.6.2 Historic Controls 267
4 Summary 267
References 267

Abstract

Many study designs and design variants have been developed in the past to either overcome or enhance drug–placebo differences in clinical trials or to identify and characterize placebo responders in experimental studies. They share many commonalities as well as differences that are discussed here: the role of deception and ethical restrictions, habituation effects and the control of the natural course of disease, assay sensitivity testing and effective blinding, acceptability and motivation of patients and volunteers, and the development of individualized medicine. These are fostered by two opposite strategies: utilizing the beneficial aspects of the placebo response—and avoiding its negative counterpart, the nocebo effect—in medical routine for the benefit of patients, and minimizing—by controlling—the negative aspects of the placebo effect during drug development.

Keywords

Trial designs • Experimental designs • Placebo effect • Control group

1 Introduction

The following chapter will present and discuss both traditional and innovative and novel approaches to study the placebo response and its underlying mechanisms in laboratory experiments and in the clinical setting, with healthy volunteers as well as with patients.

While it will discuss and exemplify the traditional randomized and placebo-controlled study design that is "gold standard" since the mid of the twentieth century, it will not go any further in history to elaborate on the origin of this concept—this has been done in the first chapter of this book.

It also will not elaborate on the ethical implications of the use of placebos in the laboratory and the clinics, as this is not the expertise of the authors and is described somewhere else in this book. However, some of the ethical implications of many of the old and as well as the new designs will be discussed where appropriate to demonstrate that new methodologies may be based on ethical grounds, but may also generate new ethical conflicts and dilemmas. Ethics is an implicit challenge in all research involving humans, healthy volunteers or patients and will never find a final solution, at least not in placebo research. Similarly, we will not discuss ethic-related aspects that refer to patient information and informed consent procedures for the same reason, and for paucity of data.

Dealing especially with experimental designs will bring this chapter close to the ones on mechanisms of the placebo responses, e.g., on learning and on expectations and conditioning, but we will not go into the details of it but restrict ourselves to issues where learning and expectations have specifically influenced design aspects. We have also excluded here studies where the purpose was dose reduction using conditioning paradigms via partial reinforcement as they are discussed elsewhere. We have finally excluded specifics of psychotherapy trials (except with respect to waiting list controls and their variants, see Sects. 3.2.1 and 3.4.2) because they represent a subset of study designs due to the fact that—different from all other nondrug interventions, e.g., surgery, physical therapy, and others—in psychotherapy the unspecific effects of drug therapy that include the placebo effects may become the specific effects of the psychological intervention (Kirsch 2005).

In the following, we will distinguish between experimental studies that are mostly performed in healthy volunteers but may also include patients, and clinical studies that are almost exclusively done in patients, at least once a drug is beyond Phase I of its development.

The latter studies are usually performed to compare a treatment (a drug, a nonpharmacological intervention, e.g., surgery) with a "sham" treatment (a placebo pill, sham-surgery, or other control procedures) to explore the benefit of the treatment above unspecific effects (often called placebo effects) that also include methodological biases, regression to the mean, and the spontaneous course of the disease (see Fig. 1).

The former are to explore mechanisms, and as such they may either explore mechanisms of action of the therapeutic intervention (drug, etc.) or of the placebo response. Only designs to explore the placebo response will be discussed here.

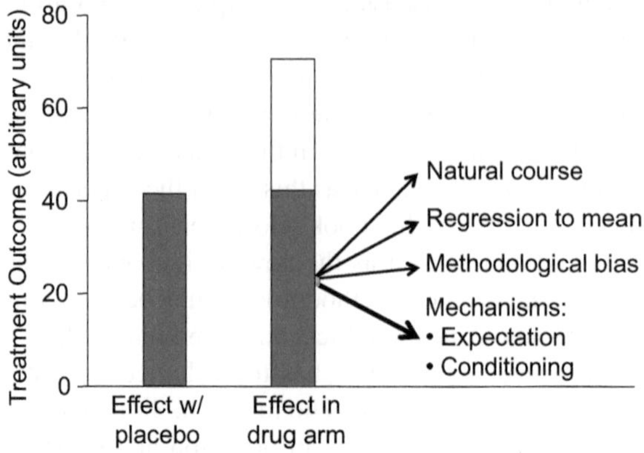

Fig. 1 The "placebo effect" in both arms of RCTs is thought to be a composite of spontaneous disease variation, regression to the mean, and specific contextual factors that represent the placebo response. It is assumed that these factors contribute to both trial arms in an equal manner [adopted from Rief et al. (2011) with permission]

2 Experimental Study Designs to Explore the Placebo Response

While in clinical studies the placebo effect is a compound effect of factors other than the placebo response of an individual (see Fig. 1), experimental designs in placebo research attempt to separate these components to—ideally—identify the "true" placebo effect. Two strategies can be singled out to do so: manipulating the timing of drug action and manipulating the information provided to the patients. The latter is much more common due to technical limitations of the first. Both carry specific ethical problems that will not be discussed here (see above).

2.1 Manipulating Timing

If placebo responses occur as an almost immediate consequence of a medical intervention intended to relief symptoms in a patient as long as the patient expects symptom improvement to occur, placebo responses may even occur before a drug action can be noted. It has in fact been noted in experimental trials that the response in the placebo arm of a drug trial may be faster than in the drug arm in depression (Petrovic et al. 2002). Responses in short-term placebo or drug run-in phases in RCTs have been used to identify placebo or drug responders (see below Sect. 3.1.2). Therefore, dissociating the act of drug application from its presumed drug action onset in the eyes of patients allows separating the true (pharmacological) drug effect from the drug-plus-placebo effect in clinical trials. Two strategies can be found in the literature, of which only one has not yet found its way into experimental placebo research.

2.1.1 Open/Hidden Treatment Paradigm

The open/hidden treatment paradigm (O/HP) was—based upon some empirical observations (Levine et al. 1978; Gracely et al. 1983)—developed by Benedetti and

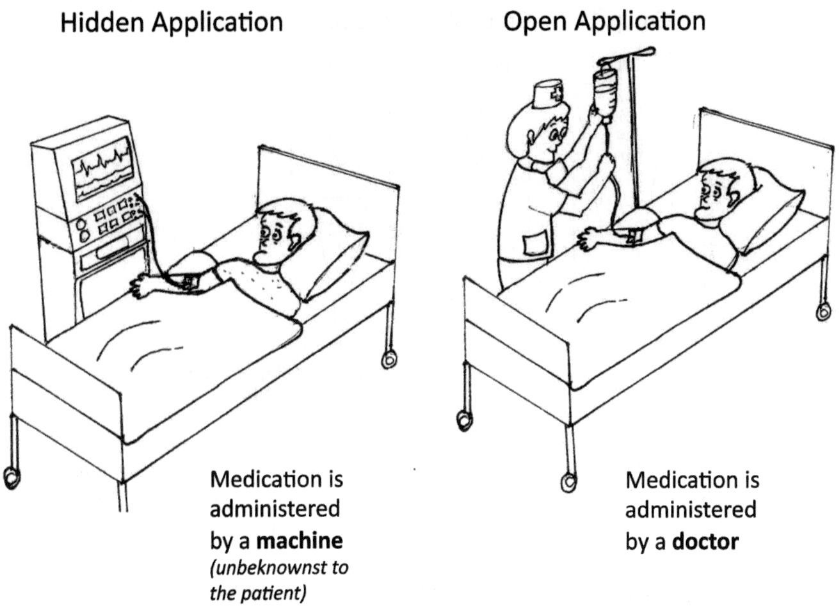

Fig. 2 Open-hidden paradigm according to Benedetti et al. (2003): In this paradigm, identical concentrations of active drugs are administered by a physician in a visible (open condition) or hidden manner, in which the patient is unaware of the timing of administration of the medication (for example, a computer is used to control infusion timing). This permits the dissociation of the pure pharmacodynamic effect of the treatment (hidden treatment) from the additional benefit of the psychological context that comes from knowing that the treatment is being administered [adopted from Enck et al. (2013) with permission]

colleagues (Colloca et al. 2004; Benedetti et al. 2011) and demonstrates an exception from rules stated earlier: that studying the placebo response needs the application of a placebo. In the O/HP, no placebo is given but the timing of drug application is hidden to the patients allowing the placebo response to occur prior to the pharmacological action of the drug (Fig. 2). At the same time, this paradigm is presumably most effective with a real medical treatment situation, e.g., in treatment of acute pain.

Benedetti et al. have applied the paradigm in a number of clinical/experimental situations and have found that many drugs carry a substantial placebo effect in a standard medical setting where the open application of a drug is the rule eliciting strong patient expectations, including opioid and nonopioid analgesics (Amanzio et al. 2001), tranquillizers (Benedetti et al. 2003), and for a nonpharmacological intervention such as deep-brain stimulation in Parkinson's Disease (Pollo et al. 2002). The paradigm has also been used in experimental settings with healthy subjects undergoing pain simulation during brain imaging (Bingel et al. 2011).

While the O/HP may not be a suitable treatment model for clinical routine situations because it discourages the use of drugs with poor or questionable pharmacology, it carries a strong message into the clinics: even poorly effective

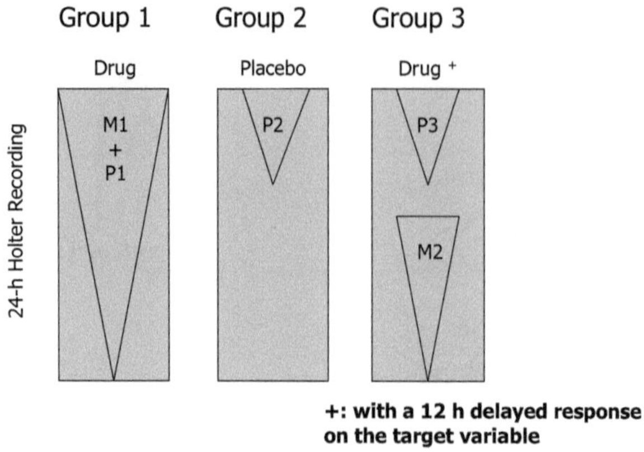

Fig. 3 The "delayed response" design; M1 and M2 stand for medication response, P1 and P2 for placebo response; the "additive model" by Kirsch (2000) assumes that P1 = P2. Under the further assumptions that M1 = M2 und P2 = P3, the hypothesis of the "additive model" is falsified if (M1+ P1 ≠ M2 + P3) [adopted from Enck et al. (2011a) with permission]

drugs can show enhanced clinical efficacy when their open application makes use of the placebo response.

2.1.2 Delayed Response Paradigm

In the O/HP, the manipulation of timing is achieved via a computer-driven drug pump that randomizes (within given limits) the medicine application. In a theoretical model, we came to a similar—though presumably less reliable—technical solution by manipulating the drug release via tablet coating technology. It would dissociate the act of medication intake (swallowing a pill) from its pharmacological action and also allow the placebo response to occur prior to the true drug response; this was called the delayed response paradigm (DRP) (Enck et al. 2011a).

Different from the O/HP, the DRP would be most suitable specifically for drug studies in healthy participants and patients, both under experimental and clinical conditions, provided the pill coating technology would allow such procedures. However, it would require more than just one treatment group; ideally it would include 3 groups (Fig. 3) to identify the true drug, the true placebo response, and to verify the "additive model" (Kirsch 2000). All participants are informed that they will receive either a drug or placebo in a double-blinded fashion. No information, however, is provided about the timing of drug response but a cover story for the potential of prolonged drug action, e.g. for 24 h.

A variation of such a design that intended to elucidate the drug response in a clinical trial in Parkinson's Disease was recently described (D'Agostino 2009): Patients in the placebo-arm are planned to switch from placebo to drug at some time point during the trial unbeknown to the patient and physician, but in this case pretreatment with placebo may affect the later drug treatment by conditioning procedures (Suchman and Ader 1992). A better way of separating drug and placebo effects may be randomized run-in and withdrawal periods (see below, Sect. 3.1.2).

2.2 Manipulating Information

Manipulation of information provided to volunteers and patients appears easier and is therefore most frequently done in placebo research—however, deception is evident in these cases and requires careful ethical consideration and approval, while with manipulation of drug timing (above), even fully informed consent may be possible.

In the majority of all experimental studies of the placebo response, the experimental group (that receives placebo) is usually provided with a 100 % security that the applied drug (pill, cream, injection, infusion, etc.) contains an effective pharmacological agent, while in fact they receive a placebo. In contrast, in clinical RCTs, patients usually receive the information that they have a 50 % (or another) chance to receive the active compound. The difference between both types of information accounts for substantially (up to sixfold) higher placebo effect size in the laboratory compared to a RCT (Vase et al. 2002), thereby allowing a better study of the underlying mechanisms. The control group serves as "no-treatment control" (see below, Sects. 2.3.1 and 3.2) and does not receive any treatment.

The downside of this common practice is the fact that the investigator is usually not blinded towards group assignment and treatment, and thereby may allow the response to be biased by implicit information and behaviors. Strictly separating data collection and data evaluation, or even using uninformed experimenters may help avoiding such bias but are not easy to establish. In the following we will present four experimental approaches to overcome these limitations.

2.2.1 The Balanced Placebo Design

The "balanced placebo design" (BPD) was traditionally used in the testing for expectancy effects of frequently consumed everyday-drugs such as caffeine, nicotine, and alcohol (Kelemen and Kaighobadi 2007), more recently also with drugs such as cocaine (Volkow et al. 2003) and marijuana (Metrik et al. 2009).

While one-half of the study sample receives placebo and the other half the drug, half of each group receives correct information while the other half receives false information on the nature of their study condition (drug or placebo) immediately prior to drug testing, thus allowing to differentiate between the "true" drug effect (those receiving the drug but are told they received placebo) and the "true" placebo effect (those receiving placebo but are told they received the drug) (Fig. 4).

The central concept of the design is—similar to the O/HP—to separate the "true" effects of drug from expectancy effects that occur when participants and patients are given a pill with the information that it may or may not contain the active compound.

A recent paper (Lund et al. 2014) used the BPD explicitly to evaluate whether the assumption of additivity that is implicitly underlying all RCT (Kirsch 2000) is correct. They found that the sum of the "true" drug effect and the "true" placebo effect is larger than the conventional "drug plus placebo" effect in trials, allowing estimating that RCTs tend to underestimate the drug effect and falsifying the additivity hypothesis.

		Information	
		Medication	**Placebo**
Application	**Medication**	*1: true positive*	*2: false negative*
	Placebo	*3: false positive*	*4: true negative*

Fig. 4 The "balanced placebo design" (BPD): All participants are told they participate in a double-blind parallel-group design study. After drug intake and immediately before testing half of the participants in each group are given false and correct information on what they received [adopted from Enck et al. (2011a) with permission]

A variant of the BPD is the "half BPD" in which all participants are given placebos, but half of them receive information that they receive the drug—this is a more common design in current placebo research, as it does not require approval for performing a drug study where the ethical and legal stakes are usually higher. However, effective double-blinding of such a study is difficult unless—as in a recent test in our laboratory (Weimer et al. 2013b)—the participants and the experimenter(s) conducting the study are made to believe that they participate in a fully BPD.

One of the pitfalls of the BPD is the fact that all participants are informed (either correctly or falsely) prior to testing whether and what they have received. In sceptical participants (especially in medical students), this may raise doubts about the truth of the information provided and may require additional measures, such as a reliable explanation why the information is given at all. This is usually done by informing them that once the drug is active, the information whether and what they received may no longer be relevant—however, the participants' acceptance of such information is difficult to prove prior to the test, and its testing afterwards may be subject to other biases.

2.2.2 The Balanced Crossover Design

In an attempt to overcome the serious limitations of the BPD, we designed another strategy that may account for some of the BDP limitations (Enck et al. 2011a). Participants are divided into four groups, and all are told they participate in a conventional trial, in which they will receive both the drug and the placebo at two different occasions in a randomized and double-blinded crossover fashion. This was called the balanced crossover design (BCD).

However, only Groups 2 and 3 will be exposed to drug and placebo in a balanced way, that is half the participants will receive the drug first and the placebo at the second occasion, while the other half will receive first placebo and then the drug. Group 1 will receive the drug twice, and Group 4 will receive placebo twice instead (Fig. 5). In this case, Groups 2 and 3 represent the conventional trial design for drug and placebo effects.

		First medication application	
		drug	**placebo**
Second medication application	**drug**	*1: drug - drug*	*2: placebo - drug*
	placebo	*3: drug - placebo*	*4: placebo placebo*

Fig. 5 The "balanced cross-over design" (BCD): All participants are told they participate in a double-blind crossover design study and will receive both drug and placebo; this is true for groups 2 and 3, while in groups 1 and 4 they receive twice the drug and the placebo, respectively [adopted from Enck et al. (2011a) with permission]

In Group 1, the minimal value of both measures represents the "true" drug effect (plus other unspecific effects), and the difference between both is the expectancy component of the drug response. In Group 4, the maximum value should represent the "true" placebo effect (plus other unspecific effects); and the difference between both values should be the expectancy component of the placebo response. Comparing these expectancy effects between groups 1 and 4 allows to test whether the expectancy component (the placebo effect) is equal under drug and placebo condition—which is the assumption of the "additive model". All other nonspecific factors are assumed to be equally effective in all groups.

The balanced crossover design (BCD) has one important methodological limitation: As with other crossover designs, interference of learning effects need to be kept in mind (Suchman and Ader 1992; Colloca and Benedetti 2006; Kessner et al. 2013), and any adaptation or habituation between measurement 1 and measurement 2 should be minimized, e.g. by increasing the time interval between the two. Its ethical limitations (deception) are similar to those of the BPD with the exception, that participants may receive a drug twice but expect it to receive only once—any risk involved in such a repetition of drug application would exclude the BCD from use, and it can only be used in patients when the deception is authorized (Miller et al. 2005).

A study in our laboratory testing the effects of a nicotine patch on cognitive performance such as reaction times and response inhibition in healthy smoking and nonsmoking volunteers (Weimer et al. 2013c) showed its applicability and limitations.

2.2.3 Modifying the Chances to Receive Drug or Placebo

It has been shown that the likelihood of receiving the active treatment determines the size of both the drug and the placebo response in RCT (Papakostas and Fava 2009): the higher the likelihood of active treatment, the higher the response to both the drug as well as the placebo, solely attributable to the increased expectancy (Rutherford et al. 2009) (see below, Sect. 3.4.1). Maximal response difference between drug and placebo is achieved with a 50 % chance when the chances to receive either drug or placebo are equalized. This is thought to be associated with

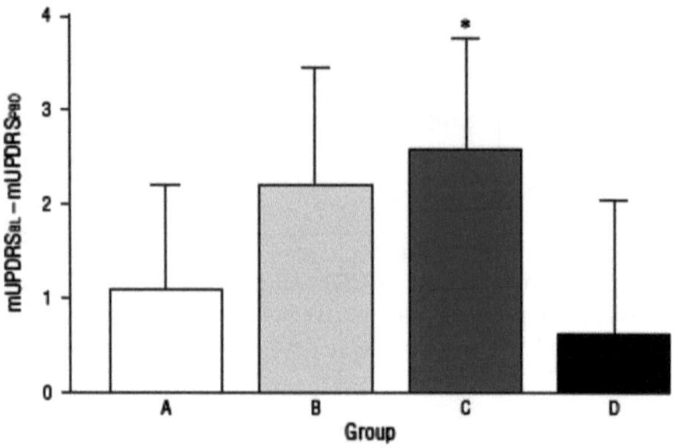

Fig. 6 Clinical response to placebo (modified Unified Parkinson Disease Rating Scale score at baseline [mUPDRSBL] – mUPDRS score following placebo [mUPDRSPBO]), adjusted for mUPDRS baseline and age. Values are given as mean (SD). There was no significant main effect of group. Only the change in group C was significant. *$p < 0.05$. In group A, subjects were told that their chances of receiving active levodopa were 25 %; group B, 50 %; group C, 75 %; and group D, 100 % [reproduced from Lidstone et al. (2010) with permission]

maximal reward activity in the brain, e.g., with maximal dopamine-release in subthalamic neurons (Fiorillo et al. 2003).

In the experimental study by Lidstone et al. (2010) only the information about the likelihood of receiving the active drug was varied while in fact all patients received placebo. This resulted in a bell-shaped curve of the placebo response with maximal efficacy in the 50–75 % range, and supports the underlying reward hypothesis (Fig. 6). Scott et al. (2007) found a strong correlation between the placebo effect and rewarding monetary responses: the larger the nucleus accumbens' responses to monetary reward, the stronger the nucleus accumbens' responses to placebos suggesting that placebo responsiveness depends on the functioning and efficiency of the reward system. In this study Scott et al. (2007) used an experimental approach that is typical of clinical trials, i.e., a 50 % chance to receive either placebo or active treatment.

This model can also be used to simulate the results of clinical trials where altered chances to receive active treatment changed the placebo response (see below, Sect. 2.2.4). In this case effective blinding of the investigator may be achieved and may secure unbiased validity of the results. However, it would require substantially more subjects and patients to be studied under both drug and placebo condition and thus may corroborate the intention to mainly study the placebo effect.

2.2.4 Inverse Enrichment

Enrichment designs in RCT (as discussed below, Sect. 3.4.1) are chosen to increase the number of patients in the drug arm of the study for ethical reasons (the Declaration of Helsinki requires the least number of patients possible to be included into the placebo arm of studies), for psychological reasons (to improve patient motivation during recruitment), or for methodological reasons (e.g., to test different drug dosages against one placebo arm). The same strategy can be applied to experimental laboratory studies to enrich the number of volunteers treated with placebo but maintaining the double blinding of the study and avoiding investigator biases.

If for instance, 90 % of volunteers are assigned to placebo and 10 % to a drug, all subjects can still receive and sign the information that they participate in a double-blinded study as long as the true ratio of drug : placebo is not disclosed. This would significantly improve the number of cases available for exploring the placebo response in comparison to a 50:50 balanced chance, and the deception of volunteers is minimized.

2.3 Habituation, Sensitization, Learning

With any repeated measure of any function or symptom in the laboratory or in a RCT, several factors may influence the outcome that are not related to the measure itself but rather to its repetition: extreme values tend to regress towards a mean value over time, participants may learn to distinguish "signals" from "noise" and thereby alter the signal-to-noise ratio of the response, volunteers may habituate to the stimulus, and systems stimulated may either sensitize or desensitize with repetitions. Patients and volunteers may also "learn" what is expected as a response and may want to please the doctor or experimenter ("placebo" in its original meaning as "it may please"). Finally, if intervals between measures are longer, interfering environmental conditions (time, circadian rhythms, other cycles or events) may directly or indirectly influence the measure differentially. In RCT, such influences are taken care of by unbiased randomization of participants into the different study arms, since this warrants an overall averaged effect of all factors in all groups. This holds true also for any spontaneous variation is clinical symptoms over time, as it is the case in many chronic medical conditions (see below, Sect. 3.2.1).

2.3.1 "No-Treatment" Controls in the Laboratory

The equivalent of a "no-treatment" control condition in laboratory experiments is the inclusion of a group in which the experimental measures are taken at the same frequency than in the experimental (placebo) group but without a placebo intervention. Such a "no-treatment" control is usually unblinded (also in RCTs), and subjects are regularly told that they belong to the control group. In RCTs this has substantial effects of the motivation of the patients to continue participation. Whether healthy volunteers in the laboratory respond differently may depend (among other) on the monetary compensation of volunteers, but other effects have never been explored.

Another open question of a "no-treatment" control group in experimental settings is whether and to what degree "no-treatment" implies that not only all timing aspects of the test, but also all experimental procedures except the presumed drug application need to be similar between the placebo and the control group. For example, in case of a (placebo = NaCl) injection of a presumed analgesic for visceral pain via a constantly running NaCl infusion line (Schmid et al. 2013), it remains to be determined whether the control condition should include the installation of the infusion line or even another NaCl injection that is labeled as placebo. As the purpose of most experiments performed is to elicit maximal placebo response in the experimental group and minimal response in the control group, this may be another source of biases that affect placebo response data as long as they are performed unblinded for the experimenter.

Similarly, the application of an inert skin cream proposed to be a powerful analgesic against experimental pain requires to apply a non-analgesic skin cream in the respective control condition to make measurements comparable otherwise, the skin may respond differentially between two measurements. However, whether volunteers truly believe that they are "controls" rather than experimental subjects has rarely been tested.

Finally, assessing the spontaneous variation of response to an experimental stimulus in "untreated" volunteers is important for the assessment of placebo responsiveness and a placebo responder analysis (as discussed below, Sect. 3.2).

2.3.2 Providing Models (Social Learning)

Another systematic way to elicit placebo responses and to control for their efficacy is to use instructed "models" that demonstrate the effectiveness of the procedure applied before the experimental subjects are tested themselves. The clinical equivalence are other patients that report effective treatment by the drug (or the doctor, or the procedure) to other patients prior to their recruitment into a study. It has been noted that "placebo by proxy" (Grelotti and Kaptchuk 2011; Whalley and Hyland 2013) is an almost completely unknown and unexplored effect in RCT, as we will discuss later (Sect. 3.5.2); in experimental settings however, a few studies have demonstrated its efficacy.

Colloca and Benedetti (2009) were the first to show that strong placebo analgesia can be elicited to the same degree than a conditioning procedure when a volunteer was allowed to observe the pain application and reduction by a presumed drug in another person, prior to being tested him- or herself. In a more recent study (Hunter et al. 2013) they also showed that this does not necessarily require the model to be present in the same room, but that a video demonstration may be sufficient, and that empathy with the patient model is not a prerequisite for its efficacy. Others (Swider and Babel 2013; Vögtle et al. 2013) have shown that also strong nocebo effects (hyperalgesia) can be elicited this way, and that (among others) the gender of the model and the experimental subjects determine the efficacy of such modeling.

This raises another relevant issue in experimental setting, especially with respect to pain and placebo analgesia: whether the gender of the experimenter and experimental volunteers play an important role in the response, and to what degree both interact. A number of studies (Aslaksen et al. 2007; Aslaksen and Flaten 2008) have pointed toward such an effect, but data are inconclusive and in part contradictory (Weimer et al. 2010).

Finally, experimental models may also operate without notice of the experimenter: recruitment of experimental subjects often runs by hear-say and subjects informing each other about the options to participate in experiments for monetary reimbursement reasons. It has never been properly assessed whether this takes influence on the experimental findings.

2.3.3 Providing Reinforcement (Instrumental Learning)

Beyond the question whether the mechanisms by which placebo responses occur include social and instrumental learning (and not only Pavlovian conditioning) (which is not the topic of our review) is the fact that providing (monetary) reinforcement for pain-suppressing behavior has been shown to elicit placebo analgesia: when healthy participants were trained to suppress painful mimic expressions during electrical stimulation, they reported lower pain levels compared to baseline stimulations with the same intensity (Kunz et al. 2011).

This calls into question whether many of the procedures installed in placebo research that operate with monetary reward for enduring painful stimuli (at an individually assessed threshold on a visual analog scale) may in fact be biased by indirect reinforcement mechanisms. This could also account for the fact that rather than pain and other sensory thresholds, cognitive assessments of standardized stimuli are responsive to placebo interventions.

2.4 Predicting Placebo Responders

The question whether "placebo responders" (patients and volunteers who reliably respond to a placebo application in a single setting) truly exists has been raised (Kaptchuk et al. 2008) but not answered. Posthoc analyses have been used both for RCT as well as for experimental studies to identify individuals who would show significant responses following a placebo application, with the prediction based on data collected prior to the intervention. The latter requirement is not always met in prediction studies: Definition of a responder based on median split (or any other separation) of the response data (Elsenbruch et al. 2012) is unacceptable, as this is a posthoc selection of the (best) predictor variables selected from a battery of tests installed in the study, thereby creating a strong publication bias. Prediction analysis instead should be based on a multifactorial regression analysis of the entire response range (rather than a dichotomous grouping) within the experimental (placebo) group compared to a "no-treatment" control group.

In a review of the respective literature we (Horing et al. 2014) identified 3 classes of predictor variables: cognitive and motivational predictors (situational optimism, self efficacy, coping strategies), other psychological predictors (suggestibility, bodily self-awareness), and symptom-related predictors (especially with respect to pain and pain control). For a retrospective analysis of own data (Horing 2013) we found the placebo response to be depending on an internal "locus of control," contrary to common belief: A higher internal locus of control was associated with lower placebo responsiveness in the experimental group, but with higher responses in the "no-treatment" control group.

However, more questions need to be answered: Are placebo responders responding to the same placebo intervention twice or more? Do placebo responders respond to different placebo interventions across modalities, e.g., in pain studies as well as in studies investigating cognitive responses? Is placebo responsiveness a stable condition over time, and how long can an experimental or clinical placebo response be observed?

Only very few studies have ever shown that placebo response in one study predicts response in a subsequent study, be it within the same domain (Whalley et al. 2008) or across modalities (Kaptchuk et al. 2008). The reason for this paucity of data is obvious: it would require investigation protocols that would exceed (by time, money, organizational efforts, and other determinants) the possibilities of most experimental laboratories.

2.5 Avoiding Ethical Conflicts

As discussed above, it cannot be the purpose of a review paper on trial designs to also review and discuss the various ethical aspects that are associated with the use of placebos in experiments, in RCT and in the clinics. However, the use of placebos in experimental research (and *not* in RCTs) raises some specific concerns that need to be addressed here as they have immediate consequences for the conductance of such experiments.

Most experiments that are performed by the majority of placebo researchers imply some type of deception of the volunteers (and in some cases also of the patients) that have stirred discussion about its acceptability (Miller et al. 2005). Different from informed consent in RCT where patients know that they may or may not receive a placebo pill or intervention, in experimental research they are incompletely informed about the purpose of the study and are told instead a "cover story" to hide that the investigation is done to induce a placebo response. Similar to research in lie detection, placebo research may not be able to generate reliable results without the use of deception.

In placebo research, two ethical principles are conflicting: autonomy which requires a fully informed patient and informed consent and assumes full autonomy of the patient, and beneficence which requires optimizing treatment effects and minimizing negative effects, including nocebo effects from informed consent. Many ethical review boards prioritize autonomy and informed consent over beneficence, although this priority should be continuously reevaluated, and new options such as "patient authorized concealments" are to consider.

For experimental research, ethicists have found a similar way out of this dilemma: the introduction of the "authorized deception" (Miller et al. 2005) whereby volunteers in experiments give written informed permission to not being fully informed about the purpose of the study prior to its conductance, to avoid challenging the entire experiment. It has been shown that in comparison to a fully deceptive study, authorized deception produces similar placebo analgesia with experimental pain in the laboratory (Martin and Katz 2010).

2.6 One Size Fits All? The "Free Choice" Paradigm

The free-choice paradigm (FCP) most radically breaks with current traditions in clinical and experimental placebo research by introducing the option to choose between drug and placebo to the patient/volunteer (Enck et al. 2012a).

The design allows volunteers/patients to choose between two pills different in colour. They receive the correct information that one contains the drug while the other contains the placebo, but that conditions are double-blinded. In this case no deception is obvious, and hence ethical limitations are minimal, and the dependent variable for measuring drug efficacy is the choice behaviour rather than reported symptoms or symptom improvement.

The design does neither manipulate the information provided to participants and patients, nor does it manipulate the timing of drug release, both of which are common when novel designs are proposed in experimental studies on the placebo effect in healthy volunteers. It thus avoids ethical concerns (deception) in case of inclusion of patients. It also increases the number of events that can be used for evaluation of drug efficacy, e.g., superiority of drug over placebo by computing.

One has, however, to make sure that patients indeed select and do not take both pills simultaneously, thus undermining the intention of the design. It further has to be made sure that technical solutions are installed to warrant appropriate compliance, to prevent over-dosage, and to monitor drug intake.

Other restrictions may be short-acting effects of the drug, the need for steady drug levels, effects on symptoms rather than biochemical disease indicators, hence symptomatic endpoints rather than disease biomarkers. In this case, the primary outcome measure of drug testing is the "selection behavior" of patients (Fig. 7).

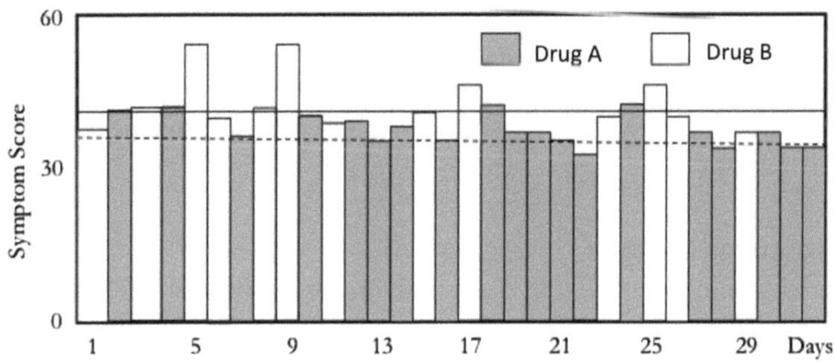

Fig. 7 The "free choice paradigm": patients can choose daily between drugs A and B. The efficacy measures are either the average symptom score with A (*solid line*) and B (*dotted line*) or the number on days with A and B were taken [adopted from Enck et al. (2012a) with permission]

The FCP may be regarded as a modification of the "adaptive response design" (Rosenberger and Lachin 1993), the "early-escape design" (Vray et al. 2004) and other adaptive strategies (Zhang and Rosenberger 2006). It may offer an alternative approach to common drug test procedures, though its statistics have still to be established.

Other requirements of such an approach may be due to the fact that the patient is allowed to switch to the other condition at any time, hence, the pharmacodynamics of the compound under investigation have to be appropriate, e.g., the speed of action, and the feasibility of on-demand medication. It would, on the other hand, allow assessment of drug efficacy via the choice behavior rather than with symptomatic endpoints.

With the FCP, no randomization is needed as all patients have the choice between drug and placebo at predefined time points. Since reasons to alter from 1 day to the next may vary within and across patients, they need to be assessed continuously, e.g., by symptom diaries, and may be taken as covariates in the efficacy analysis. Whether the FCP is suitable for clinical trials in patients needs to be shown in the future.

3 Clinical Designs to Explore the Placebo Effect

Clinical trials serve a different purpose than most experimental trials: they attempt to demonstrate clinical efficacy of a drug (or any other intervention) *against* a placebo control condition, thus attempting to prove superiority of the therapy under investigation against a placebo condition. In consequence, they try to minimize rather than to maximize (Enck et al. 2013) the placebo response in patients and volunteers. Several design variants have been developed to meet this goal.

3.1 Identifying Placebo Responders

Ideally, one would wish to identify potential responders to placebo treatment before a study starts, or at least before it is formally evaluated. Any other (posthoc) exclusion of individuals from trial evaluation would be suspected to be severely investigator-biased. Therefore, a number of study designs have been proposed to deal with this issue.

3.1.1 Crossover Designs

From the beginning of RCTs in drug trials in the early 50s and 60s of the last century, it was evident from trial statistics that within-subject variability of responses is lower than between-subject variability under most clinical conditions. In consequence, the idea of each subject providing his/her own control data is at hands and promotes the idea of crossover trials in which patients receive both the drug and placebo in separate phases (with wash-out periods in-between) and in

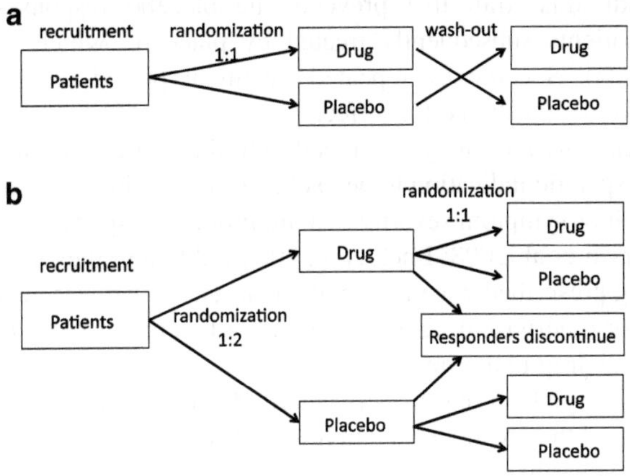

Fig. 8 The conventional cross-over design (**a**) and sequential parallel comparison design (SPCD) according to Fava et al. (2003) (**b**). Note that randomization schemes may be unbalanced in the RPCD, and that only nonresponders to drug or placebo in Phase 1 are re-randomized to drug or placebo in Phase 2 while responders discontinue. This allows merging of Phase1 and Phase 2 data in case treatment periods are equally long [see Ivanova et al. (2011) for the statistics]

completely double-blinded randomized and balanced order (Fig. 8a). This was the dominant drug study design in the second half of last century trials.

Crossover trials at the same time support patient recruitment since all patients can be confirmed that at one stage of the study they would receive active treatment. However, by the same mechanisms they encourage patients to compare both treatment phases, and may lead to increased drop-out rates in the second treatment phase if effects and side-effect profiles are so distinct that the switch from drug to placebo discourages continuation. Taken together, crossover designs do not seem to optimize assay sensitivity.

While the risk of un-blinding could be controlled for by using "active placebos" (see below, Sect. 3.3.2) that mimic side-effects of the drug under investigation, crossover trials have also been questioned because treatments in the first phase may generate conditioning effects during the second phase. This has been demonstrated in clinical and experimental studies (e.g., Suchman and Ader 1992; Colloca and Benedetti 2006; Kessner et al. 2013).

3.1.2 Placebo and Drug Run-Ins

As a further step in early identification and elimination of placebo responders in drug trials, placebo run-in phases (of days or weeks or even longer) were frequently implemented in RCTs. During this phase all patients receive placebo (and this information was usually provided in the informed-consent information), and those responding with symptom improvement were excluded from the study prior to randomization to drug or placebo.

This pragmatic way of dealing with the placebo response has however two limitations: it assumes that being a placebo responder or a placebo nonresponder is a stable individual trait that prevents the placebo responses to occur in nonexcluded patients subsequently treated by placebo—which is not the case (Lee et al. 2004). Specifically repeated treatment period designs (see below, Sect. 3.1.4) have demonstrated this effect.

Also, it carries the risk of systematically eliminating an essential subgroup of patients with a specific indication to be excluded from being studied in RCTs, e.g., patients with minor symptom severity that are prone to respond to placebo (Bridge et al. 2009; Kirsch et al. 2008; Enck et al. 2009), although they subsequently may receive the drug prescribed once it is on the market. Such a selection bias needs to be controlled for otherwise drug approval authorities may be inclined to limit the indication for the drug under investigation.

Finally, this design feature is usually nonblinded for the investigator (and maybe for some patients if they read the patient information carefully) and thus generates a bias in clinical assessment.

Drug run-in periods to identify (and exclude) patients that do not respond to the drug at all serve the same purpose of enhancing assay sensitivity, but they run a similar risk: that the drug-responders represent only a subset of all patients with this disease which may invalidate the clinical usefulness of the drug, or its general indication. In addition, especially responders during run-in will notice when they are subsequently randomized to placebo (similar to the effect in crossover trials) and will be unblinded, as will be the treating physician. Drug run-ins will therefore increase the drug effect and decrease the placebo effect, which may be helpful in early phases of drug development only, e.g., for dose-finding.

3.1.3 Randomized Run-in/Withdrawal

An elegant and unbiased way to test whether the switch from placebo to drug (run-in) and from drug to placebo (withdrawal) creates strong placebo/nocebo effects is to implement a randomized run-in and withdrawal design (Fig. 9). It is currently favored by US Food and Drug Administration (FDA) and the European Medicinal Agency (EMA), especially with patient reported outcome (PRO) measures.

Days	D1	D2	D3	D4	D5	D6	D7		Dt	Du	Dv	Dw	Dx	Dy	Dz
Subj 1	P	D	D	D	D	D	D	...	D	D	P	P	P	P	P
Subj 2	P	P	D	D	D	D	D	...	D	D	D	P	P	P	P
Subj 3	P	P	P	D	D	D	D	...	D	D	D	D	P	P	P
Subj 4	P	P	P	P	D	D	D	...	D	D	D	D	D	P	P
Subj 5	P	P	P	P	P	D	D	...	D	D	D	D	D	D	P
...															
Subj x	P	P	P	P	P	P	P	...	P	P	P	P	P	P	P
Subj y	P	P	P	P	P	P	P	...	P	P	P	P	P	P	P

Fig. 9 Schematic drawing of the randomized run-in and withdrawal: patients 1–5 start treatment at the same time but receive placebo (P) initially for a variable period of time before being switched to the drug (D) in a double-blinded manner. Similarly, at the end of a set period of the study patients are switched from the drug to placebo at variable time points. Individuals *x* and *y* receive placebo throughout the entire study [adopted from Enck et al. (2013) with permission]

Here the switches from drug to placebo and the drug withdrawal is completely blinded for patients and investigators, and as both are not standardized with respect to timing but may occur within a pre-set time window, symptom improvements (at run-in) and symptom worsening (at withdrawal) may allow the separation of "true" drug responses from drug + placebo compound effects. As this design is rather new, not many data are available to test this hypothesis (Rao et al. 2012).

3.1.4 Repetitive Drug Application Phases

A novel strategy that has recently been favored by drug approval authorities in chronic diseases in which cyclic waxing and waning of symptoms is common (such as in irritable bowel syndrome, IBS) is to implement repetitive phases of drug treatment with our without complete re-randomization of patients to drug or placebo, thus going beyond the classical crossover design (see above, Sect. 3.1.1) (Fig. 10). However, this is not primarily to distinguish between drug and placebo response *within* a patient but to demonstrate whether a drug that is taken for some time (and maybe even "on demand," given the low medication compliance in many chronic conditions) loses or maintains its efficacy during a subsequent treatment period (Rao et al. 2012).

As is evident from the example in Fig. 10, a drug may not loose its potency to improve symptoms in Phase 2, but apparently the pretreatment in Phase 1 with either drug or placebo contributes substantially but differentially to the drug efficacy in Phase 2.

An open question in such a design is whether ethical concerns prohibit a complete re-randomization for Phase 2 and allows that patients that received placebo during Phase 1 may receive placebo also during the second treatment

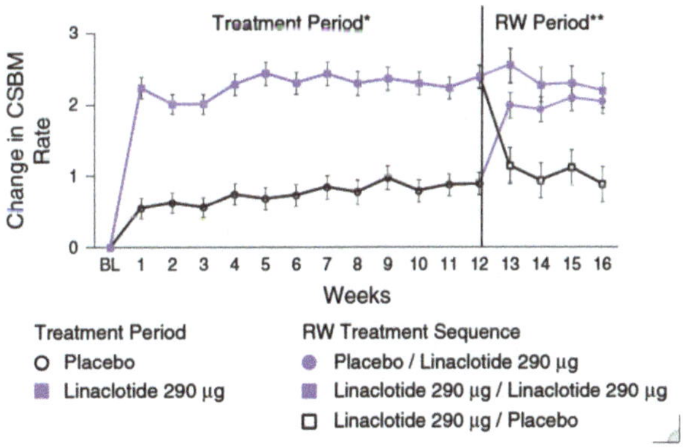

Fig. 10 Weekly results for complete spontaneous bowel movement (CSBM) frequency for linaclotide patients compared with placebo patients for each of the 12 treatment-period weeks. During the randomized withdrawal (RW) period patients that had received placebo in the treatment period were switched to linaclotide. As is evident, their symptom improvement is lower than the initial improvement seen during the treatment period, even when the initial drug-placebo difference is counted [Reproduced from Rao et al. (2012) with permission]

phase. The same applies to the following two designs that were specifically developed to overcome the high placebo response rates in recent depression RCTs.

The Sequential Parallel Comparison Design (SPCD) (Fava et al. 2003) consists of two phases: In Phase 1, patients are randomized to receive either drug or placebo in a conventional manner (RCT), but eventually with more patients randomized to placebo (Ivanova et al. 2011). For the second phase, patients in the placebo arm are screened for their response, and nonresponders to placebo will re-randomized to receive either drug or placebo during the second phase of the trial (Fig. 8b).

From the trials currently conducted according to this design (Baer and Ivanova 2013) it is evident that the placebo response is regularly lower in Phase 2 as compared to Phase 1. Statistics (Ivanova et al. 2011) allow either evaluating both phases separately or—given equal treatment duration in both phases—to merge data for a common evaluation.

The Two-way Enrichment Design (TED) (Ivanova and Tamura 2011) is similar but goes one step further: it re-randomizes not only placebo nonresponders but also drug-responders to drug or placebo in Phase 2, this way proposing to enhance the drug response and decrease the placebo response of the complete trial.

3.2 Controlling the Natural Course of Disease

Spontaneous variation of symptoms can occur with all medical conditions, and especially with chronic diseases. They are part of the "unspecific effects" seen in both arms of drug trials (Fig. 1, above). As long as the assumption of "additivity" is correct (Kirsch 2000) such variation may occur in both study arms to the same degree and may therefore be ignored for the evaluation of drug efficacy. However, with the focus on the size and mechanisms of the placebo response in RCTs, this assessment becomes essential to not overestimate the placebo response in clinical trials.

Therefore, "no-treatment" control groups have been mandated by critiques of the current placebo discussion (Hróbjartsson and Gøtzsche 2001, 2004) to account for spontaneous variation of symptoms in many clinical trials that may falsely be attributed to the placebo response. When they meta-analysed studies (Krogsbøll et al. 2009), they found that about half of the placebo response can be attributed to spontaneous remission; this was also true for included pain trials (Fig. 11). They also noted, that the number of studies that used no-treatment controls is low, they are often with benign clinical conditions (smoking cessation, insomnia), and include most often nonmedicinal interventions such as psychotherapy and acupuncture.

3.2.1 Waiting Lists, Treatment as Usual

Potential ways around the ethical issue of assigning patients to a "no-treatment" group are waiting list (WL) and "treatment as usual" (TAU) groups that are common control strategies in all nonmedication trials where an inert "placebo" treatment is difficult to provide, such as in psychotherapy, physical rehabilitation,

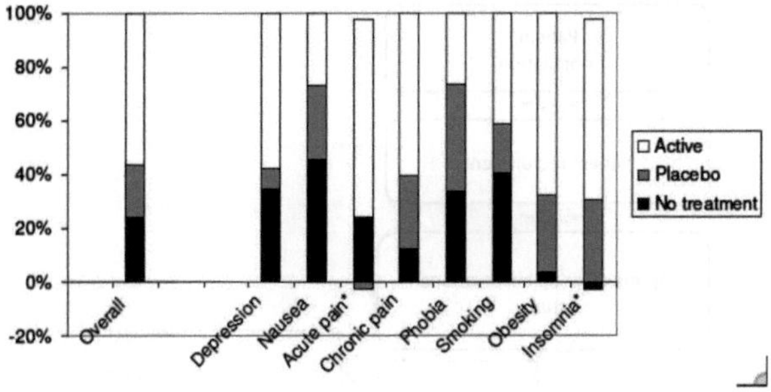

Fig. 11 Relative contributions of the spontaneous improvement, effect of placebo, and effect of active treatment to the change from baseline seen in the actively treated group in RCTs with a no-treatment control arm in different clinical conditions [reproduced from Krogsbøll et al. (2009) with permission]

surgery, and "mechanical" interventions (TENS, magnetic stimulation, laser, acupuncture). While some of these therapies have developed their own control strategy (e.g., sham surgery, sham acupuncture), others have relied on WL and TAU. Their limitations are that patients' expectation to receive effective therapy are at conflict with being randomized to routine treatment (which most of them will have experienced in the past already) and to delays in therapy onset (which may increase the placebo response, but also drop-out rates). This may significantly affect recruitment and compliance in trials, and may lead to biased patients populations in respective studies. A more advanced variant of the WL control strategy is discussed below (Sect. 3.4.3).

WL controls as well as TAU lack credibility as proper control groups in many clinical conditions, and certainly when patients with acute or chronic pain ask for therapy. According to recent meta-analyses (Saarto and Wiffen 2007; Quilici et al. 2009) many drug studies in acute and chronic pain are conducted with comparator drugs rather than with placebos for ethical reasons.

3.2.2 The "Zelen Design" or the "Cohort Multiple Randomized Controlled Trial"

A much more acceptable strategy for patients than being randomized into a "no-treatment" control group is the—classical or modified—Zelen design (Zelen 1979) (Fig. 12) that was recently "re-invented" as "cohort multiple randomized controlled trial" (CMRCT) (Relton et al. 2010). It separates recruitment for an observational study that allows assessing spontaneous symptom variation (the "no-treatment" control condition) from randomization for an interventional study, either placebo-controlled or as comparative effectiveness research (CER) study (see below, Sect. 3.6.1).

In this case, the larger the observational cohort the easier the recruitment of a subsample for a treatment study will be: patients are randomly selected from the larger cohort and can be controlled for representativeness, self-selection bias (those that agree to participate in the RCT), and other cohort descriptors.

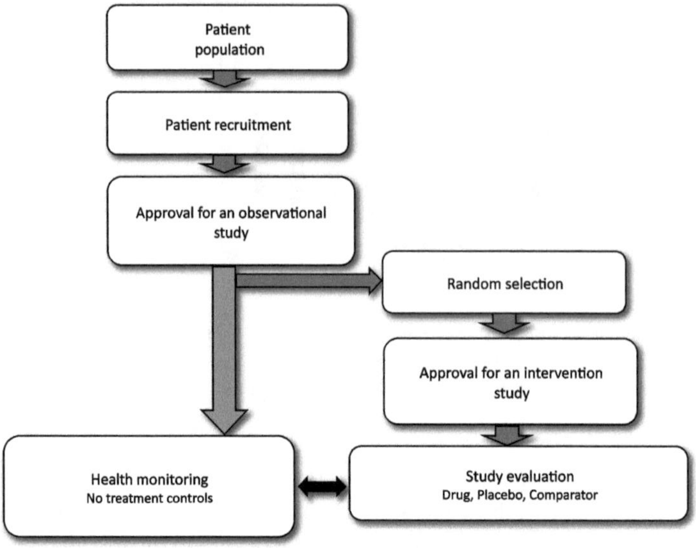

Fig. 12 Schematics of the so-called Zelen design (Zelen 1979) that separates recruitment for an observational study from recruitment for one or more intervention studies [adopted from Enck et al (2013) with permission]

However, two limitations apply: the observational cohort needs to be monitored over time (a cross-sectional sample analysis would not be sufficient to account for changes occurring over time), and it needs to be representative for the complete patient cohort affected by the diseases, both in terms of disease features (e.g., symptom severity) as well as disease management (diagnosis, TAU). Once such a cohort it established it may be used for more than one RCT.

3.2.3 Registry Trials

Instead of building up an observational cohort for one or more CMRCT, it has recently been proposed to use an already established patient registry that follows a patient cohort (Lauer and D'Agostino 2013). This may be the most elegant way to recruit patients for a trial without randomization into a "no-treatment" control group, but disease registries are only available for a few clinical conditions, e.g., in communicable, in rare, and in the more severe diseases.

3.3 Improving Assay Sensitivity

Ways to improve assay sensitivity (the distinction between drug and placebo response in RCTs) include traditional (blinding, active placebos) as well as novel strategies (adaptive designs). We will not discuss here the presumably most important factor in this respect, namely the selection of the primary outcome variable and whether this is a PRO or a disease biomarker.

3.3.1 Effective Blinding

While many studies state that they are double-blinded, they rarely report how effective the blinding actually was. In 1986, Ney et al. (1986) stated that the effectiveness of blinding was assessed in less than 5 % of studies conducted between 1972 and 1983. Twenty years later, Hróbjartsson et al. (2007) identified 1,599 blinded randomized studies and found that only 31 (2 %) reported tests for the success of blinding. Even then, only 14 of the 31 studies (45 %) reported that blinding was successful. Ineffective blinding was also noted in pain trials (Machado et al. 2008). Boutron et al. (2006) reviewed methods used in blinding of pharmacological studies and found insufficient report of the efficacy of blinding across studies and conditions. Boehmer and Yong (2009) consequently asked for inclusion of the evaluation of the effectiveness of blinding in RCTs, but this request should also be extended to experimental studies. Blinding in nondrug trials, e.g., in surgery, physical therapy, and with the use of medical devices is even more complicated and potentially costly (Boutron et al. 2007).

A metaanalysis of RCTs in IBS (Shah et al 2013) has recently shown that the drug benefit across 30 trials with 6 groups of drugs is positively and significantly correlated to the number of adverse events reported in the respective drug arm of the trial, indicating a potential un-blinding effect of the adverse events occurring during a trial that co-determines overall drug efficacy. The authors propose that at least presumed treatment allocation should be evaluated after the study.

3.3.2 Active Placebos

Active placebos mimic the side effects of a drug under investigation without inducing its main effect in clinical trials. Active placebos in experimental research induce side effects that make the volunteer believe to have received active treatment (e.g., a pain medication); this may be achieved by any perceivable effect following a placebo application, e.g., by skin, olfactory, gustatory, and other signals that are easy to induce and do not interfere with the function under test. Interestingly, active placebos have rarely been used, neither in clinical trials nor in experimental placebo research: Boutron et al. (2006) identified only 6 drug trials with active placebos. Among the few experimental studies that tested active placebos in comparison with inactive ones, Rief and Glombiewski (2012) recently showed that adding a small amount of capsaicin to an otherwise inert nasal placebo spray increased the response rate (placebo analgesia) under a 50:50 chance to that with a 100 % security.

In clinical trials, active placebos are difficult to develop and therefore used only occasionally in a few clinical conditions, e.g., in the treatment of depression (Edward et al. 2005). A Cochrane meta-analysis (Moncrieff et al. 2004) reported only 9 studies with 751 patients with depression, all conducted/ published between 1961 and 1984. In all these cases, the "active placebo" was atropine compared with amitryptilin or imipramine, and all but one study used a parallel-group design. While the overall effect size was in favour of active treatment, it was small compared with placebo-controlled trials using inactive placebos, indicating that unblinding effects may inflate the efficacy of antidepressants in trials using inert placebos.

3.4 Improving Trial Acceptability

Many design features were developed to improve patient recruitment and motivation to participate in drug studies even though they have chances to receive placebo. Patient expectations when enrolled are usually to receive active treatment, and this may lead to discontinuation when the lack of improvement may indicate randomization to placebo (Stone et al. 2005; Lindström et al. 2010).

3.4.1 Unbalanced Randomization

Unbalanced randomization can be used for different purposes: to allow more patients to receive active treatment for ethical reasons, to ease recruitment of patients for practical reasons, or to test more drug doses against a single placebo arm. In all cases, the chances of receiving drug instead of placebo improve.

Experimental evidence shows that the chance of receiving active treatment determines the response to placebo (Lidstone et al. 2010) (see above, Sect. 2.2.3). Clinical data also suggests that the number of study arms in a trial, e.g. with various dosages of the drug against placebo codetermines the size of the placebo and the drug response. In two meta-analyses of depression trials (Papakostas and Fava 2009; Sinyor et al. 2010) it was shown that the lower the likelihood of receiving active treatment (compared to placebo), the lower the response to placebo and to drug. Similar findings were made for migraine (Diener et al. 1999) earlier and for schizophrenia treatment recently (Mallinckrodt et al. 2010): with trial designs that randomized 50 % of patients to either drug or placebo (called 1:1 ratio trials here) the placebo response would be minimal compared to trials with two or more drug arms and higher numbers of patients assigned to active treatment compared to placebo (called 2:1 or $\geq$2:1 ratio trials).

Interestingly, this is not supported by data from other areas: Among more than 100 trials with various drugs in irritable bowel syndrome (IBS), 17 used a ratio of drug: placebo greater than 1:1, and these studies yielded a similar placebo response rate than 1:1 studies (Enck et al. 2012b) (Fig. 13).

The fact that maximal differences between drug and placebo is achieved with a 1:1 ratio generates an interesting ethical dilemma (Enck et al. 2011b): If exposing patients to placebo carries an ethical burden that requires the minimal number of patients to be assigned to placebo treatment (World Medical Association 2013), more active treatment arms would be in favour. On the other hand, 1:1 trials would require fewer patients to be tested to prove efficacy of the drug over placebo, and thus would claim the same ethical argument to be in favour of 1:1 trials. This dilemma becomes even more virulent with comparator trials (see below, Sect. 3.6.1).

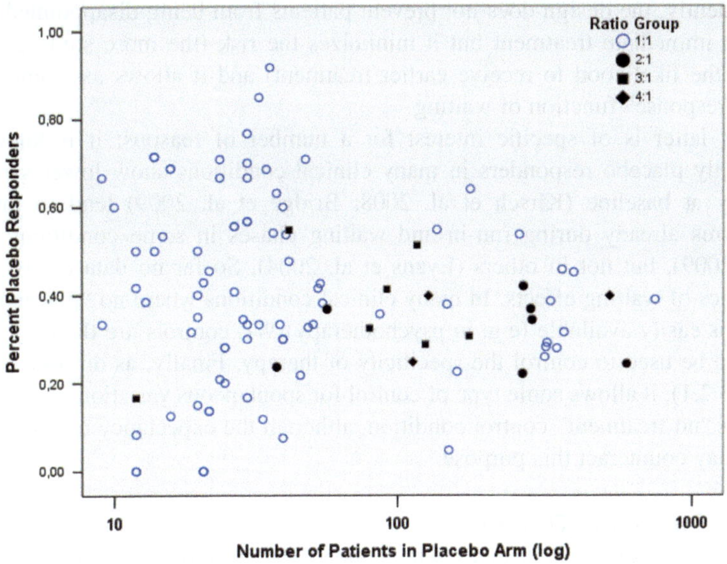

Fig. 13 Correlation between placebo response rates (%) and number of patients (log transformed) in the placebo arm of 102 randomized, double-blinded placebo-controlled irritable bowel syndrome studies. It is evident that with sample sizes of more than 100 the placebo response tends toward 40 %. Open circles indicate studies powered 1:1 and dark circles indicate studies power ≥ 2:1 drug:placebo [adopted from Enck et al. (2012b) with permission]

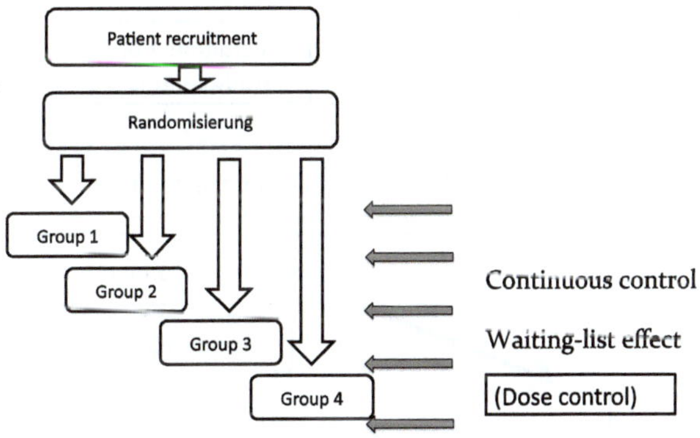

Fig. 14 The step-wedge design according to De Allegri et al. (2008) is a modified waiting-list control strategy. Patients are randomized to more than one waiting arm which increases motivation and reduces disappointment, and at the same time allows assessment of a "dose–response" function of waiting for treatment

3.4.2 Step-Wedge Design

The step-wedge design (De Allegri et al. 2008) is a modification of the WL control group and randomizes patients to different treatment groups that are stacked (immediate begin, begin after x weeks, after y weeks, etc.) so that waiting becomes less of a disappointment and waiting time allows assessment of spontaneous variation of symptoms (Fig. 14).

Evidently, the design does not prevent patients from being disappointed to not receive immediate treatment but it minimizes the risk (the more study arms the higher the likelihood to receive earlier treatment) and it allows assessment of a "dose–response" function of waiting.

This latter is of specific interest for a number of reasons: it is know that especially placebo responders in many clinical conditions show lower symptom severity at baseline (Kirsch et al. 2008; Bridge et al. 2009) tend to improve symptoms already during run-in and waiting phases in some conditions (Enck et al. 2009), but not in others (Evans et al. 2004). So far no data exists on the dynamics of waiting effects. In many clinical conditions where no "placebo treatment" is easily available (e.g. in psychotherapy) WL controls are the only option that can be used to control the specificity of therapy. Finally, as discussed above (Sect. 3.2.1), it allows some type of control for spontaneous variation of symptoms under a "no-treatment" control condition, although the expectancy of future treatment may counteract this purpose.

3.4.3 Preference Design

Especially under circumstances where more than just one treatment option is available (e.g. psychotherapy versus drug therapy for psychiatric disorders) or in comparator trials (see below, Sect. 3.6.1) where a novel drug is tested against another drug already approved for the same indication instead of being tested against placebo, the "preference design" (King et al. 2005) asks for patients' preference before patient that do not have any preference are randomized into the treatment arms (Fig. 15).

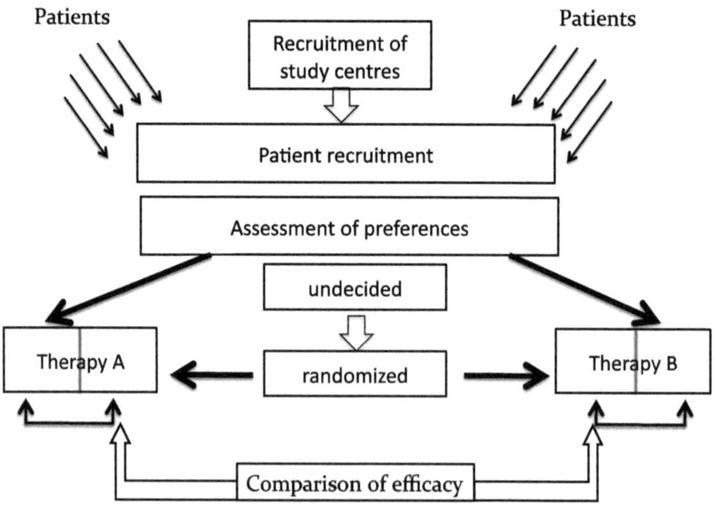

Fig. 15 The preference design (King et al. 2005) allows patients to chose between alternative treatments when available (e.g., drug vs. psychotherapy) before randomization. It also allows comparison of the efficacy in patient that preferred one arm to patients that were randomized to this arm

Assuming a nearly equal number of patients with preference for one of the two options available, and a substantial number of patients without any preference that will undergo randomization, the preference design would allow assessing whether treatment preference plays a role for treatment outcome by comparing (for each option) the patients that selected the treatment to those that were randomized to the same treatment. This information is usually not available following RCTs but hidden in the efficacy data. The role of preferences can also be included into the overall statistics of comparing both treatment effects. It needs to be shown whether preferences play a role in the placebo response, as has been speculated (Prady et al. 2013).

3.4.4 Cluster Randomization

Cluster randomization (Weijer et al. 2012) removes the randomization process further away from the patient: in this case, treatment providers (health care providers, hospitals, private practices) are grouped (clustered) and the decision which cluster provides one therapy and which the other (drug/placebo, drug A/drug B) is randomized (Fig. 16).

In consequence, the patient may not be aware that different treatment options are available, but changing to another cluster is often not feasible due to health care insurance limitations. It has been discussed (McRae et al. 2011) whether such "remote" randomization should be subject to informed consent and that patients should receive the complete information—since they are part of a RCT, ethics approval and patient consent should be identical to conventional trials.

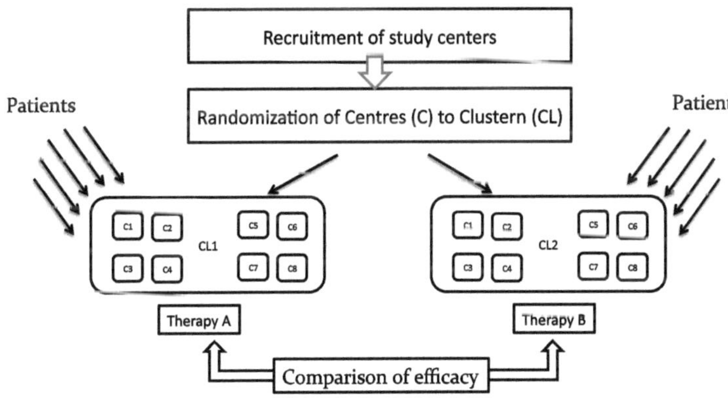

Fig. 16 Cluster randomization according to Weijer et al. (2012) randomizes treatments to different clusters (CL1, CL2) (treatment centers, hospitals, physicians), while patients are recruited by individual centers (C1 to C8). Thereby, patients have a reduced choice and may not even know that randomization has taken place. This generates ethical issue (McRae et al. 2011)

3.5 Developing Individualized Medicine

In our programmatic paper on the future of placebo effects in medicine (Enck et al. 2013) we have argued that for maximizing placebo effects in every-day medicine, individualization of responses to any treatment, including responses to treatment in a RCT—should become the standard in medicine. This includes previous drug history, previous participation in drug trials, and assessment of the role of the social environment of a patient.

3.5.1 Previous Drug History

Both positive and negative previous medical experiences co-determine whether a patient is willing to participate in a RCT, and whether or not he/she responds to drug and placebo treatment. It has been shown experimentally that a previous negative (nocebo) experience can affect the degree of placebo analgesia (and hyperalgesia) in experimental pain (Colloca and Benedetti 2006), and that with repetitive exposure to the same placebo analgesia experience can provide long-lasting efficacy (Kessner et al. 2013).

In clinical trials the situation is similar: In Parkinsons' Disease previous experience with a drug for restless leg syndrome determined similar efficacy of the same drug in a subsequent trial; unfortunately, in a second trial this was not the case but rather the opposite happened (de la Fuente-Fernández 2012).

Similar data are available for only a few other clinical conditions (Iovieno and Papakostas 2012), and the current state of knowledge is rather poor. One legal restriction that applies here is that individualized patient data that have been generated in one RCT cannot easily be transferred to another RCT especially when different investigators or drug companies are involved, for protection of the patient's anonymity. A way out of this dilemma could be the organization of a patient registry for RCTs (see below, Sect. 3.5.3).

3.5.2 "Placebo by Proxy"

The phenomenon of "placebo by proxy" has been established in assessing the determinants of placebo responses in children: While we know that placebo responses overall are larger in RCT in children and adolescents than in adults (Weimer et al. 2013a), little is known about the underlying mechanisms. Apparent mechanisms that account for high placebo response rates in adult disorders, e.g., the number and intensity of doctor visits during a RCT are not operating in children (Rutherford et al. 2011).

It has been argued (Lewis et al 2005) that placebos could operate by producing changes in how caregivers perceive children symptom changes. Placebos could also operate by producing changes in how caregivers behave toward children, which in turn produce behavioral changes in the child. The concept of "placebo by proxy" has recently received attention both from a methodological point of view (Grelotti and Kaptchuk 2011) as well as in an observational study on temper tantrums in children (Whalley and Hyland 2013).

Grelotti and Kaptchuk (2011) argued that—not only in children—the expectations of a patient towards his/her treatment is based not only on own experience and hopes, but occurs in a social context where proxies (family members, caregivers, relatives) respond to symptoms and their improvement and worsening as well. Because these can exist independently of any placebo response of the patient, their contribution to the patient's response are largely unknown and uninvestigated. One of the paradigmatic examples the authors cite refers to the fact that antibiotics are frequently overprescribed specifically in children because of parents' concerns and wishes (Mangione-Smith et al. 1999). Proxies' influences on (placebo) responsiveness may also be responsible for differences in efficacy reports seen between doctor and patient-reported outcomes, especially in depression (Rief et al. 2009).

Whalley and Hyland (2013) take the argument that placebo by proxy may play an important role especially in children one step further: They investigated whether the efficacy of an impure placebo (Bach flower therapy, a homeopathic remedy) to improve symptoms of temper tantrums in 2–5-year old children would be affected by the parents' beliefs and mood. To exclude any direct effect of physician-child and physician-parent interaction, an automated telephone system was used for symptom recording. The authors found a sustained and significant improvement of tantrum frequency and severity that was strongly correlated to parents' mood. As this was an observational study, the authors cannot conclude on the true nature of the symptomatic improvement but assume that these are "pure" placebo effects. Whether symptom improvements were mirrored in children's behavioral changes or only in parents' perception cannot be concluded from the data.

However, as discussed above not only children but most adult patients have a social environment (family, relatives, friends) that participated in the illness history, is involved in its current care and is interested in its future development. Not only the patients own experience with drugs, but also the experience of these "significant others" may co-determine responses to drug and placebo in a RCT. This field of "placebo by proxy" in adulthood is and remains vastly unexplored as long as reliable methods of assessment are missing.

3.5.3 Patient Registry

We have recently argued (Enck et al. 2013) that individualized medicine with respect to placebo responses would require some type of patient registry that serves a dual purpose: protecting patients' anonymity and data collected during one RCT but at the same time make these data available for evaluation of another RCT in which the patient may participate in the future. The legal and ethical rules of such data transfer still need to be established.

This goes far beyond what is current practice in either disease-specific databases (e.g., "... to develop a comprehensive database of individuals who are diagnosed with ...., to better understand the characteristics of these diseases, to determine areas that need further research, and to help pharmaceutical companies with the development of treatments to improve the lives of those affected"(https://connect.patientcrossroads.org/?org=apfed) or in databases for drug companies helping them to evaluate RCT outcomes (Electronic Medical Records), and it also is more than just a recruitment basis for future RCTs to ensure that only properly diagnosed patients are included into such studies.

3.6 Dismissing Placebos in RCTs

While placebo-controlled RCT are still regarded as the gold-standard in the development of novel drug treatments, they have come into question for several reasons: the recently released updated version of the Declaration of Helsinki of World Medical Association (WMA) (World Medical Association 2013) calls for an even more restrictive use of placebo controlled trials in drug development, and some countries have banned the use of such trials entirely (Ehni and Wiesing 2008). In consequence, drug approval authorities such as the FDA and the EMA favor head-to-head comparison (also called "comparator trials" or "comparative effectiveness research," CER) of novel compounds against drugs already marketed for both ethical reasons (no patient without active treatment) as well as economic reasons (novel drugs should be at least equal to what is already available).

3.6.1 Head-to-Head Trials and CER

It is said that CER trials more closely mimic the situation occurring in medical routine where several drugs are available to treat one condition, and where direct comparison of their efficacy if feasible. In contrast, the clinical equivalence of placebo treatment is said to be a "watchful waiting" decision (Hegerl and Mergl 2010) although (as we have discussed above, Sect. 3.2.1) waiting lists are inappropriate control conditions for what happened without treatment.

Because the placebo response is immanent in all medical treatments, *not* applying placebos in RCTs does not result in no placebo response at all but rather in its ignoring during evaluation of the data. As we know from the evaluation of enrichment trials and unbalanced randomization in experiments (see above, Sect. 2.2.3) and in clinical trials (see above, Sect. 3.4.1), providing a 100 % chance to receive active treatment increases the response to both drug and placebo compared to a 50:50 chance as in placebo-controlled trials. However, CER trials lack the direct possibility to assess the placebo response.

In a meta-analytic comparison of CER trials and placebo-controlled trials of the same drugs for treatment of depression it was shown that CER trials enhance the drug response (compared with placebo controlled trials of the same compounds) solely by the expectation to receive a drug by 100 %, and add another 15 % placebo response to the already established average of 40 % from placebo-controlled drug trials for depression (Rutherford et al 2009). Similar data have been shown for CER in schizophrenia (Woods et al. 2005).

This creates an ethical dilemma already discussed above (Sect. 3.4.1): CER trials need up to four times more patients for a statistical test of "noninferiority" than conventional placebo-controlled trials (Leon 2012) which contradicts the statement that the least number of patients should be included in RCT. CER trials are also associated with substantial increased costs of trials, specifically if the selected appropriate comparator drug requires it to be produced (because it is the property of a competing company), and the provision of double-dummy technology (Marušić and Ferenčić 2013). Finally, the selection of the comparator may force substantial methodological considerations and concerns, if more than one potential comparator is available on the market (Estellat and Ravaud 2012; Dunn et al. 2013).

3.6.2 Historic Controls

A completely different way of avoiding placebo-controls was recently described by a drug company (Desai et al. 2013): they screened their entire archive of previously performed RCTs (total: $n = 24{,}581$ studies) for studies where patients were recruited into a placebo arm of pain trials ($n = 3{,}119$). After screening and merging of the data (that were stored in different databases) and screening for core data available in *all* studies they were left with 203 studies with "historic" controls (called ePlacebo patients) treated with placebo. It is proposed to use these historic controls as a database rather than recruiting future patients into placebo arms of RCTs with novel compounds. Feasibility of such an approach needs however still to be verified prospectively.

4 Summary

As we have discussed, both experimental and clinical study designs have attempted to identify placebo responders, to characterize them, and to limit the effects of placebo application of primary and secondary outcome measures, with variable success. Among the different strategies chosen, early identification and exclusion of placebo responders and drug nonresponders seem most promising but carry the risk of selective indication. Enrichment strategies to enhance the placebo–drug difference are most promising for drug development, but for the purpose of characterizing mechanisms of the placebo response, it is most important to distinguish the placebo response from other influences on trial outcomes, especially of spontaneous symptom variation, statistical errors, and response biases. Novel strategies include the use of randomized run-in and withdrawal periods, historic controls, and e-patients but most of them still have to be evaluated.

Acknowledgment Supported by a grant from Deutsche Forschungsgemeinschaft

References

Amanzio M, Pollo A, Maggi G et al (2001) Response variability to analgesics: a role for non-specific activation of endogenous opioids. Pain 90:205–215

Aslaksen PM, Flaten MA (2008) The roles of physiological and subjective stress in the effectiveness of a placebo on experimentally induced pain. Psychosom Med 70:811–818

Aslaksen PM, Myrbakk IN, Høifødt RS et al (2007) The effect of experimenter gender on autonomic and subjective responses to pain stimuli. Pain 129:260–268

Baer L, Ivanova A (2013) When should the sequential parallel comparison design be used in clinical trials? Clin Invest 3:823–833

Benedetti F, Maggi G, Lopiano L et al (2003) Open versus hidden medical treatments: the patient's knowledge about a therapy affects the therapy outcome. Prevention Treatment 6(1). http://psycnet.apa.org/journals/pre/6/1/1a.html. Accessed 28 Jan 2014

Benedetti F, Carlino E, Pollo A (2011) Hidden administration of drugs. Clin Pharmacol Ther 90:651–661

Bingel U, Wanigasekera V, Wiech K et al (2011) The effect of treatment expectation on drug efficacy: imaging the analgesic benefit of the opioid remifentanil. Sci Transl Med 3:70ra14

Boehmer J, Yong P (2009) How well does blinding work in randomized controlled trials?: a counterpoint. Clin Pharmacol Ther 85:463–465

Boutron I, Estellat C, Guittet L et al (2006) Methods of blinding in reports of randomized controlled trials assessing pharmacologic treatments: a systematic review. PLoS Med 3:e425

Boutron I, Guittet L, Estellat C et al (2007) Reporting methods of blinding in randomized trials assessing nonpharmacological treatments. PLoS Med 4:e61

Bridge JA, Birmaher B, Iyengar S et al (2009) Placebo response in randomized controlled trials of antidepressants for pediatric major depressive disorder. Am J Psychiatry 166:42–49

Colloca L, Benedetti F (2006) How prior experience shapes placebo analgesia. Pain 124:126–133

Colloca L, Benedetti F (2009) Placebo analgesia induced by social observational learning. Pain 144:28–34

Colloca L, Lopiano L, Lanotte M et al (2004) Overt versus covert treatment for pain, anxiety, and Parkinson's disease. Lancet Neurol 3:679–684

D'Agostino RB (2009) The delayed-start study design. N Engl J Med 361:1304–1306

De Allegri M, Pokhrel S, Becher H et al (2008) Step-wedge cluster-randomised community-based trials: an application to the study of the impact of community health insurance. Health Res Policy Syst 6:10

de la Fuente-Fernández R (2012) The powerful pre-treatment effect: placebo responses in restless legs syndrome trials. Eur J Neurol 19:1305–1310

Desai JR, Bowen EA, Danielson MM, Allam RR et al (2013) Creation and implementation of a historical controls database from randomized clinical trials. J Am Med Inform Assoc 20:e162–e168

Diener HC, Dowson AJ, Ferrari M et al (1999) Unbalanced randomization influences placebo response: scientific versus ethical issues around the use of placebo in migraine trials. Cephalalgia 19:699–700

Dunn AG, Mandl KD, Coiera E et al (2013) The effects of industry sponsorship on comparator selection in trial registrations for neuropsychiatric conditions in children. PLoS ONE 8:e84951

Edward SJ, Stevens AJ, Braunholtz DA et al (2005) The ethics of placebo-controlled trials: a comparison of inert and active placebo controls. World J Surg 29:610–614

Ehni HJ, Wiesing U (2008) International ethical regulations on placebo-use in clinical trials: a comparative analysis. Bioethics 22:64–74

Elsenbruch S, Kotsis V, Benson S et al (2012) Neural mechanisms mediating the effects of expectation in visceral placebo analgesia: an fMRI study in healthy placebo responders and nonresponders. Pain 153:382–390

Enck P, Vinson B, Malfertheiner P et al (2009) Placebo effects in functional dyspepsia – reanalysis of trial data. Neurogastroenterol Motil 21:370–377

Enck P, Klosterhalfen S, Zipfel S (2011a) Novel study designs to investigate the placebo response. BMC Med Res Methodol 11:90

Enck P, Klosterhalfen S, Weimer K et al (2011b) The placebo response in clinical trials: more questions than answers. Philos Trans R Soc Lond B Biol Sci 366:1889–1895

Enck P, Grundy D, Klosterhalfen S (2012a) A novel placebo-controlled clinical study design without ethical concerns – the free choice paradigm. Med Hypotheses 79:880–882

Enck P, Horing B, Weimer K et al (2012b) Placebo responses and placebo effects in functional bowel disorders. Eur J Gastroenterol Hepatol 24:1–8

Enck P, Bingel U, Schedlowski M et al (2013) The placebo response in medicine: minimize, maximize or personalize? Nat Rev Drug Discov 12:191–204
Estellat C, Ravaud P (2012) Lack of head-to-head trials and fair control arms: randomized controlled trials of biologic treatment for rheumatoid arthritis. Arch Intern Med 172:237–244
Evans KR, Sills T, Wunderlich GR et al (2004) Worsening of depressive symptoms prior to randomization in clinical trials: a possible screen for placebo responders? J Psychiatr Res 38:437–444
Fava M, Evins AE, Dorer DJ et al (2003) The problem of the placebo response in clinical trials for psychiatric disorders: culprits, possible remedies, and a novel study design approach. Psychother Psychosom 72:115–127
Fiorillo CD, Tobler PN, Schultz W (2003) Discrete coding of reward probability and uncertainty by dopamine neurons. Science 299:1898–1902
Gracely RH, Dubner R, Wolskee PJ et al (1983) Placebo and naloxone can alter post-surgical pain by separate mechanisms. Nature 306:264–265
Grelotti DJ, Kaptchuk TJ (2011) Placebo by proxy. BMJ 343:d4345
Hegerl U, Mergl R (2010) The clinical significance of antidepressant treatment effects cannot be derived from placebo-verum response differences. J Psychopharmacol 24:445–448
Horing B (2013) Placebo effects and their prediction across multiple experimentally induced symptoms: motion sickness, cutaneous heat and cold pain, and rectal distension. Ph.D. thesis, University of Tübingen Medical School, Tübingen
Horing B, Weimer K, Muth ER et al (2014) Prediction of placebo responses: a systematic review of the literature. Front Psychol (in press)
Hróbjartsson A, Gøtzsche PC (2001) Is the placebo powerless? An analysis of clinical trials comparing placebo with no treatment. N Engl J Med 344:1594–1602
Hróbjartsson A, Gøtzsche PC (2004) Is the placebo powerless? Update of a systematic review with 52 new randomized trials comparing placebo with no treatment. J Intern Med 256:91–100
Hróbjartsson A, Forfang E, Haahr MT et al (2007) Blinded trials taken to the test: an analysis of randomized clinical trials that report tests for the success of blinding. Int J Epidemiol 36:654–663
Hunter T, Siess F, Colloca L (2013) Socially induced placebo analgesia: a comparison of a pre-recorded versus live face-to-face observation. Eur J Pain 18:914–922. doi:10.1002/j.1532-2149.2013.00436.x
Iovieno N, Papakostas GI (2012) Does the presence of an open-label antidepressant treatment period influence study outcome in clinical trials examining augmentation/ combination strategies in treatment partial responders/nonresponders with major depressive disorder? J Clin Psychiatry 73:676–683
Ivanova A, Tamura RN (2011) A two-way enriched clinical trial design: combining advantages of placebo lead-in and randomized withdrawal. Stat Methods Med Res. doi:10.1177/0962280211431023
Ivanova A, Qaqish B, Schoenfeld DA (2011) Optimality, sample size, and power calculations for the sequential parallel comparison design. Stat Med 30:2793–2803
Kaptchuk TJ, Kelley JM, Deykin A et al (2008) Do "placebo responders" exist? Contemp Clin Trials 29:587–595
Kelemen WL, Kaighobadi F (2007) Expectancy and pharmacology influence the subjective effects of nicotine in a balanced-placebo design. Exp Clin Psychopharmacol 15:93–101
Kessner S, Wiech K, Forkmann K et al (2013) The effect of treatment history on therapeutic outcome: an experimental approach. JAMA 173:1468–1469
King M, Nazareth I, Lampe F et al (2005) Conceptual framework and systematic review of the effects of participants' and professionals' preferences in randomised controlled trials. Health Technol Assess 9:1–186, iii–iv
Kirsch I (2000) Are drug and placebo effects in depression additive? Biol Psychiatry 47:733–735
Kirsch I (2005) Placebo psychotherapy: synonym or oxymoron? J Clin Psycholb 61:791–803

Kirsch I, Deacon BJ, Huedo-Medina TB et al (2008) Initial severity and antidepressant benefits: a meta-analysis of data submitted to the food and drug administration. PLoS Med 5:e45
Krogsbøll LT, Hróbjartsson A, Gøtzsche PC (2009) Spontaneous improvement in randomised clinical trials: meta-analysis of three-armed trials comparing no treatment, placebo and active intervention. BMC Med Res Methodol 9:1
Kunz M, Rainville P, Lautenbacher S (2011) Operant conditioning of facial displays of pain. Psychosom Med 73:422–431
Lauer MS, D'Agostino RB Sr (2013) The randomized registry trial - the next disruptive technology in clinical research? N Engl J Med 369:1579–1581
Lee S, Walker JR, Jakul L, Sexton K (2004) Does elimination of placebo responders in a placebo run-in increase the treatment effect in randomized clinical trials? A meta-analytic evaluation. Depress Anxiety 19:10–19
Leon AC (2012) Challenges in designing comparative-effectiveness trials for antidepressants. Clin Pharmacol Ther 91:165–167
Levine JD, Gordon NC, Fields HL (1978) The mechanism of placebo analgesia. Lancet 2:654–657
Lewis DW, Winner P, Wasiewski W (2005) The placebo responder rate in children and adolescents. Headache 45:232–239
Lidstone SC, Schulzer M, Dinelle K et al (2010) Effects of expectation on placebo-induced dopamine release in Parkinson disease. Arch Gen Psychiatry 67:857–865
Lindström D, Sundberg-Petersson I, Adami J et al (2010) Disappointment and drop-out rate after being allocated to control group in a smoking cessation trial. Contemp Clin Trials 31:22–26
Lund K, Vase L, Petersen GL et al (2014) Randomised controlled trials may underestimate drug effects: balanced placebo trial design. PLoS ONE 9:e84104
Machado LA, Kamper SJ, Herbert RD et al (2008) Imperfect placebos are common in low back pain trials: a systematic review of the literature. Eur Spine J 17:889–904
Mallinckrodt CH, Zhang L, Prucka WR et al (2010) Signal detection and placebo response in schizophrenia: parallels with depression. Psychopharmacol Bull 43:53–72
Mangione-Smith R, McGlynn EA, Elliott MN et al (1999) The relationship between perceived parental expectations and pediatrician antimicrobial prescribing behavior. Pediatrics 103:711–718
Martin AL, Katz J (2010) Inclusion of authorized deception in the informed consent process does not affect the magnitude of the placebo effect for experimentally induced pain. Pain 149:208–215
Marušić A, Ferenčić SF (2013) Adoption of the double dummy trial design to reduce observer bias in testing treatments. J R Soc Med 106:196–198
McRae AD, Weijer C, Binik A et al (2011) When is informed consent required in cluster randomized trials in health research? Trials 12:202
Metrik J, Rohsenow DJ, Monti PM et al (2009) Effectiveness of a marijuana expectancy manipulation: piloting the balanced-placebo design for marijuana. Exp Clin Psychopharmacol 17:217–225
Miller FG, Wendler D, Swartzman LC (2005) Deception in research on the placebo effect. PLoS Med 2:e262
Moncrieff J, Wessely S, Hardy R (2004) Active placebos versus antidepressants for depression. Cochrane Database Syst Rev:CD003012
Ney PG, Collins C, Spensor C (1986) Double blind: double talk or are there ways to do better research? Med Hypotheses 21:119–126
Papakostas GI, Fava M (2009) Does the probability of receiving placebo influence clinical trial outcome? A meta-regression of double-blind, randomized clinical trials in MDD. Eur Neuropsychopharmacol 19:34–40
Petrovic P, Kalso E, Petersson KM et al (2002) Placebo and opioid analgesia: imaging a shared neuronal network. Science 295:1737–1740
Pollo A, Torre E, Lopiano L et al (2002) Expectation modulates the response to subthalamic nucleus stimulation in Parkinsonian patients. Neuroreport 13:1383–1386

Prady SL, Burch J, Crouch S et al (2013) Insufficient evidence to determine the impact of patient preferences on clinical outcomes in acupuncture trials: a systematic review. J Clin Epidemiol 66:308–318
Quilici S, Chancellor J, Löthgren M et al (2009) Meta-analysis of duloxetine vs. pregabalin and gabapentin in the treatment of diabetic peripheral neuropathic pain. BMC Neurol 9:6
Rao S, Lembo AJ, Shiff SJ et al (2012) A 12-week, randomized, controlled trial with a 4-week randomized withdrawal period to evaluate the efficacy and safety of linaclotide in irritable bowel syndrome with constipation. Am J Gastroenterol 107:1714–1724
Relton C, Torgerson D, O'Cathain A et al (2010) Rethinking pragmatic randomised controlled trials: introducing the "cohort multiple randomised controlled trial" design. BMJ 340:c1066
Rief W, Glombiewski JA (2012) The hidden effects of blinded, placebo-controlled randomized trials: an experimental investigation. Pain 153:2473–2477
Rief W, Nestoriuc Y, Weiss S et al (2009) Metaanalysis of the placebo response in antidepressant trials. J Affect Disord 18:1–8
Rief W, Bingel U, Schedlowski M et al (2011) Mechanisms involved in placebo and nocebo responses and implications for drug trials. Clin Pharmacol Ther 90:722–726
Rosenberger WF, Lachin JM (1993) The use of response-adaptive designs in clinical trials. Contemp Clin Trials 14:471–484
Rutherford BR, Sneed JR, Rosse SP (2009) Does study design influence outcome? Psychother Psychosom 78:172–181
Rutherford BR, Sneed JR, Tandler JM et al (2011) Deconstructing pediatric depression trials: an analysis of the effects of expectancy and therapeutic contact. J Am Acad Child Adolesc Psychiatry 50:782–795
Saarto T, Wiffen PJ (2007) Antidepressants for neuropathic pain. Cochrane Database Syst Rev: CD005454
Schmid J, Theysohn N, Gaß F et al (2013) Neural mechanisms mediating positive and negative treatment expectations in visceral pain: a functional magnetic resonance imaging study on placebo and nocebo effects in healthy volunteers. Pain 154:2372–2380
Scott DJ, Stohler CS, Egnatuk CM et al (2007) Individual differences in reward responding explain placebo-induced expectations and effects. Neuron 55:325–336
Shah E, Triantafyllou K, Hana AA et al (2013) Adverse events appear to unblind clinical trials in irritable bowel syndrome. Neurogastroenterol Motil 26:482–488. doi:10.1111/nmo.12289
Sinyor M, Levitt AJ, Cheung AH et al (2010) Does inclusion of a placebo arm influence response to active antidepressant treatment in randomized controlled trials? Results from pooled and meta-analyses. J Clin Psychiatry 71:270–279
Stone DA, Kerr CE, Jacobson E et al (2005) Patient expectations in placebo-controlled randomized clinical trials. J Eval Clin Pract 11:77–84
Suchman AL, Ader R (1992) Classic conditioning and placebo effects in crossover studies. Clin Pharmacol Ther 52:372–377
Swider K, Babel P (2013) The effect of the sex of a model on nocebo hyperalgesia induced by social observational learning. Pain 154:1312–1317
Vase L, Riley JL, Price DD (2002) A comparison of placebo effects in clinical analgesic trials versus studies of placebo analgesia. Pain 99:443–452
Vögtle E, Barke A, Kröner-Herwig B (2013) Nocebo hyperalgesia induced by social observational learning. Pain 154:1427–1433
Volkow ND, Wang GJ, Ma Y et al (2003) Expectation enhances the regional brain metabolic and the reinforcing effects of stimulants in cocaine abusers. J Neurosci 23:11461–11468
Vray M, Girault D, Hoog-Labouret N et al (2004) Methodology for small clinical trials. Therapie 59:273–286
Weijer C, Grimshaw JM, Eccles MP et al (2012) The Ottawa statement on the ethical design and conduct of cluster randomized trials. PLoS Med 9:e1001346
Weimer K, Enck P, Klosterhalfen S (2010) Gender effects in placebo responses. Z Med Psychol 19:146–153

Weimer K, Gulewitsch M, Schlarb AA et al (2013a) Placebo effects in children: a review. Pediatr Res 74:96–102

Weimer K, Horing B, Stürmer J et al (2013b) Nicotine expectancy differentially affects reaction time in healthy non-smokers and smokers depending on gender. Exp Clin Psychopharmacol 21:181–187

Weimer K, Horing B, Walentin S et al (2013c) Nicotine stimulus expectancy but not nicotine affects cognitive performance in healthy smokers and non-smokers. Psychosom Med 75:A115

Whalley B, Hyland ME (2013) Placebo by proxy: the effect of parents' beliefs on therapy for children's temper tantrums. J Behav Med 36:341–346

Whalley B, Hyland ME, Kirsch I (2008) Consistency of the placebo effect. J Psychosom Res 64:537–541

Woods SW, Gueorguieva RV, Baker CB et al (2005) Control group bias in randomized atypical antipsychotic medication trials for schizophrenia. Arch Gen Psychiatry 62:961–970

World Medical Association (2013) World medical association declaration of Helsinki: ethical principles for medical research involving human subjects. JAMA 310:2191–2194

Zelen M (1979) A new design for randomized clinical trials. N Engl J Med 300:1242–1245

Zhang L, Rosenberger WF (2006) Response-adaptive randomization for clinical trials with continuous outcomes. Biometrics 62:562–569

Lessons to be Learned from Placebo Arms in Psychopharmacology Trials

Bettina K. Doering, Winfried Rief, and Keith J. Petrie

Contents

1 Introduction 274
2 The Placebo Effect in Psychopharmacology Trials 275
  2.1 The Relevance of Spontaneous Remission 275
  2.2 The Increasing Power of Placebo 277
  2.3 The Impact of Trial Design on Placebo Responses 278
  2.4 Open-Label Placebo Application 281
3 Side Effects in Psychopharmacology Trials 283
  3.1 The Nocebo Effect 283
  3.2 Onset Sensations 284
4 Implications for Drug Trials: Possible Interaction Effects 285
References 287

Abstract

Large placebo effects are typically reported in clinical drug trials and evidence suggests placebo effects have increased over time. The diminishing drug–placebo difference calls into question the effectiveness of pharmacological treatments and provides a challenge to prove the effectiveness of new medications. This chapter discusses explanations for the increasing placebo effect. It highlights the contribution of spontaneous remission to the improvement in placebo groups, but focuses particularly on the role of patient and clinician expectations. Certain characteristics of the trial design can influence the formation of patient expectations and, subsequently, true placebo responses. Side effects in clinical trials may also contribute inadvertently to placebo

B.K. Doering (✉) • W. Rief
University of Marburg, Marburg, Germany
e-mail: doeringb@staff.uni-marburg.de

K.J. Petrie
University of Auckland, Auckland, New Zealand

F. Benedetti et al. (eds.), *Placebo*, Handbook of Experimental Pharmacology 225,
DOI 10.1007/978-3-662-44519-8_15

responses. Side effects after starting medication can inform participants about their allocation to an active treatment group. Thus, they may enhance expectations of improvement and contribute to nonspecific effects in clinical trials. It is argued that specific and nonspecific effects interact in drug groups of clinical trials. This interaction influences drug–placebo differences in clinical trials (i.e., trial sensitivity). Future research should aim to identify which patients will respond best to drugs and those who may be better treated with placebos.

Keywords
Clinical trials • Placebo • Psychopharmacology • Additive model • Onset sensation • Nocebo

1 Introduction

Double-blind randomized placebo-controlled trials (RCT) have been the standard research design to investigate the effect of a new pharmacological substance on a medical condition since the 1950s (Hill 1990). Placebo interventions may consist of pharmacologically inactive pills or other sham treatments. In RCTs, patients are randomized to either an active drug arm or a placebo arm, and patient outcomes in both study arms are contrasted. Thus, RCTs seek to disentangle the specific effect of the pharmacological substance under investigation from nonspecific effects of the treatment. Nonspecific effects manifest themselves as an improvement in the placebo arm. This improvement is partly due to phenomena such as symptom fluctuation or statistical artifacts. According to Enck and colleagues (2013), the term "placebo effect" will be used in this chapter to denote all symptom changes in the placebo group, irrespective of their origin. There are different mechanisms underlying this phenomenon, including spontaneous remission, regression to the mean, natural course of a disease, biases, and true placebo responses. The term placebo response, therefore, will be reserved for the neurobiological and psycho-physiological response of an individual to an inert substance or sham treatment that is mediated by factors of the treatment context.

The double-blind RCT design makes several basic assumptions (Enck et al. 2013). First, nonspecific effects should be identical in placebo and drug arms. True placebo responses due to expectancy and learning mechanisms should therefore be equally present in placebo and active drug arms. Secondly, the nonspecific effect in the placebo group should be independent of the drug. Thirdly, the improvement in the placebo group should be constant, i.e., it should not change over the course of the trial, or, at least, changes of nonspecific effects over the course of treatment should be parallel in drug and placebo groups. Lastly, the outcome in the active drug group is thought to indicate clinical relevance, i.e., to mirror the drug's effectiveness in clinical practice. Specific and nonspecific effects must be additive in order to identify the drug-specific effect by means of comparing the symptom change in the active drug group with the change in the placebo group.

This chapter discusses empirical evidence for placebo and nocebo phenomena that challenges these assumptions. It leads to the question of whether we are drawing the right conclusions from the placebo groups of clinical trials. It focuses on placebo arms in psychopharmacology trials, since notably strong placebo effects have been observed in clinical trials involving psychiatric disorders (Kirsch et al. 2008; Price et al. 2008; Rief et al. 2011b). The discussion is therefore of particular relevance to psychopharmacology trials.

2 The Placebo Effect in Psychopharmacology Trials

Psychopharmacological drug trials often report significant symptom improvement in their respective placebo arms. This is especially true for antidepressant pharmacological treatment. Based on the results of published antidepressant drug trials, 30 % of patients in the placebo group respond to treatment, compared to 50 % of patients in the active medication arm (Walsh et al. 2002). This may still underestimate the prevalence of the placebo effect, since serious concerns about a publication bias in the antidepressant trial literature have been raised. Among 74 antidepressant trials registered with the Food and Drug Administration (FDA), 31 % of the trials, accounting for 3,449 study participants, were not published (Turner et al. 2008). Publication was associated with study outcome, with lower probability of publication for studies that were viewed by the FDA as having unfavorable results for the investigational treatment.

Kirsch and colleagues analyzed both published and unpublished data from the Food and Drug Administration (FDA) for a subgroup of four new-generation antidepressant drugs (Kirsch et al. 2008). They reported a strong placebo effect that questioned the clinical effectiveness of antidepressant treatment. When changes in the Hamilton Depression Rating Scale (Hamilton 1960) were considered as primary outcome, patients in active drug groups demonstrated a weighted mean improvement of 9.6 points on the scale, while patients assigned to placebo groups reported 7.8 points improvement. The mean drug–placebo difference therefore amounts to only 1.8 points. This has led researchers to claim that up to 75 % of the positive effect of antidepressant medication is accounted for by placebo effects (Kirsch et al. 2008). Reanalyses of the FDA data (Fountoulakis et al. 2013; Horder et al. 2010) have questioned the statistical approach of Kirsch and colleagues and have argued in favor of drug-specific antidepressant effects. However, these new analyses have not reached substantially different conclusions and have inadvertently corroborated the substantial magnitude of nonspecific effects.

2.1 The Relevance of Spontaneous Remission

The improvement seen in placebo arms of clinical trials only partially represents a true placebo response. A portion of change over the study course is likely to be caused by symptom fluctuation, i.e., spontaneous improvement or worsening in a patient's disease. Epidemiologic surveys report high spontaneous remission rates for depression (Rhebergen et al. 2009, 2011; Wells et al. 1992). However, data from these naturalistic study designs cannot evaluate the effect of treatment on the observed course of depressive symptoms and the proportion of treated and untreated depressed study participants in the sample. In order to assess the true placebo response, a comparison of placebo groups with untreated control groups that demonstrate the natural course of the disease is needed. Unfortunately, data from no-treatment control groups from antidepressant treatment trials are scarce: in psychopharmacological trials, no-treatment control groups are not considered a valid and necessary control condition (Laughren 2001). Additionally, the inclusion of a no-treatment control group raises ethical concerns, since patients are left without treatment. The scarcity of natural course data from psychopharmacological trials is illustrated by a recent meta-analysis (Krogsboll et al. 2009). The meta-analysis attempted to quantify the spontaneous improvement in RCTs, based on a Cochrane review of the effect of placebo interventions across different medical conditions (Hrobjartsson and Gotzsche 2004). Only three-armed trials with no-treatment groups, placebo groups, and active treatment groups were included. Across all medical conditions, only 5 of 37 trials with this design employed a pharmacological treatment. For antidepressant treatment, only three 3-armed trials could be identified, and two of these trials used non-pharmacological interventions. Based on the paucity of data, it is very difficult to draw definite conclusions about the contribution of spontaneous remission to the observed symptom changes in placebo groups of antidepressant pharmacological trials.

Data concerning spontaneous improvement come primarily from trials of psychotherapeutic interventions for depression. Waitlist-controlled trials offer the treatment under investigation to patients of the control group only after a fixed waiting time, and observe the natural course of the disease during the wait. However, change in a waitlist group may be caused by various factors. While symptom fluctuation, spontaneous remission, and regression to the mean are obvious factors of influence, it is also necessary to consider other explanations (Arrindell 2001). Patients randomized to a waitlist may be disappointed about the wait, potentially resulting in an exacerbation of symptoms and an underestimation of spontaneous remission. On the other hand, diagnostic assessments during the wait may exert therapeutic benefit, and a guaranteed treatment option may induce hope and thus lead to additional improvement above natural course. It is also unclear whether patients who are enrolled in psychotherapy trials and patients enrolling in psychopharmacology trials are comparable regarding for example symptom severity or other disease-specific characteristics: if not, their spontaneous symptom change may not be identical. In spite of the limited explanatory power of waitlist control groups, they provide the best estimate of spontaneous remission effects. Therefore, focusing on psychotherapy trials, a recent meta-analytic review has attempted to investigate the contribution of spontaneous improvement to the symptom changes in placebo groups of antidepressant trials (Rutherford et al. 2012a). The authors report a medium effect size for the change in depression scores in the waitlist group. This translates to a mean improvement of four points on the Hamilton Rating Scale for Depression (Hamilton 1960) in waitlist control groups. In placebo groups of antidepressant drug trials, an average improvement of 8 points on the scale has been reported (Kirsch et al. 2008). Since these results are based on different data sets, they cannot be compared directly. However, the estimated improvement in waitlist control groups is unlikely to account for the full magnitude of the placebo effect seen in antidepressant trials. This highlights the relevance of true placebo responses. To summarize, preliminary evidence argues that placebo effects in antidepressant clinical trials are substantially more than only spontaneous remission. Due to the paucity of data, however, additional explanatory factors need to be taken into account when interpreting the symptom change in placebo arms.

2.2 The Increasing Power of Placebo

An increasing number of clinical trials in psychopharmacology fail to demonstrate the superiority of active medication over placebo. Substantial improvement in their respective placebo arms is considered an important explanatory factor for the high failure rate in clinical trials, including antidepressant medication in both adult populations (Khin et al. 2011) and pediatric populations (Bridge et al. 2009), and antipsychotic drugs (Kemp et al. 2010). A recent meta-analysis investigated the effectiveness of second-generation antipsychotic drugs in placebo-controlled RCTs (Leucht et al. 2009). Thirty-seven RCTs representing data from over seven thousand patients diagnosed with schizophrenia were included and analyzed concerning 13 different outcome measures. Forty-one percent of patients responded to the drug compared with 24 % of patients who responded to placebo. Effect sizes varied across the treatment outcome, but they were all of only moderate size (standardized mean difference −0.51 for “overall symptoms” as predefined primary outcome). Meta-regression showed a decline in drug–placebo differences over time and a funnel plot suggested the possibility of publication bias. This bias indicates a selective publication of trials that demonstrate a significant superiority of the active medication and possibly report larger drug–placebo differences. This could mean that the already substantial placebo effect observed may be only a conservative estimate.

A decline in drug–placebo differences has already been reported in other meta-analyses of antipsychotic trials (Chen et al. 2010; Kemp et al. 2010; Potkin et al. 2011), and it is mirrored in antidepressant trials (Khin et al. 2011; Rief et al. 2009b; Walsh et al. 2002). The so-called “publication year effect” describes that the reported magnitude of placebo effects over the years has grown steadily, while experimental design has fundamentally stayed the same. Various explanations have been proposed for this effect: changes in the populations included in the trials or decreasing quality in the implementation of recent trials could contribute to this finding. To cite one example, the number of trials conducted outside the United States of America has increased. Region of data acquisition (U.S. trials versus non-U.S. trials) has been implicated as a factor of influence for diminished drug–placebo differences in antipsychotic drug trials (Chen et al. 2010), but not in antidepressant trials (Khin et al. 2011). From a statistical point of view, larger effect sizes may originate from increased sample homogeneity in clinical trials. More homogeneous samples artificially inflate the effect size since effect sizes are calculated by dividing the mean difference by the pooled standard deviation. Indeed, evidence for increased sample homogeneity over time has been reported, for example, in a moderate association of the standard deviation of baseline depression scores with publication year (Mora et al. 2011). Nevertheless this finding can only partially explain the magnitude of the publication year effect.

Therefore, our meta-analysis investigated alternative explanations focusing on methods of assessing treatment outcome and their potential role in the publication year effect. Like Walsh and colleagues (2002) we found that effect sizes based on observer ratings in antidepressant trials correlate significantly and substantially with publication year (Rief et al. 2009b). If, however, effect sizes in placebo groups based on patient self-ratings were considered, these ratings demonstrated no significant association with publication year. Thus, while observer ratings demonstrate an increasing placebo effect, this trend is not apparent in the patients' self-ratings. To explain this surprising finding it is helpful to consider not only the role of patient expectation for placebo responses but also the role of clinician expectation about the trial. Trials of an investigational treatment that is likely to be perceived as ineffective by the study personnel report extremely low placebo effects (Shelton et al. 2001). In line with Fava and colleagues (2003), we would argue that clinician expectations about the effectiveness of antidepressant medication have probably increased over time, for example, through positive clinical experience with antidepressant pharmacotherapy. Clinician expectation may therefore be more positive than patient expectation and thus contribute to the increase of the placebo effect. However, this hypothesis awaits further investigation. Nevertheless, the diverging pattern of effect sizes in placebo groups based on observer ratings and patient self-ratings certainly questions the exclusive role of observer ratings as the gold standard of outcome assessment.

2.3 The Impact of Trial Design on Placebo Responses

While the impact of different assessment methods has already been discussed in the previous section, there are other additional characteristics of the trial design that influence the magnitude of the placebo response. Among these, characteristics that have been investigated in psychopharmacology trials (Alphs et al. 2012; Enck et al. 2013; Papakostas and Fava 2009) are:

- The duration of the clinical trial
- The number of active treatment groups or presence of a placebo group
- The number of study visits
- The use of placebo run-in phases
- Crossover design

- Flexible or fixed dosing regimes

The next section will discuss two trial characteristics that may result in changes in patient expectations based on examples from psychopharmacology trials: the number of treatment arms and effects of blinding/concealment.

An important factor of the trial design is the blinding. Double-blind design involves the blinding of study personnel and raters who evaluate the outcome. Additionally, it pertains to the blinding of patients since absent or deficient patient blinding may confound the trial outcome. While most psychopharmacological trials are designed as double-blind randomized controlled trials, a minority of trials are conducted with an open-label design, i.e., both study participants and study personnel are informed about the individual allocation to treatment arms. Additionally, some trials may be conceptualized as double blind but blinding may be broken inadvertently (cf. onset sensations). In a double-blind comparison of alprazolam, imipramine, and placebo for panic disorder, the majority of both patients and physicians were able to correctly guess the assignment to active treatment and placebo arm, respectively. Additionally, physicians were also able to accurately guess the type of active treatment that a patient had been assigned to (Margraf et al. 1991).

The influence of blinding has been demonstrated impressively in a meta-analysis of antipsychotic drugs versus placebo for relapse prevention in schizophrenia (Leucht et al. 2012). The analysis included randomized trials of patients with schizophrenia who were continued or withdrawn from antipsychotic medication after an initial stabilization period. Relapse between 7 and 12 months was defined as primary outcome and assessed by clinical judgment, e.g., need for medication or rating scales. As anticipated, all antipsychotic drugs were more successful at preventing relapse than placebo. Additionally, however, a significant difference emerged between blinded and unblinded studies. The proportion of patients in the drug groups of unblinded trials who relapsed was only 17 % compared to 28 % in blinded trials, while the proportion of patients who relapsed in the respective placebo groups was practically identical in blinded and unblinded trials (64 and 65 %, respectively). This translates to a significantly reduced risk ratio of relapse in the drug groups of unblinded trials (RR = 0.26) compared to blinded trials (RR = 0.42). Thus, antipsychotic drugs are apparently more effective in open-label trials. This finding is important, because open-label conditions mimic clinical practice more closely that double-blind trials.

This result also leads to the question of whether the increase in effectiveness with open-label use is caused by nonspecific treatment factors, i.e., a placebo mechanism such as expectancy. Patients who knew that they were certainly receiving the active medication may have developed more positive expectations that in turn may have resulted in a better treatment outcome. However, patient expectations are not routinely assessed in clinical trials; therefore, this explanation remains hypothetical. Another explanation could focus on clinician expectation. Since the study personnel also knew about individual allocation, this may have biased their rating of symptom severity and stability in the open-label trials and led to an overestimation of the effectiveness of the antipsychotic drugs. However, both open-label studies used the criterion "hospital admission" in addition to more subjective data like rating scales in order to define the occurrence of "relapse." Nevertheless, without further data, both explanations are possible. Again, they call our attention to the need for refined assessment methods on a multimodal level.

Another important characteristic of clinical trial designs is the number of active treatment arms and the definition of the control group. Adequate and well-controlled trials are needed to provide evidence for a drug's effectiveness. The use of both placebo control groups and active medication control groups is considered to meet this requirement. Comparative effectiveness research conducts trials that employ active medication control groups: the investigational product is tested against an established standard treatment, so that all patients receive active therapy. Active comparators can also be used in combination with an additional placebo control group to result in a three-armed clinical trial design (investigational treatment, active comparator, and placebo). Obviously, these designs vary with regard to the likelihood of receiving active medication or placebo.

A recent meta-analysis of atypical antipsychotic trials in schizophrenia examined whether the investigational active treatments performed equally well in active-controlled or low-dose controlled trials compared to placebo-controlled trials (Woods et al. 2005). Based on published and unpublished data, it demonstrated that the effectiveness of investigational treatments depended on trial design: all investigational treatments were associated with greater symptom improvement in active-controlled designs. The same drugs and doses were almost twice as effective when employed in an active-controlled design compared to placebo-controlled studies. Similar results have been reported for antidepressant trials. The response rate to antidepressants is higher in trials that do not include a placebo arm (65.4 %) than in placebo-controlled trials (57.7 %) (Sinyor et al. 2010). However, active-controlled trials and placebo-controlled trials may vary with regard to additional characteristics, e.g., different completion rates and study sample selection. These differences could partially account for the design-specific placebo effect. In antidepressant trials, however, different dropout rates in active-controlled designs and placebo-controlled design do not seem to add to this effect (Rutherford et al. 2012b). A convincing explanation for this differential improvement is an expectancy effect: patients who are enrolled in an active-controlled trial know that they will definitively receive active medication after the informed consent procedure. This knowledge engenders positive treatment expectations. These expectations in turn act as nonspecific treatment factors (i.e., a placebo mechanism) that contribute to the symptom improvement observed in both active treatment groups. In addition to patient expectations, expectations of study personnel will probably also differ in active-controlled and placebo-controlled trials for the same reason. The clinician expectations may also impact ratings of improvement in the placebo groups. Since the definition of treatment response in the meta-analysis was based on observer ratings (Sinyor et al. 2010), concurrent influences of patient and clinician expectations cannot be quantified.

In either case, the improvement in active drug arms varies as a function of control group. This finding is complemented by varying response rates in placebo groups in antidepressant trials with one or more active medication arms (Sinyor et al. 2010). Trials that include only the investigational treatment as active medication and a placebo treatment as control group yield lower response rates in placebo groups (34 %) than trials that include at least a second active treatment arm (46 %). Thus, depressed patients respond better to placebo in trials that offer a higher likelihood of receiving active medication than in trials that offer only a 50 % chance of active treatment. In consequence, the trial design can lead to increased placebo responses (i.e., nonspecific treatment factors) that may not only impact the improvement in the placebo group but also in the active medication group. In a meta-regression of antidepressant trials a greater probability of receiving placebo predicted a better efficacy separation of drug and placebo (Papakostas and Fava 2009). This association remained significant independent of a simultaneous consideration of publication year and baseline depression severity as additional predictors.

The meta-analytical evidence that the number of treatment arms can impact the placebo response is corroborated by preliminary evidence from a pilot study. In this trial, assignment to placebo-controlled or active-controlled trial, respectively, directly influenced treatment expectation (Rutherford et al. 2013). Depressed patients were randomly allocated to either a placebo-controlled trial or a comparative effectiveness trial of two active antidepressant treatments. Expectancy of improvement was assessed once before randomization and at beginning of the trial. Group assignment led to the hypothesized changes in expectancy: patients in the active-controlled trial reported significantly greater expectancy of improvement than patients in the placebo-controlled trial. Importantly, baseline depression, which may be a source of more negative expectations, was not associated with this expectancy score. Additionally, higher expectancy scores were associated longitudinally with lower depression scores at the end of the study and a greater improvement in depressive symptoms over time. The mean difference between active medication groups in the placebo-controlled and active controlled trials, however, was not statistically significant. While these results should certainly be interpreted with caution, due to the limited sample size and minor methodological concerns, they illustrate the importance of accounting for patient expectation when assessing clinical trial outcome. The study also offers preliminary evidence that trial design may exert its influence on trial outcome in placebo and active medication groups through changes in patient expectation.

2.4 Open-Label Placebo Application

A special case of placebo use in clinical trials is open-label placebo application. Open label in this context means that patients are correctly informed that the pill they are receiving contains no pharmacologically active ingredient. However, positive treatment expectancies are formed through additional information, e.g., referring to large effects that placebo pills have demonstrated in other clinical trials. This is a novel approach to the research of placebo effects since deception has long been regarded a prerequisite for placebo responses by both healthcare professionals and laypeople. Early proof-of-principle experiments employed methodologically weak research designs (Aulas and Rosner 2003; Park and Covi 1965) and are therefore of limited internal validity. Recently, open placebo application has also been investigated in pilot RCTs. A groundbreaking study in the treatment of Irritable Bowel Syndrome (Kaptchuk et al. 2010) contrasted open-label application of a placebo pill with a natural history control group. The open-label placebo condition introduced the pill truthfully as pharmacologically inactive but also as known to result in significant improvement in Irritable Bowel Syndrome through mind–body self-healing processes. Results demonstrated clinically meaningful improvements: participants of the open-label placebo application reported significantly greater global improvement, reduced symptom severity, and increased relief.

Based on the substantial improvements in the placebo arms of psychopharmacological trials that have been reported in previous sections of this chapter, the identical rationale has been applied to the treatment of Major Depressive Disorder. Kelley and colleagues (2012) conducted a pilot waitlist-controlled RCT in 20 patients. Placebo pills were correctly introduced as pharmacologically inactive but also with regard to their substantial positive effects in clinical trials of depression and with additional explanations for their use. Patients were assessed at baseline before treatment and again after 2 weeks with the Hamilton Rating Scale for Depression (Hamilton 1960). The experimental group and the control group demonstrated no significant differences. However, preliminary data show an interesting trend: the improvement in Hamilton Rating scores was of medium effect size in the open-label placebo group ($d = 0.53$). Notably, this trend emerged in spite of a minimal sample size ($n = 11$) and a very limited observation period. A replication investigating a larger sample over a longer period of time is desirable, before any conclusions about the efficiency of open-label placebo application in the treatment of depression can be drawn.

A different approach to an open-label placebo application has been investigated in the treatment of Attention Deficit Hyperactivity Disorder (Sandler and Bodfish 2005; Sandler et al. 2010). The design combined pharmacologically active drugs and open-label placebo application in a classical conditioning paradigm using two control conditions. In the experimental group, mixed amphetamine salts were paired in the acquisition period with a visually distinct placebo capsule that was truthfully specified as a placebo. Additionally, the placebo pill was also referred to as a "dose extender" that could generate positive effects on ADHD based on mind–body interactions and the placebo mechanisms of learning and expectancy. After 1 month of acquisition, the dose of amphetamines was reduced for 1 month to 50 % of the original amount and again paired with the placebo. Outcomes were contrasted with two control groups. The first control group received their original dose continuously. The second control group received only 50 % of the original dose (similar to the experimental group) but without the placebo application. Compared to the simple dose reduction group, the open-label placebo group demonstrated better outcomes such as lower side effect rates and maintained ADHD symptom control.

Evidence for the effectiveness of open-label placebo application is still sparse. The few studies suffer from weaknesses such as small sample sizes or the inherently impossible double-blind masking in open-label applications. Special attention must also be paid to the role of the patient–provider relationship and the instructions about the placebo pill in the respective contexts (Kaptchuk et al. 2010). Nevertheless, this innovative approach has yielded first encouraging results. Open-label placebo application may be of special interest to medical conditions that demonstrate substantial placebo effects in clinical trials and that involve a treatment that is associated with severe side effects.

3 Side Effects in Psychopharmacology Trials

Adverse events that occur in the placebo group of a clinical trial have been termed nocebo effects (Barsky et al. 2002). Like placebo responses, nocebo responses are induced by patient's response expectations about the treatment outcome and the medication under investigation. The nocebo phenomenon is of great relevance to clinical practice (Doering and Rief 2013) but also to clinical trials. The next section discusses evidence that nocebo effects may lead to an increase in the symptom burden and may distress the patient. Nocebo-induced side effects significantly influence a patient's decision to adhere to a prescribed treatment and may ultimately lead to the decision to discontinue participation in a clinical trial. The second section elaborates how sensations or minor symptoms that patients associate with study medication intake may inadvertently contribute to placebo responses.

3.1 The Nocebo Effect

Nocebo research requires systematic assessment of adverse events in clinical trials, preferably both on objective and on subjective level (Rief et al. 2011a). Unfortunately, this issue is not routinely addressed in psychopharmacological trials: in clinical studies of antipsychotic medication, only a minority of studies investigated subjectively experienced side effects and standardized, systematic assessment methods were rarely used (Pope et al. 2010). Our knowledge about nocebo responses in clinical trials is therefore limited and has to be interpreted in the context of differing and mostly unsystematic assessment methods.

A convincing example for the relevance of the nocebo phenomenon comes from a review of statin drug trials (Rief et al. 2006): in these trials, a comparable number of patients from both active treatment and placebo groups discontinued trial participation, with dropout rates varying from 10 to 28 %. Of note, a considerable number of patients from the placebo group discontinued treatment specifically because of side effects that they had experienced (4–26 %). A meta-analysis of antidepressant drug trials (including only tricyclic antidepressants and selective serotonin reuptake inhibitors) reports comparable results: discontinuation rates were nearly identical for placebo groups and corresponding drug groups, 24.7 and 24.8 %, respectively (Rief et al. 2009a). Similar results have been reported for clinical trials in the pharmacological treatment of fibromyalgia, investigating drugs including the antidepressants duloxetine and milnacipran and the anticonvulsant gabapentine (Mitsikostas et al. 2012). Thus, adverse events in placebo groups of psychopharmacology trials are relatively frequent, can even lead to trial discontinuation, and must certainly be taken into account when interpreting clinical trial data.

Recently, research has focused on the comparison of the side effect profile that is reported in the placebo group with the side effect profile that is reported in the respective active drug group. Adverse events are assumed to originate from the pharmacological profile of the drug. In the case of antidepressant drug trials for example, tricyclic antidepressants (TCA) would be expected to produce more adverse events than serotonin reuptake inhibitors (SSRI) due to their differential pharmacological mode of action. Interestingly, the placebo groups mirror this expectation: placebo groups from TCA trials report significantly more side effects than placebo groups from SSRI trials (Rief et al. 2009a). In a similar vein, adverse events in placebo groups of clinical trials of drug treatment for fibromyalgia mirrored quantitatively and qualitatively the side effects of the respective active drug arm (Mitsikostas et al. 2012). A meta-analysis of clinical trials of various anti-migraine medications (nonsteroidal anti-inflammatory drugs, triptans, anticonvulsants) reports the same drug-specific nocebo effects in the placebo arm (Amanzio et al. 2009): only placebo groups of anticonvulsant trials report anticonvulsant-specific side effects, e.g., memory difficulties and anorexia, while patients in placebo groups of nonsteroidal anti-inflammatory drug trials report more gastrointestinal symptoms. This is an important finding that again demonstrates how symptom changes of patients in placebo groups mirror those of patients in the respective active drug arm. Both the improvement of symptoms and the development of side effects in placebo groups can only be understood within the context of the individual study. This illustrates that a pooling of placebo groups derived from different clinical trials may lead to false conclusions.

3.2 Onset Sensations

Minor bodily symptoms associated with medication intake are not necessarily only considered in the context of adverse events, but also conceptualized as "onset sensations." In clinical trials these onset sensations may occasionally be experienced in placebo groups as nocebo phenomena, but they occur primarily in the drug group. Onset sensations have been discussed as a confounding influence that can unblind trial participants and raters to the randomization, and thus endanger the internal validity of clinical trials (Fava et al. 2003; Margraf et al. 1991; Rief et al. 2011b). Therefore, active placebos have been proposed as an alternative; these placebos induce minor side effects that mimic those of the active drug.

However, the placebo contains no active ingredient with specific therapeutic benefit to the medical condition under investigation. In antidepressant research atropine has been used as an active placebo in several clinical trials of TCA. A review of these studies (Moncrieff et al. 2004) concludes that the drug–placebo difference for trials using active placebos is reduced below any clinical relevance: the pooled effect size for antidepressants over placebo was 0.17, and the 95 % confidence interval ranged from 0.00 to 0.37. The review can be criticized since it only included a limited number of relatively old studies, focusing only on TCAs. However, the findings suggest that drug–placebo differences become less evident when active placebos are used as a control condition. This could be caused by a rather unlikely decrease in drug effectiveness, though it seems more likely that active placebos may be more powerful than "inert" placebos.

This hypothesis was tested empirically in the domain of placebo analgesia in healthy volunteers (Rief and Glombiewski 2012). In an experimental study inert placebos were compared with active placebos in combination with different instructions about group allocation (probability of receiving drug: 0, 50, 100 %). Participants were informed that they either had a 50 % chance of receiving the active drug (to mirror a clinical trial) or that they had a 100 % chance of receiving active medication (to mirror clinical practice). In reality, all volunteers received only placebo. Pain thresholds were assessed before and after placebo treatment. In inert placebo conditions, the well-known expectancy effect of placebo analgesia was replicated: participants who believed they had received an active drug reported the highest pain thresholds. Pain thresholds in the active placebo group differed substantially from the inert placebo group in the 50 % chance condition. Compared to participants who noted no bodily symptoms after "medication" intake, participants with minor onset sensations from active placebo intake demonstrated a greater placebo effect. It can be hypothesized that these onset sensations convinced participants that they were receiving the active medication. Increased placebo analgesia was then triggered by this expectancy effect. Since the 50 % condition most closely resembles clinical trial design, the results argue that minor onset conditions serve to strengthen nonspecific effects in clinical studies. The placebo effect observed in experimentally induced pain in healthy participants is not necessarily identical to placebo effects observed in patients who suffer from a chronic disease. In combination with data from clinical trials using active placebos (Moncrieff et al. 2004), however, these results question the relevance of drug–placebo differences stemming from inert placebos.

4 Implications for Drug Trials: Possible Interaction Effects

The evidence presented in this chapter argues strongly for the consideration of interactions between drug-specific and nonspecific effects in clinical trials, as illustrated in Fig. 1. Before a trial starts, patients will form outcome expectancies, based for example on their individual chance of receiving the active treatment in the respective trial design. Moreover, patients will probably hold expectations about

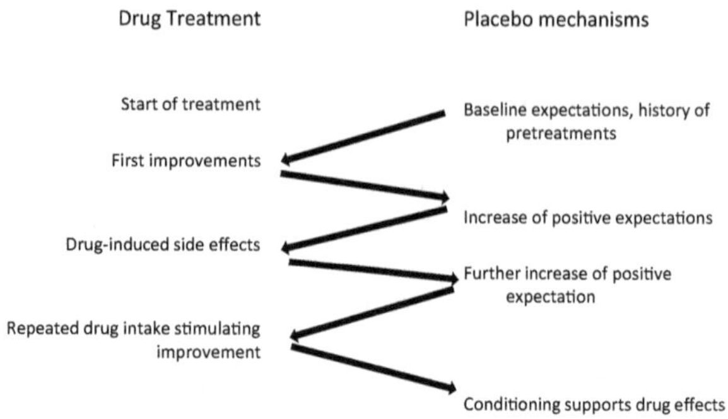

Fig. 1 Complex interaction of placebo mechanisms with specific treatment effects. Therefore, in this example, nonspecific effects in placebo and drug groups can differ

the respective treatment in general or have previous experience with the treatment in the case of chronic medical conditions. This may also influence their response to placebo and medication, and possibly to a varying degree: in depression, previous treatment experience has been reported to have a negative impact on symptom change in placebo groups, but not in active treatment groups (Hunter et al. 2010). Furthermore, the magnitude of nonspecific effects varies not only with patient expectation but also with clinician expectation, as the publication year effect suggests.

During a clinical trial, onset sensations may unblind patients to their treatment allocation and trigger expectancy effects that in turn lead to more positive treatment outcomes. However, these nonspecific effects are more likely to occur in the active treatment arm, since most clinical trials employ only inert placebos. Additionally, associative learning processes (i.e., conditioning) that link the ritual of medication intake with the experience of symptom alleviation may occur and support the drug effect.

These considerations challenge the basic assumptions of the additive model in RCTs. If nonspecific effects interact with specific effects and are strengthened by onset sensations, then nonspecific effects are not identical in placebo and drug groups. If the nonspecific effects in the placebo group vary with regard to the treatment under investigation, as demonstrated by the drug specificity of nocebo effects, then nonspecific effects can no longer be considered independent of the drug. If drug-specific effects and nonspecific effects interact and reinforce each other, true placebo responses (as a portion of the improvement observed in the drug group) will not remain constant but change over the course of the trial. Thus, an interactive model of RCTs is proposed (Enck et al. 2013) that accounts for these interaction effects in the drug group of clinical trials. This new model should guide our interpretation of clinical trial results.

Conclusion: Lessons to Be Learned

The accumulated evidence demonstrates that placebo effects are substantial, even when accounting for methodological bias and spontaneous remission. Large placebo effects challenge the development of new drugs due to diminished drug–placebo differences. Various explanations have been proposed for this phenomenon, both pointing to methodological biases and increasing expectancy effects. Placebo effects and, in consequence, drug–placebo differences in clinical trials must be interpreted within the context of the RCT design. For example, placebo-controlled clinical trials with a second active comparator (three-armed RCTs) may yield different drug–placebo differences for a given drug than a two-armed, placebo-controlled trial of the same drug. The large placebo response in psychopharmacological trials needs to be investigated in more detail and with more suitable assessment methods. Patient and clinician expectation should be considered, and side effects assessed more carefully, in order to advance our understanding of placebo and nocebo responses in clinical trials. In the context of clinical research, alternative trial designs that are better suited to evaluate the true efficacy of the investigational drug should be employed.

Nevertheless, the large placebo response in psychopharmacology should also be a warning: for at least 25 % of depressed patients receiving antidepressants, placebos may be better options (Gueorguieva et al. 2011). For some patients, no treatment could be the recommendation of choice (i.e., spontaneous remission in natural course), especially when considering potential unwanted consequences of antidepressant treatment. At present, physicians have no empirically founded guidelines to help them to determine which depressed patient should receive no treatment, placebo treatment, or active drug treatment. Thus, refined treatment guidelines for the use of psychopharmacological medication are clearly needed, both to reduce overtreatment and to prevent under-treatment. Special attention must be paid to ethical concerns in informed consent procedures, both when using placebo interventions in clinical trials and when using verum treatments with a considerable placebo or nocebo component in clinical practice (Blease 2010; Miller and Colloca 2009; Wells and Kaptchuk 2012). The findings should also be a motivation to harness nonspecific effects and maximize them in clinical practice.

References

Alphs L, Benedetti F, Fleischhacker WW, Kane JM (2012) Placebo-related effects in clinical trials in schizophrenia: what is driving this phenomenon and what can be done to minimize it? Int J Neuropsychopharmacol 15:1003–1014

Amanzio M, Corazzini LL, Vase L, Benedetti F (2009) A systematic review of adverse events in placebo groups of anti-migraine clinical trials. Pain 146:261–269

Arrindell WA (2001) Changes in waiting-list patients over time: data on some commonly-used measures. Beware! Behav Res Ther 39:1227–1247

Aulas JJ, Rosner I (2003) Efficacy of a non blind placebo prescription. Encéphale 29:68–71

Barsky AJ, Saintfort R, Rogers MP, Borus JF (2002) Nonspecific medication side effects and the nocebo phenomenon. JAMA 287:622–627

Blease C (2010) Deception as treatment: the case of depression. J Med Ethics 37:13

Bridge JA, Birmaher B, Iyengar S, Barbe RP, Brent DA (2009) Placebo response in randomized controlled trials of antidepressants for pediatric major depressive disorder. Am J Psychiatry 166:42–49

Chen Y-F, Wang S-J, Khin NA, Hung HMJ, Laughren TP (2010) Trial design issues and treatment effect modeling in multi-regional schizophrenia trials. Pharm Stat 9:217–229

Doering BK, Rief W (2013) Nocebos in daily clinical practice: the potential side effects of the treatment context and the patient–doctor interaction on pain in clinical populations. In: Colloca L, Flaten MA, Meissner K (eds) Placebo and pain. Academic, San Diego, pp 257–266

Enck P, Bingel U, Schedlowski M, Rief W (2013) The placebo response in medicine: minimize, maximize or personalize? Nat Rev Drug Discov 12:191–204

Fava M, Evins AE, Dorer DJ, Schoenfeld DA (2003) The problem of the placebo response in clinical trials for psychiatric disorders: culprits, possible remedies, and a novel study design approach. Psychother Psychosom 72:115–127

Fountoulakis K, Veroniki A, Siamouli M, Moller H-J (2013) No role for initial severity on the efficacy of antidepressants: results of a multi-meta-analysis. Ann Gen Psychiatry 12:26

Gueorguieva R, Mallinckrodt C, Krystal JH (2011) Trajectories of depression severity in clinical trials of duloxetine: insights into antidepressant and placebo responses. Arch Gen Psychiatry 68:1227–1237

Hamilton M (1960) A rating scale for depression. J Neurol Neurosurg Psychiatry 23:56–62

Hill AB (1990) Suspended judgment. Memories of the British streptomycin trial in tuberculosis. The first randomized clinical trial. Control Clin Trials 11:77–79

Horder J, Matthews P, Waldmann R (2010) Placebo, Prozac and PLoS: significant lessons for psychopharmacology. J Psychopharmacol 25:1277–1288

Hrobjartsson A, Gotzsche PC (2004) Placebo interventions for all clinical conditions. Cochrane Database Syst Rev:Cd003974

Hunter AM, Cook IA, Leuchter AF (2010) Impact of antidepressant treatment history on clinical outcomes in placebo and medication treatment of major depression. J Clin Psychopharmacol 30:748–751

Kaptchuk TJ, Friedlander E, Kelley JM, Sanchez MN, Kokkotou E, Singer JP, Kowalczykowski M, Miller FG, Kirsch I, Lembo AJ (2010) Placebos without deception: a randomized controlled trial in irritable bowel syndrome. PLoS ONE 5:e15591

Kelley JM, Kaptchuk TJ, Cusin C, Lipkin S, Fava M (2012) Open-label placebo for major depressive disorder: a pilot randomized controlled trial. Psychother Psychosom 81:312–314

Kemp AS, Schooler NR, Kalali AH, Alphs L, Anand R, Awad G, Davidson M, Dube S, Ereshefsky L, Gharabawi G, Leon AC, Lepine JP, Potkin SG, Vermeulen A (2010) What is causing the reduced drug-placebo difference in recent schizophrenia clinical trials and what can be done about it? Schizophr Bull 36:504–509

Khin NA, Chen YF, Yang Y, Yang P, Laughren TP (2011) Exploratory analyses of efficacy data from major depressive disorder trials submitted to the US Food and Drug Administration in support of new drug applications. J Clin Psychiatry 72:464–472

Kirsch I, Deacon BJ, Huedo-Medina TB, Scoboria A, Moore TJ, Johnson BT (2008) Initial severity and antidepressant benefits: a meta-analysis of data submitted to the Food and Drug Administration. PLoS Med 5:e45

Krogsboll LT, Hrobjartsson A, Gotzsche PC (2009) Spontaneous improvement in randomised clinical trials: meta-analysis of three-armed trials comparing no treatment, placebo and active intervention. BMC Med Res Methodol 9:1

Laughren TP (2001) The scientific and ethical basis for placebo-controlled trials in depression and schizophrenia: an FDA perspective. Eur Psychiatry 16:418–423

Leucht S, Arbter D, Engel RR, Kissling W, Davis JM (2009) How effective are second-generation antipsychotic drugs? A meta-analysis of placebo-controlled trials. Mol Psychiatry 14:429–447

Leucht S, Tardy M, Komossa K, Heres S, Kissling W, Salanti G, Davis JM (2012) Antipsychotic drugs versus placebo for relapse prevention in schizophrenia: a systematic review and meta-analysis. Lancet 379:2063–2071

Margraf J, Ehlers A, Roth WT, Clark DB, Sheikh J, Agras WS, Taylor CB (1991) How "blind" are double-blind studies? J Consult Clin Psychol 59:184–187

Miller FG, Colloca L (2009) The legitimacy of placebo treatments in clinical practice: evidence and ethics. Am J Bioeth 9:39–47

Mitsikostas DD, Chalarakis NG, Mantonakis LI, Delicha E-M, Sfikakis PP (2012) Nocebo in fibromyalgia: meta-analysis of placebo-controlled clinical trials and implications for clinical practice. Eur J Neurol 19:672–680

Moncrieff J, Wessely S, Hardy R (2004) Active placebos versus antidepressants for depression. Cochrane Database Syst Rev:CD003012

Mora MS, Nestoriuc Y, Rief W (2011) Lessons learned from placebo groups in antidepressant trials. Philos Trans R Soc Lond B Biol Sci 366:1879–1888

Papakostas GI, Fava M (2009) Does the probability of receiving placebo influence clinical trial outcome? A meta-regression of double-blind, randomized clinical trials in MDD. Eur Neuropsychopharmacol 19:34–40

Park LC, Covi L (1965) Nonblind placebo trial: an exploration of neurotic patients' responses to placebo when its inert content is disclosed. Arch Gen Psychiatry 12:36–45

Pope A, Adams C, Paton C, Weaver T, Barnes TR (2010) Assessment of adverse effects in clinical studies of antipsychotic medication: survey of methods used. Br J Psychiatry 197:67–72

Potkin S, Agid O, Siu C, Watsky E, Vanderburg D, Remington G (2011) Placebo response trajectories in short-term and long-term antipsychotic trials in schizophrenia. Schizophr Res 132:108–113

Price DD, Finniss DG, Benedetti F (2008) A comprehensive review of the placebo effect: recent advances and current thought. Annu Rev Psychol 59:565–590

Rhebergen D, Beekman ATF, Rd G, Nolen WA, Spijker J, Hoogendijk WJ, Penninx BWJH (2009) The three-year naturalistic course of major depressive disorder, dysthymic disorder and double depression. J Affect Disord 115:450–459

Rhebergen D, Batelaan NM, de Graaf R, Nolen WA, Spijker J, Beekman ATF, Penninx BWJH (2011) The 7-year course of depression and anxiety in the general population. Acta Psychiat Scand 123:297–306

Rief W, Glombiewski JA (2012) The hidden effects of blinded, placebo-controlled randomized trials: an experimental investigation. Pain 153:2473–2477

Rief W, Avorn J, Barsky AJ (2006) Medication-attributed adverse effects in placebo groups. Arch Intern Med 166:155–160

Rief W, Nestoriuc Y, von Lilienfeld-Toal A, Hofmann SG, Barsky AJ, Avorn J (2009a) Differences in adverse effect reporting in placebo groups in SSRI and tricyclic antidepressant trials: a systematic review and meta analysis. Drug Saf 32:1041–1056

Rief W, Nestoriuc Y, Weiss S, Welzel E, Barsky AJ, Hofmann SG (2009b) Meta-analysis of the placebo response in antidepressant trials. J Affect Disord 118:1–8

Rief W, Barsky AJ, Glombiewski JA, Nestoriuc Y, Glaesmer H, Braehler E (2011a) Assessing general side effects in clinical trials: reference data from the general population. Pharmacoepidem Drug Saf 20:405–415

Rief W, Bingel U, Schedlowski M, Enck P (2011b) Mechanisms involved in placebo and noceco responses and implications for drug trials. Clin Pharmacol Ther 90:722–726

Rutherford BR, Mori S, Sneed JR, Pimontel MA, Roose SP (2012a) Contribution of spontaneous improvement to placebo response in depression: a meta-analytic review. J Psychiatr Res 46:697–702

Rutherford BR, Sneed JR, Roose SP (2012b) Does differential drop-out explain the influence of study design on antidepressant response? A meta-analysis. J Affect Disord 140:57–65

Rutherford BR, Marcus SM, Wang P, Sneed JR, Pelton G, Devanand D, Duan N, Roose SP (2013) A randomized, prospective pilot study of patient expectancy and antidepressant outcome. Psychol Med 43:975–982

Sandler AD, Bodfish JW (2005) Open-label use of placebos in the treatment of ADHD: a pilot study. Child Care Hlth Dev 34:104–110

Sandler AD, Glesne CE, Bodfish JW (2010) Conditioned placebo dose reduction: a new treatment in attention-deficit hyperactivity disorder? J Dev Behav Pediatr 31:369–375

Shelton RC, Keller MB, Gelenberg A, Dunner DL, Hirschfeld R, Thase ME, Russell J, Lydiard RB, Crits-Christoph P, Gallop R, Todd L, Hellerstein D, Goodnick P, Keitner G, Stahl SM, Halbreich U (2001) Effectiveness of St. John's worth in major depression: a randomized controlled trial. JAMA 285:1978–1986

Sinyor M, Levitt AJ, Cheung AH, Schaffer A, Kiss A, Dowlati Y, Lanctot KL (2010) Does inclusion of a placebo arm influence response to active antidepressant treatment in randomized controlled trials? Results from pooled and meta-analyses. J Clin Psychiatry 71:270–279

Turner EH, Matthews AM, Linardatos E, Tell RA, Rosenthal R (2008) Selective publication of antidepressant trials and its influence on apparent efficacy. New Engl J Med 358:252–260

Walsh BT, Seidman SN, Sysko R, Gould M (2002) Placebo response in studies of major depression: variable, substantial, and growing. JAMA 287:1840–1847

Wells R, Kaptchuk TJ (2012) To tell the truth, the whole truth, may do patients harm: the problem of the nocebo effect for informed consent. Am J Bioeth 12:22–29

Wells KB, Burnam MA, Rogers W, Hays R, Camp P (1992) The course of depression in adult outpatients. Results from the medical outcomes study. Arch Gen Psychiatry 49:788–794

Woods SW, Gueorguieva RV, Baker CB, Makuch RW (2005) Control group bias in randomized atypical antipsychotic medication trials for schizophrenia. Arch Gen Psychiatry 62:961–970

The Emperor's New Drugs: Medication and Placebo in the Treatment of Depression

Irving Kirsch

Contents

1 The Emperor's New Drugs: Medication and Placebo in the Treatment of Depression .. 292
2 Severity of Depression and Antidepressant Effectiveness 295
3 Predicting Response to Treatment 296
4 How Did These Drugs Get Approved? 296
5 The Serotonin Myth 298
6 Antidepressants as Active Placebos 299
7 What to Do? 300
References 301

Abstract

Antidepressants are supposed to work by fixing a chemical imbalance, specifically, a lack of serotonin in the brain. Indeed their supposed effectiveness is the primary evidence for the chemical imbalance theory. But analyses of the published data and the unpublished data that were hidden by the drug companies reveal that most (if not all) of the benefits are due to the placebo effect. Some antidepressants increase serotonin levels, some decrease it, and some have no effect at all on serotonin. Nevertheless, they all show the same therapeutic benefit. Even the small statistical difference between antidepressants and placebos may be an enhanced placebo effect, due to the fact that most patients and doctors in clinical trials successfully break blind. The serotonin theory is as close to any theory in the history of science having been proved wrong. Instead of curing depression, popular antidepressants may induce a biological vulnerability making people more likely to become depressed in the future.

I. Kirsch (✉)
Harvard Medical School, Beth Israel Deaconess Medical Center, Boston, MA, USA
e-mail: irvkirsch@gmail.com

F. Benedetti et al. (eds.), *Placebo*, Handbook of Experimental Pharmacology 225,
DOI 10.1007/978-3-662-44519-8_16

Keywords
Antidepressants • Placebos • Expectancy • Depression

1 The Emperor's New Drugs: Medication and Placebo in the Treatment of Depression

On February 26, 2008, an article about antidepressants that my colleagues and I wrote was published in the journal PLoS Medicine (Kirsch et al. 2008). That morning, I awoke to find that our paper was the front-page story in all of the leading national newspapers in the United Kingdom. A few months later, Random House invited me to expand the article into a book, entitled *The Emperor's New Drugs: Exploding the Antidepressant Myth*, which has since been translated into French, Italian, Japanese, Polish, and Turkish (Kirsch 2009). Two years later, the book, and the research reported in it, was the topic of a five-page cover story in the influential American news magazine, *Newsweek*. And 2 years after that, it was the focus of a 15 min segment on *60 Minutes*, America's top-rated television news program. Somehow, I had been transformed, from a mild-mannered university professor into a media superhero—or super villain, depending on whom you asked. What had my colleagues and I done to warrant this transformation?

To answer that question, we have to go back to 1998, when a former graduate student, Guy Sapirstein, and I published a meta-analysis on antidepressants in an online journal of the American Psychological Association (Kirsch and Sapirstein 1998). When they were new, meta-analyses were somewhat controversial and our article was accompanied by an editorial warning to that effect—not unlike the suicide warning that the U.S. Food and Drug Administration (FDA) requires for antidepressants. But now meta-analyses are published in all of the major medical journals, where they are widely considered to be the best and most reliable way of making sense of the data from studies with different and sometimes conflicting results.

When Sapirstein and I began our analysis of the antidepressant clinical trial data, we were not particularly interested in antidepressants. Instead, we were interested in understanding the placebo effect. I have been fascinated by the placebo effect for my entire academic career. How is it, I wondered, that the belief that one has taken a medication can produce some of the effects of that medication?

It seemed to Sapirstein and me that depression was a good place to look for placebo effects. After all, one of the prime characteristics of depression is the sense of hopelessness that depressed people feel. If you ask depressed people to tell you what the worst thing in their life is, many will tell you that it is their depression. The British psychologist John Teasdale called this being depressed about depression (Teasdale 1985). If that is the case, then the mere promise of an effective treatment should help to alleviate depression, by replacing hopelessness with hopefulness—

the hope that one will recover after all. It was with this in mind that we set out to measure the placebo effect in depression.

We searched the literature for studies in which depressed patients had been randomized to receive an inert placebo or no treatment at all. The studies we found also included data on the response to antidepressants, because that was the only place one finds data on the response to placebo among depressed patients. I was not particularly interested in the drug effect. I assumed that antidepressants were effective. As a psychotherapist, I sometimes referred my severely depressed clients for prescriptions of antidepressant drugs. Sometimes the condition of my clients improved when they began taking antidepressants; sometimes it did not. When it did, I assumed it was the effect of the drug that was making them better. Given my long-standing interest in the placebo effect, I should have known better, but back then I did not.

Analyzing the data we had found, we were not surprised to find a substantial placebo effect on depression. What surprised us was how small the drug effect was. Seventy-five percent of the improvement in the drug group also occurred when people were given dummy pills with no active ingredient in them. Needless to say, our meta-analysis proved to be very controversial. Its publication led to heated exchanges (e.g., Klein 1998; Beutler 1998; Kirsch 1998). The response from critics was that these data could not be accurate. Perhaps our search had led us to analyze an unrepresentative subset of clinical trials. Antidepressants had been evaluated in many trials, the critics said, and their effectiveness had been well established.

In an effort to respond to these critics, we decided to replicate our study with a different set of clinical trials (Kirsch et al. 2002). To do this, we used the Freedom of Information Act to request that the Food and Drug Administration (FDA) send us the data that pharmaceutical companies had sent to it in the process of obtaining approval for six new-generation antidepressants that accounted for the bulk of antidepressant prescriptions being written at the time. There are a number of advantages to the FDA dataset. Most important, the FDA requires that the pharmaceutical companies provide information on all of the clinical trials that they have sponsored. Thus, we had data on unpublished trials as well as published trials. This turned out to be very important. Almost half of the clinical trials sponsored by the drug companies have not been published (Turner et al. 2008; Melander et al. 2003). The results of the unpublished trials were known only to the drug companies and the FDA, and most of them failed to find a significant benefit of drug over placebo. A second advantage of the FDA trials in the FDA dataset is that they all used the same primary measure of depression—the Hamilton depression scale (HAM-D). That made it easy to understand the clinical significance of the drug–placebo differences. Finally, the data in the FDA files were the basis upon which the medications were approved. In that sense they have a privileged status. If there is anything wrong with those trials, the medications should not have been approved in the first place.

In the data sent to us by the FDA, only 43 % of the trials showed a statistically significant benefit of drug over placebo. The remaining 57 % were failed or negative trials. Similar results have been reported in other meta-analyses (Turner et al. 2008), including one conducted by the FDA on the clinical trials of all antidepressants that it had approved between 1983 and 2008 (Khin et al. 2011). The results of our analysis indicated that the placebo response was 82 % of the response to these antidepressants. Subsequently, my colleagues and I replicated our meta-analysis on a larger number of trials that had been submitted to the FDA (Kirsch et al. 2008). With this expanded dataset, we found once again found that 82 % of the drug response was duplicated by placebo, with an effect size (*d*) of 0.32. More important, in both analyses, the mean difference between drug and placebo was less than two points on the HAM-D. The HAM-D is a 17-item scale on which people can score from 0 to 53 points, depending on how depressed they are. A 6-point difference can be obtained just by changes in sleep patterns, with no change in any other symptom of depression. So the 1.8 difference that we found between drug and placebo was very small indeed—small enough to be clinically insignificant. But you don't have to take my word for how small this difference is. The National Institute for Health and Clinical Excellence (NICE), which drafts treatment guidelines for the National Health Service in the United Kingdom, has established a drug–placebo effect size (*d*) of 0.50 or a 3-point difference between drug and placebo on the HAM-D as criteria of clinical significance (NICE 2004). Thus, when published and unpublished data are combined, they fail to show a clinically significant advantage for antidepressant medication over inert placebo.

I should mention here the difference between *statistical* significance and *clinical* significance. Statistical significance concerns how reliable an effect is. Is it a real effect, or is it just due to chance? Statistical significance does not tell you anything about the size of the effect. Clinical significance, on the other hand, deals with the size of an effect and whether it would make any difference in a person's life. Imagine, for example, that a study of 500,000 people has shown that smiling increases life expectancy—by 5 min. With 500,000 subjects, I can virtually guarantee you that this difference will be statistically significant, but it is clinically meaningless.

The results of our analyses have since been replicated repeatedly (Turner et al. 2008; NICE 2004; Fournier et al. 2010; Fountoulakis and Möller 2011). Some of the replications used our data; others analyzed different sets of clinical trials. The FDA even did its own meta-analysis on all of the antidepressants that they have approved (Khin et al. 2011). But and despite differences in the way the data have been spun, the numbers are remarkably consistent (Table 1). Differences on the HAM-D are small—always below the criterion set by NICE. Thomas P. Laughren, the director of the FDA's psychiatry products division, acknowledged this on the American television news program *60 Minutes*. He said, "I think we all agree that the changes that you see in the short term trials, the difference in improvement between drug and placebo, is rather small."

And it is not only the short-term trials that show a small, clinically insignificant difference between drug and placebo. In their meta-analysis of published clinical trials, NICE (2004) found that the differences between drug and placebo in the long-term trials were no larger than those in short-term trials.

Table 1 Drug–placebo effect sizes and HAM-D difference scores in meta-analyses of antidepressant trials

Meta-analysis	Effect size	HAM-D difference	Data source
Kirsch and Sapirstein (1998)	0.39	NA	Published trials
Kirsch et al. (2002, 2008)	0.32	1.80	FDA files
NICE (2004)[a]	0.34	NA	Published trials
NICE (2004)[b]	0.28	NA	Published trials
Turner et al. (2008)[c]	0.31	NA	FDA files
Fournier et al. (2010)[d]	0.30	1.94	Published trials
Fountoulakis and Möller (2011)	0.32	2.68	FDA files
Khin et al. (2011)	NA	2.50	FDA files

[a]All trials
[b]Long trials (≥8 weeks) only
[c]Fournier et al. (2010) was a pooled analysis of patient level data
[d]Fountoulakis and Möller (2011) weighted studies by 1/var, instead of the conventional n/var, thus failing to adjust for differences in sample size (see Huedo-Medina et al. 2012)

2 Severity of Depression and Antidepressant Effectiveness

Critics of our 2002 meta-analysis argued that our results were based on clinical trials conducted on subjects who were not very depressed (e.g., Thase 2002; Hollon et al. 2002). In more depressed patients, they argued, a more substantial difference might be found. This criticism led my colleagues and I to reanalyze the FDA data in 2008 (Kirsch et al. 2008). We categorized the clinical trials in the FDA database according to the severity of the patients' depression at the beginning of the trial, using conventionally used categories of depression. As it turns out, all but one of the trials were conducted on moderately depressed patients, and that trial failed to show any significant difference between drug and placebo. Indeed, the difference was virtually nil (0.07 points on the HAM-D). All of the rest of the trials were conducted on patients whose mean baseline scores put them in the "very severe" category of depression, and even among these patients, the drug–placebo difference was below the level of clinical significance.

Still, severity did make a difference. Patients at the very extreme end of depression severity, those scoring at least 28 on the HAM-D, showed an average drug–placebo difference of 4.36 points. To find out how many patients fell within this extremely depressed group, I asked Mark Zimmerman from the Brown University School of Medicine to send me the raw data from a study in which he and his colleagues assessed HAM-D scores of patients who had been diagnosed with unipolar major depressive disorder (MDD) after presenting for an intake at a psychiatric outpatient practice (Zimmerman et al. 2005). Patients with HAM-D scores of 28 or above represented 11 % of these patients. This suggests that 89 % of depressed patients are not receiving a clinically significant benefit from the antidepressants that are prescribed for them.

Yet this 11 % figure may overestimate the number of people who benefit from antidepressants. Antidepressants are also prescribed to people who do not qualify for the diagnosis of major depression. My neighbor's pet dog died; his physician prescribed an antidepressant. A friend in the United States was diagnosed with lumbar muscle spasms and was prescribed an antidepressant. I have lost count of the number of people who have told me they were prescribed antidepressants when complaining of insomnia—even though insomnia is a frequently reported side effect of antidepressants. About 20 % of patients suffering from insomnia in the United States are given antidepressants as a treatment by their primary care physicians (Simon and VonKorff 1997), despite the fact that "the popularity of antidepressants in the treatment of insomnia is not supported by a large amount of convincing data, but rather by opinions and beliefs of the prescribing physicians" (Wiegand 2008, p. 2411).

3 Predicting Response to Treatment

Severity of depression is one of the few predictors of response to treatment. Type of antidepressant has little if any impact on treatment response. As summarized in a 2011 meta-analysis of studies comparing one antidepressant to another:

> On the basis of 234 studies, no clinically relevant differences in efficacy or effectiveness were detected for the treatment of acute, continuation, and maintenance phases of MDD. No differences in efficacy were seen in patients with accompanying symptoms or in subgroups based on age, sex, ethnicity, or comorbid conditions... Current evidence does not warrant recommending a particular second-generation antidepressant on the basis of differences in efficacy. (Gartlehner et al. 2011, p. 772)

Although type of medication does not make a clinically significant difference in outcome, response to placebo does. Almost all antidepressant trials include a placebo run-in phase. Before the trial begins, all of the patients are given a placebo for a week or two. After this run-in period, the patients are reassessed, and anyone who has improved substantially is excluded from the trial. That leaves patients who have not benefitted at all from placebo and those who have benefited only a little bit. These are the patients who are randomized to be given drug or kept on placebo. As it turns out, the patients who show at least a little improvement during the run-in period are the ones most likely to respond to the real drug, as shown not only by physician ratings, but also by changes in brain function (Hunter et al. 2006; Quitkin et al. 1998).

4 How Did These Drugs Get Approved?

How is it that medications with such weak efficacy data were approved by the FDA? The answer lies in an understanding of the approval criteria used by the FDA. The FDA requires two adequately conducted clinical trials showing a significant difference between drug and placebo. But there is a loophole: There is no limit to the number of trials that can be conducted in search of these two significant trials. Trials showing negative results simply do not count. Furthermore, the clinical significance of the findings is not considered. All that matters is that the results are statistically significant.

The most egregious example of the implementation of this criterion is provided by the FDA's approval of vilazodone in 2011 (http://www.accessdata.fda.gov/drugsatfda_docs/nda/2011/022567Orig1s000StatR.pdf). Seven controlled efficacy trials were conducted. The first five failed to show any significant differences on any measure of depression, and the mean drug–placebo difference in these studies was less than ½ point on the HAM-D, and in two of the three trials, the direction of the difference actually favored the placebo. The company ran two more studies and managed to obtain small but significant drug–placebo differences (1.70 points). The mean drug–placebo difference across the seven studies was 1.01 HAM-D points. This was sufficient for the FDA to grant approval, and the information approved by the FDA for informing doctors and patients reads, "The efficacy of VIIBRYD was established in two 8-week, randomized, double-blind, placebo-controlled trials." No mention is made of the five failed trials that preceded the two successful ones.

The failure to mention the unsuccessful trials was not merely an oversight; it reflects a carefully decided FDA policy dating back for decades. To my knowledge, there is only one antidepressant in which the FDA included information on the existence of negative trials. The exception is citalopram, and the inclusion of the information followed an objection raised by Paul Leber, who was at the time the director of the FDA Division of Neuropharmacological Drug Products. In an internal memo dated May 4, 1998, Leber wrote:

> One aspect of the labelling deserves special mention. The [report] not only describes the clinical trials providing evidence of citalopram's antidepressant effects, but make mention of adequate and well controlled clinical studies that failed to do so. . .The Office Director is inclined toward the view that the provision of such information is of no practical value to either the patient or prescriber. I disagree. I believe it is useful for the prescriber, patient, and 3rd-party payer to know, without having to gain access to official FDA review documents, that citalopram's antidepressants effects were not detected in every controlled clinical trial intended to demonstrate those effects. I am aware that clinical studies often fail to document the efficacy of effective drugs, but I doubt that the public, or even the majority of the medical community, is aware of this fact. I am persuaded that they not only have a right to know but that they should know. Moreover, I believe that labeling that selectively describes positive studies and excludes mention of negative ones can be viewed as potentially 'false and misleading'. (Leber 1998).

Hooray for Paul Leber. I have never met or corresponded with this gentleman, but because of this courageous memo, he is one of my heroes.

5 The Serotonin Myth

Over the years, I have noticed something very strange in the antidepressant literature. When different antidepressants are compared with each other, their effects are remarkably similar. I first noticed this when Guy Sapirstein did our 1998 meta-analysis of the published literature. When we first saw how small the actual drug effect was, we thought we might have done something wrong. Perhaps we had erred by including trials that had evaluated different types of antidepressants. Maybe we were underestimating the true effectiveness of antidepressants by including clinical trials of drugs that were less effective than others.

Before submitting our paper for publication, we went back to the data and examined the type of antidepressant used each trial. Some were selective serotonin reuptake inhibitors (SSRIs), others were tricyclic medications, and there were trials on antidepressant drugs that were neither SSRIs nor tricyclics. And then we noticed that there was a fourth category of drugs in the trials we had analyzed. These were trials in which drugs that are not thought to be antidepressants at all—tranquilizers and thyroid medications, for example—were given to depressed patients and evaluated for their effect on depression.

When we analyzed the drug and placebo response for each type of drug, we found another surprise awaiting us. It did not matter what kind of drug the patients had been given in the trial. The response to the drug was always the same, and 75 % of that response was also found in the placebo groups. I recall being impressed by how unusual the similarity in results was, but I have since learned that they are not unusual at all. I have since encountered this phenomenon over and over again. In the STAR*D trial, which, at a cost of $35,000,000, is the most costly clinical trial of antidepressants ever conducted, patients who did not respond to the prescribed SSRI were switched to a different antidepressant (Rush et al. 2006). Some were switched to an SNRI, a drug that is supposed to increase norepinephrine as well as serotonin in the brain. Others were switched to an NDRI, which is supposed to increase norepinephrine and dopamine, without affecting serotonin at all. And still others were simply given a different SSRI. About one out of four patients responded clinically to the new drug, but it did not matter which new drug they were given. The effects ranged from 26 to 28 %; in other words, they were exactly the same regardless of type of drug.

The most commonly prescribed antidepressants are SSRIs, drugs that are supposed to selectively target the neurotransmitter serotonin. But there is another antidepressant that has a very different mode of action. It is called tianeptine, and it has been approved for prescription as an antidepressant by the French drug regulatory agency. Tianeptine is an SSRE, a selective serotonin reuptake enhancer. Instead of increasing the amount of serotonin in the brain, it is supposed to decrease it. If the theory that depression is caused by a deficiency of serotonin were correct, we would expect to make depression worse. But it doesn't. In clinical trials comparing the effects of tianeptine to those of SSRIs and tricyclic antidepressants, 63 % of patients show significant improvement (defined as a 50 % reduction in symptoms), the same response rate that is found for SSRIs, NDRIs, and tricyclics, in this type of trial (Wagstaff et al. 2001). It simply does not matter what is in the medication—it might increase serotonin, decrease it, or have no effect on serotonin at all. The effect on depression is the same.

What do you call pills, the effects of which are independent of their chemical composition? I call them "placebos."

6 Antidepressants as Active Placebos

All antidepressants seem to be equally effective, and although the difference between drug and placebo is not clinically significant, it is significant statistically. This leads to the obvious question: What do all of these active drugs have in common that make their effect on depression slightly, but statistically significantly, better than placebo?

One thing that antidepressants have in common is that they all produce side effects. Why is that important? Imagine that you are a subject in a clinical trial. You are told that the trial is double blind and that you might be given a placebo. You are told what the side effects of the medication are. The therapeutic effects of the drug may take weeks to notice, but the side effects might occur more quickly. Would you not wonder to which group you had been assigned, drug or placebo? And noticing one of the listed side effects, would you not conclude that you had been given the real drug? In one study, 89 % of the patients in the drug group correctly "guessed" that they had been given the real antidepressant, a result that is very unlikely to be due to chance (Rabkin et al. 1986). In a more recent study (Chen et al. 2011), actual treatment assignment (sertraline, hypericum, or placebo) did not affect treatment outcome, but patients' guesses about which treatment they were getting did.

In other words, clinical trials are not really double blind. Many patients in clinical trials realize that they have been given the real drug, rather than the placebo, most likely because of the drug's side effects. What effect is this likely to have in a clinical trial? We do not have to guess at the answer to this question. Bret Rutherford and his colleagues at Columbia University have provided the answer. They examined the response to antidepressants in studies that did not have a placebo group with those in studies where they did have a placebo group (Rutherford et al. 2009). The main difference between these studies is that in the first case, the patients were certain they were getting an active antidepressant, whereas in the placebo-controlled trials, they knew that they might be given a placebo. Knowing for sure that they were getting an active drug boosted the effectiveness of the drug significantly. This supports the hypothesis that the relatively small differences between drug and placebo in antidepressant trials are at least in part due to "breaking blind" and discerning that one is in the drug group, because of the side effects produced by the drug.

7 What to Do?

To summarize, there is a strong therapeutic response to antidepressant medication. But the response to placebo is almost as strong. In the FDA files my colleagues and I analyzed (Kirsch et al. 2008), the response to antidepressants was a mean improvement of 10.1 points on the HAM-D, whereas the response to placebo was an improvement of 8.3 points. Furthermore, meta-analyses of published trials reveal that the response to placebos is mostly a true placebo effect; it is not due to spontaneous remission, the natural history of depression, or regression toward the mean (Khan et al. 2012; Kirsch and Sapirstein 1998). This presents a therapeutic dilemma. The drug effect of antidepressants is not clinically significant, but the placebo effect is. What should be done clinically in light of these findings?

One possibility would be to use antidepressants as active placebos. But the risks involved in antidepressant use render this alternative problematic (Andrews et al. 2012; Domar et al. 2013; Serretti and Chiesa 2009). Among the side effects of antidepressants are sexual dysfunction (which affects 70–80 % of patients on SSRIs), long-term weight gain, insomnia, nausea, and diarrhea. Approximately 20 % of people attempting to quit taking antidepressants show withdrawal symptoms. Antidepressants have been linked to increases in suicidal ideation among children and young adults. Older adults have increased risks of stroke and death from all causes. Pregnant women using antidepressants are at increased risk of miscarriage, and if they don't miscarry, their offspring are more likely to be born with autism, birth malformations, persistent pulmonary hypertension, and newborn behavioral syndrome. Furthermore, some of these risks have been linked to antidepressant use during the first trimester of pregnancy, when women may not be aware that they are pregnant. Perhaps the most surprising health consequence of antidepressant use is one that affects people of all ages. Antidepressants increase the risk of relapse after one has recovered. People are more likely become depressed again after treatment by antidepressants than after treatment by other means—including placebo treatment (Andrews et al. 2012; Babyak et al. 2000; Dobson et al. 2008). Furthermore, the degree to which the risk of relapse increases depends on the degree to which the particular antidepressant used changes neurotransmission in the brain. Given these health risks, antidepressants should not be used as a first-line treatment for depression.

Another possibility is to prescribe placebos. They are almost as effective as antidepressants, but elicit far fewer side effects. Surveys indicated that many physicians do in fact prescribe placebos (Tilburt et al. 2008; Raz et al. 2011). The conventional wisdom is that for a placebo to be effective, patients must believe they are receiving active medication, which entails deception. Besides being ethically questionable, the practice of deceiving patients runs the risk of undermining trust, which may be one of the most important clinical tools that clinicians have at their disposal. But is the conventional wisdom correct? My colleagues and I have tested and confirmed the hypothesis that placebos can be effective even when given openly, without deception, when given in the context of a warm therapeutic relationship and with an honest but convincing rationale as to why they should be effective (Kaptchuk et al. 2010). Our study targeted irritable bowel syndrome, rather than depression, but a small pilot study suggests that it might work also in the treatment of depression (Kelley et al. 2012). Until this is confirmed, however, placebo treatment is not a viable option.

Fortunately, placebos are not the only alternative to antidepressant treatment. My colleagues and I have conducted a meta-analysis of various treatments for depression, including antidepressants, psychotherapy, the combination of psychotherapy and antidepressants, and "alternative" treatments, which included acupuncture and physical exercise (Khan et al. 2012). We found no significant differences between these treatments or within different types of psychotherapy. When different treatments are equally effective, choice should be based on risk and harm, and of all of these treatments, antidepressant drugs are the riskiest and most harmful. If they are to be used at all, it should be as a last resort, when depression is extremely severe and all other treatment alternatives have been tried and failed.

References

Andrews PW, Thomson JA, Amstadter A, Neale MC (2012) Primum non nocere: an evolutionary analysis of whether antidepressants do more harm than good. Front Psychol 3:117. doi:10.3389/fpsyg.2012.00117

Babyak MA, Blumenthal JA, Herman S, Khatri P, Doraiswamy PM, Moore KA et al (2000) Exercise treatment for major depression: maintenance of therapeutic benefit at 10 months. Psychosom Med 62:633–638

Beutler LE (1998) Prozac and placebo: there's a pony in there somewhere. Prevention Treatment 1, 0003c

Chen JA, Papakostas GI, Youn S, Baer L, Clain AJ, Fava M, Mischoulon D (2011) Association between patient beliefs regarding assigned treatment and clinical response: reanalysis of data from the Hypericum Depression Trial Study Group. J Clin Psychiatry 72(12):1669–1676

Dobson KS, Hollon SD, Dimidjian S, Schmaling KB, Kohlenberg RJ, Gallop RJ et al (2008) Randomized trial of behavioral activation, cognitive therapy, and antidepressant medication in the prevention of relapse and recurrence in major depression. J Consult Clin Psychol 76(3):468–477

Domar AD, Moragianni VA, Ryley DA, Urato AC (2013) The risks of selective serotonin reuptake inhibitor use in infertile women: a review of the impact on fertility, pregnancy, neonatal health and beyond. Hum Reprod 28(1):160–171. doi:10.1093/humrep/des383

Fountoulakis KN, Möller HJ (2011) Efficacy of antidepressants: a re-analysis and re-interpretation of the Kirsch data. Int J Neuropsychopharmacol 14(3):405

Fournier JC, DeRubeis RJ, Hollon SD, Dimidjian S, Amsterdam JD, Shelton RC et al (2010) Antidepressant drug effects and depression severity: a patient-level meta-analysis. J Am Med Assoc 303(1):47–53

Gartlehner G, Hansen RA, Morgan LC, Thaler K, Lux L, Van Noord M et al (2011) Comparative benefits and harms of second-generation antidepressants for treating major depressive disorder. Ann Intern Med 155(11):772–785. doi:10.1059/0003-4819-155-11-201112060-00009

Hollon SD, DeRubeis RJ, Shelton RC, Weiss B (2002) The emperor's new drugs: effect size and moderation effects. Prevention Treatment 5:Article 27. Available on the World Wide Web: http://www.journals.apa.org/prevention/volume5/pre0050027c.html

Huedo-Medina TB, Johnson BT, Kirsch I (2012) Kirsch et al.'s (2008) calculations are correct: reconsidering Fountoulakis & Möller's re-analysis of the Kirsch data. Int J Neuropsychopharmacol 1(1):1–6

Hunter AM, Leuchter AF, Morgan ML, Cook IA (2006) Changes in brain function (quantitative EEG cordance) during placebo lead-in and treatment outcomes in clinical trials for major depression. Am J Psychiatr 163(8):1426–1432

Kaptchuk TJ, Friedlander E, Kelley JM, Sanchez MN, Kokkotou E, Singer JP et al (2010) Placebos without deception: a randomized controlled trial in irritable bowel syndrome. PLoS ONE 5 (12):e15591. doi:10.1371/journal.pone.0015591

Kelley JM, Kaptchuk TJ, Cusin C, Lipkin S, Fava M (2012) Open-label placebo for major depressive disorder: a pilot randomized controlled trial. Psychother Psychosom 81(5):312–314

Khan A, Faucett J, Lichtenberg P, Kirsch I, Brown WA (2012) A systematic review of comparative efficacy of treatments and controls for depression. PLoS ONE 7(7):e41778. doi:10.1371/journal.pone.0041778

Khin NA, Chen YF, Yang Y, Yang P, Laughren TP (2011) Exploratory analyses of efficacy data from major depressive disorder trials submitted to the US Food and Drug Administration in support of new drug applications. J Clin Psychiatry 72(4):464

Kirsch I (1998) Reducing noise and hearing placebo more clearly. Prevention Treatment 1(2):7r. doi:10.1037/1522-3736.1.1.17r

Kirsch I (2009) The emperor's new drugs: exploding the antidepressant myth. The Bodley Head, London

Kirsch I, Sapirstein G (1998) Listening to prozac but hearing placebo: a meta-analysis of antidepressant medication. Prevention Treatment 1, 0002a. doi:10.1037/1522-3736.1.1.12a

Kirsch I, Moore TJ, Scoboria A, Nicholls SS (2002) The emperor's new drugs: an analysis of antidepressant medication data submitted to the U.S. Food and Drug Administration. Prevention Treatment 5, 23

Kirsch I, Deacon BJ, Huedo-Medina TB, Scoboria A, Moore TJ, Johnson BT (2008) Initial severity and antidepressant benefits: a meta-analysis of data submitted to the Food and Drug Administration. PLoS Med 5(2):e45. doi:10.1371/journal.pmed.0050045

Klein DF (1998) Listening to meta-analysis but hearing bias. Prevention Treatment 1, 0006c

Leber, P (1998) NDA 20-822 Celexa (citalopram HBr) for the management of depression. Memorandum to the Department of Health and Human Services, Public Health Service, Food and Drug Administration, Center for Drug Evaluation and Research. Approvable action on Forrest Laboratories, Inc, Washington, DC

Melander H, Ahlqvist-Rastad J, Meijer G, Beermann B (2003) Evidence b(i)ased medicine–selective reporting from studies sponsored by pharmaceutical industry: review of studies in new drug applications. Br Med J 326:1171–1173

NICE (2004) Depression: management of depression in primary and secondary care. Clinical practice guideline No 23. www.nice.org.uk/page.aspx?o=235213. Accessed 24 May 2005

Quitkin FM, McGrath PJ, Stewart JW, Ocepek-Welikson K, Taylor BP, Nunes E et al (1998) Placebo run-in period in studies of depressive disorders: clinical, heuristic and research implications. Br J Psychiatry 173:242–248

Rabkin JG, Markowitz JS, Stewart JW, McGrath PJ, Harrison W, Quitkin FM et al (1986) How blind is blind? Assessment of patient and doctor medication guesses in a placebo-controlled trial of imipramine and phenelzine. Psychiatry Res 19:75–86

Raz A, Campbell N, Guindi D, Holcroft C, Déry C, Cukier O (2011) Placebos in clinical practice: a pan-Canadian review of attitudes and patterns of use between academic psychiatrists and non-psychiatrists. Can J Psychiatry 56(4):198–208

Rush AJ, Trivedi MH, Wisniewski SR, Nierenberg AA, Stewart JW, Warden D et al (2006) Acute and longer-term outcomes in depressed outpatients requiring one or several treatment steps: a STAR*D report. Am J Psychiatr 163:1905–1917

Rutherford BR, Sneed JR, Roose SP (2009) Does study design influence outcome? Psychother Psychosom 78(3):172–181

Serretti A, Chiesa A (2009) Treatment-emergent sexual dysfunction related to antidepressants: a meta-analysis. J Clin Psychopharmacol 29(3):259–266

Simon GE, VonKorff M (1997) Prevalence, burden, and treatment of insomnia in primary care. Am J Psychiatr 154(10):1417–1423

Teasdale JD (1985) Psychological treatments for depression: how do they work? Behav Res Ther 23:157–165

Thase ME (2002) Antidepressant effects: the suit may be small, but the fabric is real. Prevention Treatment 5(32)

Tilburt JC, Emanuel EJ, Kaptchuk TJ, Curlin FA, Miller FG (2008) Prescribing "placebo treatments": results of national survey of US internists and rheumatologists. Br Med J 337:1097–1100. doi:10.1136/bmj.a1938

Turner EH, Matthews AM, Linardatos E, Tell RA, Rosenthal R (2008) Selective publication of antidepressant trials and its influence on apparent efficacy. N Engl J Med 358:252–260

Wagstaff AJ, Ormrod D, Spencer CM (2001) Tianeptine: a review of its use in depressive disorders. CNS Drugs 15:231–259

Wiegand MH (2008) Antidepressants for the treatment of insomnia: a suitable approach? Drugs 68 (17):2411–2417

Zimmerman M, Chelminski I, Posternak MA (2005) Generalizability of antidepressant efficacy trials: differences between depressed psychiatric outpatients who would or would not qualify for an efficacy trial. Am J Psychiatr 162(7):1370–1372. doi:10.1176/appi.ajp.162.7.1370

Index

A

Activation Likelihood Estimation (ALE), 41, 62
"Additive model", 10–11, 242, 245, 286
Altered sensory processing, 76–77
Analgesic placebo effects
- attention, pain reducing effects, 229
- components, 229–230
- current expectation, 224
- description, 218–219
- detoxification, 228
- drug, 225, 226
- external and internal situation, 224
- medication, intake, 223–224
- model, 222, 223
- negative expectancies, 226
- pain reduction, 229
- pain relief, 228–229
- pharmacological component, 225, 229–230
- positive effect, 226
- postoperative pain management, 227
- power of instructions, 225

Attention deficit hyperactivity disorder, 23, 282
Autonomic nervous system (ANS)
- bowel motility, 190
- organs and brain, 186
- premotor neurons, 186
- smooth muscles regulation, 193

B

Balanced crossover design (BCD)
- drug and placebo, 244
- expectancy component, 245
- limitations, 244

Balanced placebo design (BPD)
- Participants' acceptance, 244
- variant, 243–244

BCD. *See* Balanced crossover design (BCD)
Behavioral conditioning model
- afferent pathways, 165
- cyclosporine, 164
- efferent neural pathway, 165
- insular cortex, 165
- neuroanatomical pathway, 165
- rats, 164

Biomarker, fMRI, 6, 8–10
Blood pressure (BP)
- and heart rate, 189
- hypertensive and normotensive, 187
- placebo effects, 188
- systolic, 187
- "verum", 188

BPD. *See* Balanced placebo design (BPD)
Brain imaging studies
- fMRI and PET, 6
- placebo and nocebo effect, 7
- subjective pain intensity ratings, 6

C

CAN. *See* Central autonomic network (CAN)
CCRs. *See* Conditioned compensatory responses (CCRs)
Central autonomic network (CAN), 186–187
CER. *See* Comparative effectiveness research (CER)
Classical conditioning
- autonomic placebo responses, 185
- conditioned taste aversion (CTA), 162
- fear conditioning, 109
- mechanisms, 231
- and negative expectations, 192
- and observational learning, 30–31
- posttreatment nausea, 192
- shaping expectancy, 219–220
- steps, 163

F. Benedetti et al. (eds.), *Placebo*, Handbook of Experimental Pharmacology 225,
DOI 10.1007/978-3-662-44519-8

Clinical trial design
  blinding, 279
  expectancy, 279–280
  meta-analytical evidence, 281
  open-label placebo application, 281–282
  placebo control groups, 280
  psychopharmacology trials, 278–279
"Cohort multiple randomized controlled trial" (CMRCT), 257–258
Comparative effectiveness research (CER), 257–258, 266–267
Complete spontaneous bowel movement (CSBM), 255
Conditioned compensatory responses (CCRs), 21
Conditioning. *See also* Classical conditioning
  and cognitive expectancy
    adaptive process, 4
    automatic instinctual process, 4
    healthy and patient populations, 5
    placebo and nocebo effects, 4–5
  non-pharmacological, 24–26
  pharmacological (*see* Pharmacological conditioning and placebo responses)
Conditioning-like expectancy manipulation model, 10
Consciousness and cognitive expectancy, 4–6
Context learning
  nocebo effect (*see* Nocebo effect)
  placebo effect (*see* Placebo effect)
Coronary arteries
  angiograms, 189
  mechanisms, 188
  vasoconstriction, 189
CSBM. *See* Complete spontaneous bowel movement (CSBM)

D
Delayed response paradigm (DRP), 242
Depression
  and antidepressant effectiveness, 295
  drug-placebo difference, 297
  drug-placebo effect sizes, 295
  drug's side effects, 299
  FDA, 293
  HAM-D, 293–294
  insomnia, 296
  MDD, 295
  medication and placebo, 292
  NICE, 294
  placebo effect, 293
  placebo run-in phase, 296
  pregnant women, antidepressants, 300
  psychotherapist, 293
  regression, 300
  SSIs, 298
  therapeutic effects, 298
  treatment response, 296
Diffuse noxious inhibitory control (DNIC), 76
Dopamine reward system, 7, 8
DRP. *See* Delayed response paradigm (DRP)

E
Emotion. *See also* Expectations
  fatigue, 153–155
  and homeostasis, 91–92
  individual differences, 91
  modulation, 8
  pain-related negative (*see* Pain-related negative emotions)
  trait and state emotions, 110–111
Expectancy-based pain modulation
  expectancy-induced increases, 49, 51
  expectancy-induced reductions, 49, 50
  meta-analyses advantages, 40–42
Expectancy effect
  and conditioning, 5
  conditioning-like, 10
  conscious cognitive, 4–6
  and lidocaine, 11
  positive and negative, 9, 10
  and remifentanil, 10–11
  reward system, 7
Expectations
  dimensions, 82
  and emotional feelings
    IBS patients, 125
    selfreinforcing placebo analgesia effect, 126
    verbal suggestions, pain relief, 125, 126
  motor performance, 150
  Parkinson's disease
    clinical motor performance, 144
    [^{11}C] raclopride PET, 144
    definition, 143
  positive and negative, 150–152
  response, 82

F
Fatigue
  central governor, 155
  The Hill's original peripheral model, 153–154

increased thalamo-insular activation, 154
physical, 154
FCP. *See* Free-choice paradigm (FCP)
FDA. *See* Food and Drug Administration (FDA)
FGDs. *See* Functional gastrointestinal disorders (FGDs)
fMRI. *See* Functional magnetic resonance imaging (fMRI)
Food and Drug Administration (FDA), 275
Free-choice paradigm (FCP)
"adaptive response design", 252
"early-escape design", 252
"selection behavior", patients, 251
Functional gastrointestinal disorders (FGDs)
etiological concepts, 99
IBS (*see* Irritable bowel syndrome (IBS))
Functional magnetic resonance imaging (fMRI)
cervical spinal cord, 77
emotion-related brain regions, 10
IBS patients, 127
meta-analysis, 41
pain
actual experience, 127
anticipation, 126
and PET, 39
placebo analgesia, 39
task-based studies, 59

G
Gastrointestinal motility, 190–191

H
Heart rate variability, 88, 171, 172, 188–190, 194, 196
The Hill's original peripheral model, 153, 154
Homeostasis, 91–92, 154, 186
Hospital Anxiety and Depression Scale, 111

I
IBS. *See* Irritable bowel syndrome (IBS)
Idiopathic pain
expectations and emotional feelings, 125–126
IBS patients, 123–124
neurobiological underpinnings, 126–127
rectal lidocaine and rectal placebo, 124
Immune responses
behavioral conditioning model (*see* Behavioral conditioning model)
behavior and neuroendocrine, 161
chemotherapy agents, 168
classical conditioning protocol, 163
conditioned stimulus (CS), 163
conditioned taste aversion (CTA) paradigm, 163–164
CsA-saccharin paradigm, 169
in humans, 166–168
humoral and neural pathway, 162–163
lipopolysaccharide (LPS), 169–170
MS, 166
placebo, 161
SLE, 166
T-cell function, 169
T-cell mitogens, 168
Insular cortex, 165
International Affective Picture System (IAPS), 88, 123
Irritable bowel syndrome (IBS)
analgesia responses, 104
fMRI study, 103
gastrointestinal conditions, 108
painful visceral stimuli, 99
placebo analgesia effects, 123
placebo response, 100
in RCTs, 259
symptom, 113
treatment approaches, 102
and visceral pain, 98–99
in VLPFC, 104
women, 98
Itch and pain
central and peripheral sensitization mechanisms, 206
contagion, 207–208
neurophysiological processing, 206
nocebo effects, 208–209
placebo effects, 207, 208

L
Learning/conditioning
classical conditioning, 110
fear conditioning, 109
IBS patients, 110
pharmacological preconditioning, 152
visual cues, 109
Lung function and respiratory activity, 193

M
Major depressive disorder (MDD), 295
Meta-analysis
advantages
ALE and MKDA, 41
different pain modalities, 40

Meta-analysis (*cont.*)
neuroimaging placebo paradigm, 40
qualitative and quantitative meta-analysis, 41–42
expectancy-based pain modulation, 49, 61
extending, 62–64
MKDA approach, 48
modulatory mechanisms, 60
placebo effect, brain regions, 58–59
prior meta-analyses, placebo analgesia, 61–62
reductions during pain, amygdala, 60
stimulus expectancies *vs.* treatment expectancies (*see* Stimulus expectancies *vs.* treatment expectancies)
study selection and coordinate identification
details, 48
expectancy-based pain modulation., 47
expectancy-related increases, 47
expectancy-related reductions, 47
experiments, 43–46
MKDA. *See* Multilevel kernel density analysis (MKDA)
Motor performance. *See* Physical performance
MS. *See* Multiple sclerosis (MS)
Multilevel kernel density analysis (MKDA), 41, 48
Multiple sclerosis (MS), 166, 169

N

National Institute for Health and Clinical Excellence (NICE), 294
Neuroendocrine functions
description, 169
glucose, 172
and immune functions, 172–173
insulin injections, 172
pharmacological treatments, 173
placebo effects, in humans, 170–171
psychological/biological variables, 173
in schizophrenic patients, 170
Neuropathic pain
hyperalgesia area, 129
open *vs.* hidden lidocaine administration, 128
placebo effect, 129
psychological factors, 129–130
NICE. *See* National Institute for Health and Clinical Excellence (NICE)
Nocebo effect
brain imaging studies, 6–7
duloxetine and milnacipran, 284
experimental studies, 106–108
and gender, 30
hyperalgesia, 5, 9, 93
hyperalgesic effects, 28
mass psychogenic illness, 30
negative emotions, 110
observational learning, 30
physical performance (*see* Physical performance)
and placebo effect, 9
statin drug trials, 283
TCA, 284
Nocebo responses
in animals and humans, 18
behavioral and clinical placebo, 26
on blood pressure, 188
in clinical trials, 283
learning mechanisms, 31
non-painful and painful stimuli, 109
placebo, 25
in women, 192
Non-pharmacological conditioning and placebo responses
CS-US pairings, 25
intervention, fMRI study, 26
painful stimuli, 24
placebo and nocebo responses, 25
simulated effective intervention, 25–26

O

Observational learning
aversive learning, 31
gender effects, 30
nocebo mass psychogenic illness, 30
placebo analgesic responses, 28
video, 28–29
O/HP. *See* Open/hidden treatment paradigm (O/HP)
Open/hidden treatment paradigm (O/HP)
clinical routine situations, 241–242
pharmacological action, 241

P

Pain. *See* Visceral pain
Pain-related negative emotions
active placebo, 85–86
compensatory reactions, 93
emotional valence and arousal, 88
emotions and homeostasis, 91–92

emotions and placebo response, 88–90
individual differences, emotions, 91
pharmacology implications, 93–94
placebo analgesic response, 82–83
placebo effects, 87
stress, 82
PANAS. *See* Positive Affect Negative Affect Schedule (PANAS)
Parkinson's disease
conditioned placebo responses, 22
[^{11}C] raclopride PET, 141
dopamine
replacement, treatment, 140
reward processing, 142–143
dopaminergic neurons, 140
"effector" physiological response, 145
endogenous dopamine release, 140
expectation, 143–144
neurochemical/neurophysiological response, placebo, 140
placebo response, 72
placebos as rewards, 142–143
reward system, 7
STN-DBS surgery, 141, 142
subthalamic nucleus deep-brain stimulators, 141
surgical trials, 141
Pharmacological conditioning and placebo responses
animals
CCRs, 21
conditioned stimulus, hypoglycemic reflex, 19
cyclosporin A, 20
hypoglycemic conditioning, 19
immune system, 19–20
opioid and non-opioid exposure, 20
"placebo effect in rat", 19
rACC and placebo analgesia, 20
humans
analgesic responses, 22
apomorphine dopaminergic agonist, 22
children, ADHD, 23
corticosteroids, 23
immune system, 21
morphine, 22
placebo-induced GH, 22
serotonin, 22
side effects, 24
single neuronal activity, 22, 23
Physical performance
central governor of fatigue, 153–155
learning, 152–153
positive and negative expectations, 150–152
Placebo analgesia
altered sensory processing, 76–77
anticipatory responses, 74–75
assessment, 84
biological mechanisms, 39
endorphins, 83
individual differences, emotions, 91
naloxone, 83, 92
opioids, 75–76
prefrontal regions loss, 77–78
prior expectations, 72–73
reduced evoked potentials, 84
reduced stress and negative emotions, 88
sex/gender differences, 112–13
signal detection errors and cognitive bias, 73–74
Placebo and nocebo effects, itch
behavioral scratching responses, 209–210
conditioning and verbal suggestion, 210
individual predictors, 211
QST, 209
treatment strategies, 211–212
verbal suggestion, 209
Placebo effect. *See also* Pharmacological conditioning and placebo responses
additivity, 10–11, 131–132
analgesia, 8–9, 20
brain imaging, 6–7
brain regions, pain processing, 58–59
classical conditioning, 219
clinical application, 219, 231–232
expectancy processes, 221
and expectations, 87
idiopathic pain
expectations and emotional feelings, 125–126
IBS patients, 123–124
neurobiological underpinnings, 126–127
rectal lidocaine and placebo, 124
"instruction", 220
neurobiological basis, 221–222
neuropathic pain
hyperalgesia area, 129
open *vs.* hidden lidocaine administration, 128
psychological factors, 129–130
nocebo, 9, 219–220
Parkinson's disease (*see* Parkinson's disease)
and patient information, 230

Placebo effect. (*cont.*)
physical performance (*see* Physical performance)
predicting responses, 11–12
RCTs, 131–133
reward system, 7
"social learning", 220
visceral pain, 99–101
Placebo response
"active placebos", 253, 259–260
analgesia, 197
assay sensitivity improvement, 258
"authorized deception", 250–251
autonomic nervous system (ANS), 186
BCD, 244–245
blood pressure, 187–188
BPD, 243–244
caffeine-related stimuli, 195
CAN, 186–187
"cephalic phase response", 197–198
CER, 266–267
classical conditioning, 185
clinical trials, 246, 252
cluster randomization, 263
cognition, 78
on coronary arteries, 188–189
cross-over design, 252–253
description, 184, 238
DRP, 242
and drug run-ins, 253–254
effective blinding, 259
and emotions, 88–90
ethical principles, 250
FCP, 251–252
on gastrointestinal motility, 190–191
habituation, sensitization and learning, 247
heart rate, 189–190
immune responses (*see* Immune responses)
instrumental learning, 249
inverse enrichment, 246–247
lung function and respiratory activity, 193
manipulation of information, 243
on nausea and gastric activity, 191–192
neuroendocrine functions (*see* Neuroendocrine functions)
"no-treatment" controls, laboratory, 247–248
O/HP, 240–242
"open–hidden paradigm", 185
Parkinson's disease, 72
participation, drug trials, 264
patient registry, 265–266
peripheral placebo and nocebo responses, 195–196
"placebo by proxy", 264–265
"placebo responders", 249–250, 252
preference design, 262–263
prior expectation, 72
randomized run-in/withdrawal, 254–255
RCT (*see* Randomized clinical trial (RCT))
registry trials, 258
social learning, 248–249
stress and sympathetic activity, 194
study designs, 240
timing manipulation, 240
treatment, 239
trial acceptability, 260
unbalanced randomization, 260–261
verbal suggestions, 185
Positive Affect Negative Affect Schedule (PANAS), 130
Positron emission tomography (PET)
EEG system, 6
and fMRI, 39
meta-analysis, 47
placebo analgesia and opioid analgesia, 39
technical advances, 6
Psychopharmacology
clinical trial design (*see* Clinical trial design)
clinician expectation, 278
drug–placebo differences, 277–278
drug-specific and nonspecific effects, 286
drug trials, 275, 285–286
FDA, 275
nocebo effect, 283–284
placebo group, 274
RCT, 274
sensations, 284–285
spontaneous remission, 276
treatment effects, 286
waitlist-controlled trials, 276–277

Q

QST. *See* Quantitative sensory testing (QST)
Quantitative sensory testing (QST), 128, 209, 210

R

Randomized clinical trial (RCT)
in children and adolescents, 264
designs, 246

double-blind design, 274
and placebo mechanisms studies
additivity assumption, 131–132
design, 131
expectations and emotional feelings, 131
meta-analyses, 133
total treatment effect, 132
placebo response, 245
treatment and placebo arms, 10–11
RCT. *See* Randomized clinical trial (RCT)
Reward system, 7

S
Selective serotonin reuptake inhibitors (SSRIs), 298–300
Sequential parallel comparison design (SPCD), 253, 256
Signal detection errors, 73–74
SLE. *See* Systemic lupus erythematosus (SLE)
SPCD. *See* Sequential parallel comparison design (SPCD)
SSRIs. *See* Selective serotonin reuptake inhibitors (SSRIs)
Step-wedge design, 261–262
Stimulus expectancies *vs.* treatment expectancies
expectancy-induced increases, 55–57
expectancy-induced reductions, pain, 52, 55
placebo-induced increases, 51–53
placebo-induced reductions, 49, 52
stimulus expectancy-induced increases, 52, 53
stimulus expectancy-induced reductions, 50
Systemic lupus erythematosus (SLE), 165–166, 170

T
"Treatment as usual" (TAU), 256–258
Tricyclic antidepressants (TCA), 284–285
Two-way enrichment design (TED), 256

U
Unconsciousness and conscious cognitive expectancy, 4–6

V
Vasoconstriction, 189
Verbal communication
airway resistance, 27
aversive and fear models, 31
gender effects, 30
iontophoretic pain stimulation, 26
mass psychogenic illness, 30
N1 and biphasic N2-P2 components, LEPs, 27
negative outcomes, 27–28
observationally induced placebo analgesia and empathy, 28–29
observing others, 28
placebo responses, 18
reverse verbal suggestions, 27
Verbal suggestions, 185
Visceral pain
and IBS, 98–99
learning/conditioning, 109–110
mechanistic studies, 101–108
placebo and nocebo effects, 99–101
sex/gender differences, 112–113
trait and state emotions, 110–111

W
Waiting list (WL), 256, 257, 261, 262
World Anti-Doping Agency (WADA), 152

Z
"Zelen design", 257–258

MIX
Papier aus verantwortungsvollen Quellen
Paper from responsible sources
FSC® C105338

If you have any concerns about our products,
you can contact us on
ProductSafety@springernature.com

In case Publisher is established outside the EU,
the EU authorized representative is:
Springer Nature Customer Service Center GmbH
Europaplatz 3, 69115 Heidelberg, Germany

Printed by Libri Plureos GmbH
in Hamburg, Germany